World Survey of Climatology Volume 13

CLIMATES OF AUSTRALIA AND NEW ZEALAND

World Survey of Climatology

Editor in Chief:

H. E. LANDSBERG, Washington, D.C. (U.S.A.)

Editors:

H. ARAKAWA, Tokyo (Japan)
R. A. BRYSON, Madison, Wisc. (U.S.A.)
H. FLOHN, Bonn (Germany)
J. GENTILLI, Nedlands, W.A. (Australia)
J. F. GRIFFITHS, College Station, Texas (U.S.A.)
P. E. LYDOLPH, Milwaukee, Wisc. (U.S.A.)
S. ORVIG, Montreal, Que. (Canada)
D. F. REX, Boulder, Colo. (U.S.A.)
W. SCHWERDTFEGER, Madison, Wisc. (U.S.A.)
H. THOMSEN, Charlottenlund (Denmark)
C. C. WALLÉN, Geneva (Switzerland)

World Survey of Climatology Volume 13

Climates of
Australia and New Zealand

edited by J. GENTILLI

University of Western Australia
Nedlands, W.A. (Australia)

ELSEVIER PUBLISHING COMPANY Amsterdam-London-New York 1971

ELSEVIER PUBLISHING COMPANY
335 Jan van Galenstraat
P.O. Box 211, Amsterdam, The Netherlands

ELSEVIER PUBLISHING COMPANY LTD.
Barking, Essex, England

AMERICAN ELSEVIER PUBLISHING COMPANY, INC.
52 Vanderbilt Avenue
New York, New York 10017

Library of Congress Card Number: 71-103354

ISBN 0-444-40827-4

With 141 illustrations and 113 tables

Printed in The Netherlands

World Survey of Climatology

Editor in Chief: H. E. LANDSBERG

List of Contributors to this Volume:

J. Gentilli
Department of Geography
University of Western Australia
Nedlands, W.A. (Australia)

W. J. Maunder
Department of Geography
University of Otago
Dunedin (New Zealand)

U. Radok
Department of Meteorology
University of Melbourne
Melbourne, Vic. (Australia)

Contents

Contents

Chapter 8. THE CLIMATE OF NEW ZEALAND—PHYSICAL AND DYNAMIC
           FEATURES
by W. J. MAUNDER

Chapter 9. ELEMENTS OF NEW ZEALAND'S CLIMATE
by W. J. MAUNDER

Chapter 10. CLIMATIC AREAS OF NEW ZEALAND
by W. J. MAUNDER

Appendix. CLIMATIC TABLES OF AUSTRALIA AND NEW ZEALAND

Chapter 1

Introduction

J. GENTILLI

When Professor Landsberg offered me the opportunity to edit this volume and to write any part of it I chose, I was well aware of the difficulties of the task. Now, having completed the manuscript, I am more than ever aware of the gaps in our knowledge of Australian climates. I have endeavoured to do justice to the painstaking, assiduous and penetrating records left by the early observers who are too readily forgotten, and to the immense amount of data published by the Australian Bureau of Meteorology. On the other hand, most of these data have been published only recently and, therefore, are still readily obtainable in official publications; it would have been out of place to burden this volume with numerous tables reproducing such data, which should be sought at the source. The same policy was followed in the editing of the New Zealand section.

The problem which posed itself from the very beginning was whether it was preferable to write a detailed static description of the various climates, full of factual information, or whether a dynamic account of the spatial relationships of climate was more desirable. The latter way was chosen, but the provision of numerous maps still makes it possible to obtain a large amount of detailed, factual knowledge by interpolation.

The first unexpected difficulty was the lack of historical studies in this field, with the notable exception of the work by RUSSELL (1889) on the early astronomers and meteorologists of New South Wales. This work, however, was mainly biographical. I am convinced that further research will disclose details and data long forgotten, but I had to call an early end to my work in this particular aspect in order to avoid further delays in the preparation of the main part of the book. The chronology which opens the volume gives the essential facts.

Use has been made of daily weather maps in order to illustrate the genesis of some climatic features or manifestations. This is not general practice in works of this kind, but I decided to do it where the frequency, repetition or magnitude of certain weather situations resulted in a definite and characteristic aspect of the climate in this or that area or at this or that time of the year. It was only possible to make a first approach and it is hoped that more research will lead to further extensions and interpretations.

The tables of the Appendix at the end of the volume provide a fuller climatography of Australia than had ever been assembled before. The many aspects of climate had been recorded over different periods; a near completeness of the climatic record has been preferred to a narrow homogeneity. Each one of the tables (or each climatic element taken through many tables) warrants statistical and graphic analysis to discover correlations, seasonal variations, geographic patterns which could not be shown in this volume without increasing its size beyond any reasonable limits.

While there was a clear editorial directive as to the number of pages in each volume, the arrangement of the subject matter was left to the editors and authors of each volume. This is likely to bring forth a refreshing variety, but may also produce some striking contrasts. In this volume, for instance, there is no chapter on comparative regional methodology, in which the subdivision of the area into climatic regions could be carried out according to each one of the principal methods of climatic classification. On the one hand, such a chapter might have lacked appeal for the non-specialist. On the other hand it might perhaps have arrived at some evaluation of the many methods available. A great deal of space has been devoted to the Waite Index and to its use for the purpose of classification. This was done because the index itself is little known outside Australia, notwithstanding its very great merits, and because its originators did not use it for classificatory purposes, having always considered that such classifications may too easily become codified and be used uncritically or, in any case, too rigidly. The fact remains that other classifications, which are based on much less sound methods of estimating the hydric effectiveness of the various climates, are very widely used.

Some effort has been made throughout the work to establish priorities. It is quite understandable that in the pioneering conditions of early Australian settlement, avenues of publication may have been few and far between, so that early works may have been published late, or only in part, or not at all, and very often not in Australia, but in Great Britain. Hence the frequent ignorance of prior work in one's own field, resulting in duplication and waste of effort, ill afforded in a pioneering community. In fact, the very existence of separate colonial (later state) communities caused this fragmentation, so that the man in Melbourne totally ignored what his colleague in Brisbane or in Hobart may have been doing in that very same field of research.

Some of the early pioneers had some happy intuitions, many of them were keen and perceptive observers, a few had a precision and conciseness of style which modern writers might well envy and try to emulate, but unfortunately, far too few had adequate training. But the happy intuitions and the precise descriptions deserve a place in the ever widening sequence of Australia's progress towards a fuller understanding of the realities of climatic environment.

Dr. Radok, who kindly accepted the difficult task of writing the chapter on the general atmospheric circulation in the Australian region, confided to me that, since we are now acquiring knowledge by new and more powerful means, and at a much faster rate than a few years ago, we could have found many more answers if this volume could have delayed by three or four years. I am in full sympathy with these feelings, because I have been painfully aware of the reality of such a situation throughout my own writing. But I am also firmly convinced that the ever widening frontier of knowledge is, like a tempting rainbow, ever shifting further away with our every move forwards.

My thanks are due to Dr. Radok and Mr. Maunder for writing their respective chapters, and my sincere editorial apologies go to Mr. Maunder for the ruthless cuts applied to the statistical details of his contribution.

It is a pleasure to express my appreciation to the University of Western Australia, which granted me sabbatical leave during 1966 and to Professor H. Flohn, who welcomed me at the Meteorologisches Institut of the University of Bonn, and placed at my disposal a most attractive and well equipped study where I could give the manuscript its first shape. Professor Flohn also kindly read through the whole manuscript and suggested several improvements. Professor H. Riehl read the first chapters and his remarks were also gratefully accepted although, of course, the responsibility for the final form of the manuscript is entirely my own.

My greatest debt is to the director and staff of the Australian Bureau of Meteorology, Melbourne, and to the regional director for Western Australia, always most generous with publications and data, and to the many professional and voluntary observers, past and present. Without their indefatigable work, my task would not have been feasible.

The weather maps used were originally drawn at the Regional Weather Bureau, Perth, and published in the *West Australian*.

I wish to thank Miss Margaret Kennedy and Miss Sandra Jenkins, who typed most of the manuscript, and Mr. R. Reid, who drew the base map of Australia used for many of the illustrations. Mr. P. Barrett kindly read the galley proofs, and Mr. D. I. Milton some of the page proofs.

Reference

RUSSELL, H. C., 1889. Astronomical and meteorological workers in New South Wales, 1778–1860. In: A. LIVERSIDGE and R. ETHERIDGE (Editors), *Report of the First Meeting of the Australasian Association for the Advancement of Science*. Government Printer, Sydney, pp.45–76.

Chapter 2

A Chronology of Climatic Work in Australia

J. GENTILLI

Early climatological observations have more than a merely historical interest; early observers were keen to learn about their new world, they had few distractions, and were usually punctilious in their use of words. However, they were often misled by insufficient knowledge or by prejudice. Early data may seldom be used for further research because the quality of the instruments, their exposure, and the mode of reading are unknown. And yet, if one is aware of the pitfalls and limitations, some very good use can be made of this early work, the study of which often gives the background for a better understanding of present-day developments. While further details more properly belong to the field of historical research, a succinct chronology of climatic work in Australia provides the best introduction to the present work.

1788: Pressure and temperature observations on board ship anchored in Botany Bay (WHITE, 1790).

1815: Publication of BAUDIN's *Observations physiques et météorologiques faites pendant le voyage aux Terres Australes, 1801–1803*.

1821: Beginning of pressure and temperature (BRISBANE, 1824) and rainfall (DOVE, 1837) observations at Governor Brisbane's residence at Paramatta (now Parramatta, New South Wales).

1822: Temperature observations five times a day at Hobart's Town (now Hobart) and Macquarie Harbour in Van Diemen's Land (now Tasmania). With Capetown, perhaps the only sites of standard climatic observations in the Southern Hemisphere at that time (BRISBANE, 1825).

1824: Temperature observations inside a well at Sydney and at sea-level and on a hill at Port Macquarie (now Macquarie Harbour, Tasmania) to measure thermal lapse rate (BRISBANE, 1824a, 1827).

1825: Publication of FIELD's *Geographical Memoirs on New South Wales*, containing F. Goulburn's and T. M. Brisbane's meteorological observations.

1829: Governor J. Stirling's order to the colonial surgeon of the future Swan River Colony (now Western Australia) to keep a weather journal.

1830: First records begun at Perth (manuscript at Australian Weather Bureau; CROSS, 1833; LANDOR, 1847).

1831: Detailed and accurate meteorological journal kept at Albany, Western Australia (CROSS, 1833).

1832: Sydney observations by Dunlop and King (RUSSELL, 1877).

1833: Publication of Breton's excursions in New South Wales, Western Australia and Van Diemen's Land, with notes on climate—pressure, temperature, wind and rainfall—recorded at York, Western Australia (OMMANEY, 1839).

1835: First Tasmanian rainfall records at Hampshire Hills, 34 km south-southwest of Burnie.

1836: Kingston's private rainfall readings in Grote Street, Adelaide (KINGSTON, 1861, 1874, 1879).

1837: MEINICKE's *Das Festland Australien*, containing 20 pages on the climates of Australia, is published in Germany.

1838: Records begun at Woolnorth, Circular Head, Port Arthur and Launceston, Tasmania (DE STRZELECKI, 1845).

1839: P. P. King begins observations at Tahlee near Port Stephens, New South Wales.

1840: Records of pressure, temperature, humidity and rainfall begun at the South Head (8 km from Sydney) and Port Macquarie, New South Wales, and at Port Phillip (now Melbourne, Victoria) (RUSSELL, 1877); J. C. Wickham's weather journal begun at Brisbane; Ross Bank Observatory established at Hobart, with J. H. Kay as director.

1843: Essay by POWER on "The Climate of South Australia" in the *Royal South Australia Almanack*.

1845: Publication of DE STRZELECKI's *Physical Description of New South Wales and Van Diemen's Land*, with original climatic observations.

1846–1852:
End of observations at Parramatta (1846), Brisbane (1850), Melbourne (1851) and Port Macquarie (1852).

1852: Rainfall records begun at Fremantle Signal Station, Western Australia.

1853: Temperature and hours of rain observed at Sydney University (JEVONS, 1859).

1854: End of official observations at Hobart.

1855: Beginning of F. Abbott's private records at 82 Murray Street, Hobart (ABBOTT, 1861, 1866, 1872); new observatory at Williamstown, 7 km southwest of Melbourne, and rainfall gauge at Yan Yean Reservoir, Victoria (BARACCHI, 1919); L. L. Becker's observations at Bendigo, Victoria (BECKER, 1855); end of observations at the South Head, New South Wales.

1856: W. B. Scott appointed Government Astronomer, New South Wales; records begun at Rockhampton, Brisbane, Casino, Armidale, Maitland, Bathurst, Parramatta, Sydney, Goulburn, Deniliquin, Albury and Cooma, New South Wales; private records at Bukalong, New South Wales (RUSSELL, 1877); C. Todd appointed Superintendant of Telegraphs and Government Astronomer in South Australia, begins private observations to be continued officially at Adelaide (TODD, 1861).

1856: Observations begun at Ballarat, Geelong, Mount Buninyong, Port Albert and (1857) Bendigo and Portland, Victoria; W. S. Jevons begins private observations at Petersham near Sydney with weekly reports to the newspaper *Empire*.

1858: Flagstaff Hill Observatory built at Melbourne, with G. Neumayer as director; records begin at Beechworth, Camperdown, Castlemaine, Heathcote in Victoria (NEUMAYER, 1861, 1864, 1867). Flagstaff Hill Observatory built at Sydney (RUSSELL, 1877). A. C. Gregory advocates the use of the telegraph to transmit weather data (GREGORY, 1859).

1859: W. S. JEVONS's paper *Some data concerning the climate of Australia and New Zealand* published in *Waugh's Almanac*. Sydney; first evaporation tank installed at Sydney.

1861: Weather records from the "Dolphin" at anchor in Nickol Bay, 18 km northwest of Roebourne, Western Australia.

1862: Pressure recorded four times daily at King Point lighthouse and five times daily at Breaksea Island lighthouse, near Albany, Western Australia (SCOTT, 1876).

1863: R. L. J. Ellery appointed Government Astronomer, Victoria; new observatory at St. Kilda, Melbourne, to replace Williamstown and Flagstaff Hill; first pluviograph installed at Sydney; J. Tebbutt begins private observations at Windsor, New South Wales.

1865: First Queensland country records from Warwick (continuing) and Somerset (near Cape York).

1868: Rainfall records begun at Stanley, Tasmania.

1870: H. C. Russell appointed Government Astronomer in New South Wales.

1871: Installation of numerous rain gauges in Tasmania, including several by the Marine Board at lighthouses; beginning of annual publication of results of observations in New South Wales, and abstract of earlier data by Russell.

1876: Publication of LORENZ and ROTHE's *Lehrbuch der Klimatologie* containing brief notes on Australian climates.

1876: Meteorological branch added to the Surveyor General's Department in Perth, new records begun; Eucla, Western Australia linked to Adelaide by telegraph, begins observations; paper on "Meteorological Periodicity" by RUSSELL.

1877: Publication of daily weather maps in *Sydney Morning Herald*.

1878: Beginning of records at several country centres, Western Australia (COOKE, 1901).

1879: End of Kingston's series (1839–1879) at Adelaide.

1880: Inter-colonial meteorological conference at Sydney; J. Thorpe advocates a weather map for all Australia; end of the Abbot series (1855–1880) at Hobart; beginning of records at Rottnest Island, off Fremantle, Western Australia.

1881: First Western Australian tropical station at Cossack; suspension of records in Tasmania.

1882: Captain J. Shortt appointed first Government Meteorologist in Tasmania, and new observatory built.

1883: Derby, Esperance and other stations established in Western Australia; BARKER's first description of southern mid-latitude depressions.

1884: Publication of VOEIKOV's book on *The Climates of the Terrestrial Globe* at St. Petersburg (now Leningrad) with account of Australian climates and map.

1885: Transfer of Perth's site to Botanical Gardens.

1887: C. L. Wragge appointed Government Meteorologist in Queensland, and new meteorological office opened; expansion of service.

1893: H. C. Kingsmill appointed in Hobart; RUSSELL's study of anticyclones published in London.

1895: C. Wragge inspects facilities and sites in Tasmania and reports to Tasmanian government on reorganization of service; initiates observations on Mount Wellington, Tasmania, and Mount Kosciusko, New South Wales, to obtain high-altitude atmospheric data, endowing Kosciusko station at personal expense; begins practice of naming cyclones.

TABLE I

NUMBER OF RECORDING STATIONS

Year	W.A.	N.T.	S.A.	Qld.	N.S.W.	Vic.	Tasmania
All stations							
1860	–	–	12	2	10	15	1
1870	–	1	53	36	35	24	2
1880	8	12	159	76	174	122	7
1890	80	28	353	422	589	465	56
1900	295	34	447	609	821	660	82
1910	580	41	524	627	1115	928	161
1920	922	43	512	966	1254	916	240
1930	1018	58	581	1075	1436	906	278
1940	1106	85	641	1126	1601	984	283
1950	1119	100	664	1149	1580	998	303
1960	1124	141	771	1304	1851	906	298
1970	1243	346	>900	1785	2416		431
Climatological stations[1]							
1860	–	–	1	1	1	1	1
1870	1	–	3	1	8	3	1
1880	1	–	5	1	18	7	1
1890	1	1	11	1	20	12	3
1900	36	3	15	1	25	33	7
1910	52	4	19	64	79	54	13
1920	58	5	23	89	114	67	17
1930	60	5	28	90	116	71	19
1940		6					
1950	85	27	55	46	196	>96	>23
1960	92	38	62	52	209		48
1970	136	46	85	156	216	113	61
Pluviographs							
1920	1	–	1	2	2	1	1
1930	1	–	1	2	3	13	1
1940	3	–	1	5	9	16	3
1950	3	–	2	6	23	22	4
1960	24	18	12	41	160	69	49
1970	74	27	16	131	300	91	71

[1] Observing rainfall, temperature, in some cases humidity, evaporation, etc. (includes 3 automatic stations in Western Australia, 3 in the Northern Territory, 5 in Queensland).

1896: Publication in Sydney of *Three Essays on Australian Weather*, edited by ABERCROMBY, containing the prize-winning essay by H. A. Hunt on southerly bursters, an essay by H. C. Russell and a report on Australian weather types, also by H. A. Hunt, based on research privately financed by Abercromby.

1897: End of J. Tebbutt's series (1863–1897) at Windsor, New South Wales.

1901: Publication of observations (1876–1899) in Western Australia (COOKE, 1901).

1903: H. A. Hunt appointed Government Meteorologist, New South Wales.

1906: Meteorology Act by Federal Parliament amalgamating services.

1907: Creation of Commonwealth Bureau of Meteorology, with H. A. Hunt first Director; central office in Melbourne, State Observatories named Divisional Observatories.

1908: Melbourne observation site transferred to corner of Latrobe and Victoria Streets, near Central Weather Bureau, Royal Society Grounds.

1911 : Publication of first volume of *Results of Rainfall Observations* (ANONYMOUS, 1911–1954); J. Watt, Divisional Meteorologist in Tasmania.

1919 : Publication by BARACCHI of the complete set of Victorian meteorological statistics, 1856–1907.

1920 : TAYLOR's *Australian Meteorology* published.

1922 : New Divisional Bureau building at Sydney; cyclone warning station established at Willis Island, Queensland.

1932 : TAYLOR's *Climatology of Australia* published in vol. 4, part S, of Köppen's *Handbuch der Klimatologie*.

1935 : Paper by KIDSON on frontal analysis.

1937 : Beginning of regular three hourly observations and expansion of service to meet civil aviation needs; teaching of frontal and upper-air analysis to trainee meteorologists by H. M. Treloar.

1938 : New Weather Bureau building at Brisbane.

1939 : Extensions to Central Bureau building in Melbourne; new building at Adelaide; creation of Division of Radiophysics (Council for Scientific and Industrial Research) at Sydney.

1940 : Meteorological Service transferred to Royal Australian Air Force for the duration of the war.

1941 : First issue of monthly *Meteorological Summary* (now *Statistical Summary*).

1942 : Publication of *Weather on the Australia Station* by the ROYAL AUSTRALIAN AIR FORCE.

1945 : Use of radar for storm detection and tracking; 27 aerodromes with forecasting facilities, 13 with observation equipment; Department of Meteorology at University of Melbourne, with F. Loewe as Head.

1946 : Creation of Section of Meteorological Physics of the Council for Scientific and Industrial Research (C.S.I.R.) with C. H. B. Priestley as chief; Division of Radiophysics begins work on stimulation of clouds to produce rain.

1947 : First man-made rain reaches the ground (KRAUS and SQUIRES, 1947); aerological flights to sub-Antarctic from Perth and Melbourne.

1951 : *Book of Normals–Rainfall (1911–1940)* published (ANONYMOUS, 1951).

1952 : First issue of *Australian Meteorological Magazine*.

1953 : KRAMER's Australian bibliography in the *Meteorological Abstracts and Bibliography* with 281 entries.

1954 : Last volume of *Results of Rainfall Observations* published (ANONYMOUS, 1911–1954); meteorological stations at Heard and Macquarie Islands and at Mawson, Antarctica.

1954 : Section of Meteorological Physics expanded to Division of Commonwealth Scientific and Industrial Research Organisation (C.S.I.R.O.).

1955 : Meteorology Act by Federal Parliament to bring service into line with W.M.O. and I.C.A.O. requirements.

1956 : Book of *Climatic Averages* published (ANONYMOUS, 1956); UNESCO Symposium on Arid Zone Climatology (microclimatology) held in Canberra.

1958 : Creation of Hydro-Meteorological Service within Weather Bureau; publication of booklets, *50 Years of Weather*.

1960 : Use of televised data from "Tiros II"; setting up of 165 temporary rainfall stations for Darling Downs cloud-seeding experiment, Queensland; Creasi's bibliographies for U.S. Weather Bureau with 305 entries (CREASI, 1960a, 1960b).

1962: Automatic weather station on Ashmore Island (Sahul Shelf).

1963: Astronaut Cooper warns Perth Bureau of thunderstorm activity far over Indian Ocean; front reaches coast two days later.

1965: Commonwealth Bureau of Meteorology (Director: W. J. Gibbs) has staff of 893, including 113 professional meteorologists, 10 engineers, 423 weather officers and observers, with 7,000 voluntary observers cooperating. Branches are: Research, Climatology, Hydro-Meteorology, Forecasting, Recording (with computing).

1967: Establishment of World Meteorological Centre at Melbourne, to gather and analyze data from the Southern Hemisphere as part of the World Weather Watch. New Weather Bureau building at Perth. History of meteorological and climatological studies in Australia published (GENTILLI, 1967)—36 pages, 140 bibliographical entries.

1968: Installation of a computer at the Bureau of Meteorology. Publication of *Monthly Rainfall and Evaporation* data (volume with lowest, highest, 10, 30, 50, 70, 90 percentile monthly rainfall values for nearly 1,400 stations, and atlas with maps of monthly and annual 10, 50 and 90 percentile rainfall, and mean evaporation).

1969: Paper by E. C. Barrett on the estimation of rainfall in data-remote regions from satellite data (with tropical Australian examples) read at the Adelaide meeting of the Australian and New Zealand Association for the Advancement of Science. Commonwealth Meteorology Research Centre established in Melbourne (Officer-in-charge: G. B. Tucker) jointly sponsored by Bureau of Meteorology and C.S.I.R.O. Division of Meteorological Physics.

1970: Division of Meteorological Physics in C.S.I.R.O. (Chief: C. H. B. Priestley), at Aspendale near Melbourne, has a staff of 81, of whom 20 are research scientists. Division of Radiophysics (Chief: E. G. Bowen), at Epping near Sydney, has a numerous staff, among whom 11 research scientists, 6 other professional staff and 20 technicians are mostly working in the field of cloud physics.

References

ABBOTT, F., 1861. *Results of Meteorological Observations for Hobart Town, January 1841 to December 1860*. Mercury Office, Hobart, 24 pp.

ABBOTT, F., 1866. *Results of Twenty-Five Years of Meteorological Observations for Hobart Town*. Mercury Office, Hobart, 48 pp.

ABBOTT, F., 1872. *Results of Five Years' Meteorological Observations for Hobart Town* (with which are incorporated the results of twenty-five years' observations previously published (1866) and completing a period of thirty years). Mercury Office, Hobart, 377 pp.

ABERCROMBY, R., (Editor), 1896. *Three Essays on Australian Weather*. White, Sydney, 104 pp.

ANONYMOUS, 1911–1954. *Results of Rainfall Observations made in* (last volumes for each state appeared for/in:) *Western Australia (1929), Tasmania (1936), Victoria (1937), Queensland (1940), New South Wales (1948), South Australia and Northern Territory (1954)*. Australian Bureau of Meteorology. (The series is now discontinued.)

ANONYMOUS, 1951. *Book of Normals, 1. Rainfall. Standard Period 30 Years, 1911–1940*. Australian Bureau of Meteorology, Melbourne, 169 pp.

ANONYMOUS, 1956. *Climatic Averages Australia—Temperature, Relative Humidity, Rainfall*. Australian Bureau of Meteorology, Melbourne, 107 pp.

ANONYMOUS, 1963. *Lists of Official Rainfall Stations—Australia* (alphabetical and numerical). Australian Bureau of Meteorology, Melbourne, 256 pp.

ANONYMOUS, 1965. *Fitzroy Region Resources Series—Climate*. Department of National Development, Canberra, 28 pp.

ANONYMOUS, 1966. *Rainfall Statistics—Western Australia, Northern Territory, South Australia, Queensland, New South Wales, Victoria, Tasmania—Long Period Averages (All Years), Standard Period Normals of Rainfall and Raindays (1931–1960)*. Australian Bureau of Meteorology, Melbourne, 312 pp.

ANONYMOUS, 1968. *Monthly Rainfall and Evaporation*. Bureau of Meteorology for Australian Water Resources Council, Melbourne, I: 682 pp.; II: 55 maps.

ASHTON, H. T., 1964. Meteorological data for air conditioning in Australia. *Bur. Meteor. Bull.*, 47: 15 pp.

BARACCHI, P., 1919. *Victorian Meteorological Statistics, 1856–1907*. Government Printer, Melbourne, 114 pp.

BARKER, D. W., 1883. Notes on the storms of high south latitudes. *Trans. Proc. Roy. Soc. Victoria*, 19: 139–143.

BAUDIN, N., 1815. Observations physiques et météorologiques faites pendant le voyage aux Terres Australes, 1801–1803. En: L. C. DE S. DE FREYCINET (Rédacteur), *Voyage de découvertes aux terres australes, exécuté sur les corvettes le Géographe, le Naturaliste, et la goélette le Casuarina... sous le commandement du Capitaine de Vaisseau N. Baudin*. Imprimerie Royale, Paris, 576 pp.

BECKER, L., 1855. Meteorological observations at Bendigo. *Transactions Phil. Soc. Victoria*, 1: 87–91.

BRISBANE, T. M., 1824a. Observations on the temperature of the earth at Paramatta, New South Wales. *Edinburgh Phil. J.*, 10: 219–222.

BRISBANE, T. M., 1824b. Mean of twelve months' meteorological observations in the years 1822–23, made at Paramatta, New South Wales. *Edinburgh Phil. J.*, 11: 119–120.

BRISBANE, T. M., 1825. On the meteorological tables kept in 1822 at Macquarie Harbour and Hobart's Town in Van Diemen's Land. *Edinburgh J. Sci.*, 2: 75–77.

BRISBANE, T. M., 1827. Observations on the mean temperature of the earth at Sydney, made in the years 1824–1825. *Edinburgh J. Sci.*, 6: 226–228.

COOKE, W. E., 1901. *The Climate of Western Australia from Meteorological Observations made during the Years 1876–1899*. Government Printer, Perth, 126 pp.

CREASI, V. J., 1960a. *Bibliography of Climatic Maps for Australia*. U.S. Weather Bureau, Washington, D.C., 42 pp.

CREASI, V. J., 1960b. *Selected Bibliography on the Climate of Australia*. U.S. Weather Bureau, Washington, D.C., 109 pp.

CROSS, J. (Editor), 1833. *Journals of Several Expeditions made in Western Australia... under the Sanction of the Governor, Sir James Stirling*. Cross, London, 264 pp.

DE STRZELECKI, P. E., 1845. *Physical Description of New South Wales and Van Diemen's Land*. Longmans, London, 462 pp. (Facsimile edition, 1967, Library Board of South Australia, Adelaide.)

DOVE, H. W., 1837. *Meteorologische Untersuchungen*. Sander, Berlin, 344 pp.

FIELD, B. (Editor), 1825. *Geographical Memoirs on New South Wales, by various hands... together with... the meteorology of New South Wales and Van Diemen's Land*. Murray, London, 504 pp.

GENTILLI, J., 1967. A history of meteorological and climatological studies in Australia. In: *University Studies in History*, 5: 54–88.

GIBBS, W. J. and MAHER, J. V., 1967. Rainfall deciles as drought indicators. *Bur. Meteorol. Bull.*, 48: 33 pp., 85 maps.

GREGORY, A. C., 1859. Some interesting facts founded on barometrical observations. *Trans. Phil. Inst. Victoria*, 4: 14–15.

JEVONS, W. S., 1859. Some data concerning the climate of Australia and New Zealand. In: *Waugh's Almanac*. Waugh, Sydney, pp.XV–XVI; 47–98.

KIDSON, E., 1935. The analysis of weather charts. *Australian Geograph.*, 2(5): 3–16.

KINGSTON, G. S., 1861. *Register of the Rain Gauge kept in Grote Street, Adelaide, January 1st, 1839 to December 31st, 1860*. Government Printer, Adelaide, 17 pp.

KINGSTON, G. S., 1874. *Tables showing the Monthly and Yearly Depth of Rain in Adelaide, Melbourne and Sydney*. Government Printer, Adelaide, 27 pp.

KINGSTON, G. S., 1879. *Register of the Rainfall kept in Grote Street, Adelaide, January 1st to December 16th, 1879*. Government Printer, Adelaide, 29 pp.

KRAMER, H. P., 1953. Selective annotated bibliography on the climate of Australia and New Zealand. *Meteorol. Abstr. Bibl.*, 4: 481–528.

KRAUS, E. B. and SQUIRES, P., 1947. Experiments on the stimulation of clouds to produce rain. *Nature*, 159: 489–491.

LANDOR, E. W., 1847. *The Bushman; or, Life in a New Country*. Bentley, London, 355 pp.

LORENZ, J. R. und ROTHE, C., 1876. *Lehrbuch der Klimatologie*. Braumüller, Wien, 481 pp.

MEINICKE, C. E., 1837. *Das Festland Australien—eine geographische Monographie*. Kalbersberg, Prenzlau, 437 pp.

NEUMAYER, G., 1861. Climatological Outlines for the Colony of Victoria. In: *Catalogue, Victorian Exhibition, 1861*. Government Printer, Melbourne, pp.131–158.

NEUMAYER, G., 1864. *Results of the Meteorological Observations taken in the Colony of Victoria, 1859– 1862*. Government Printer, Melbourne, 392 pp.

NEUMAYER, G., 1867. *Discussion on the Meteorological and Magnetical Observations made at the Flagstaff Observatory, Melbourne, during the Years 1858–1863*. Akademische Verlag, Mannheim, 306 pp.

NEWMAN, B. W. and DEACON, E. L., 1956. A "dynamic" meteorologist—Clement Wragge, 1852–1922. *Weather*, 11: 3–7.

OMMANEY, H., 1839. Meteorological summary for 1833, at York, Western Australia. *Meteor. Soc. Trans.*, 1: 136.

POWER, W. J., 1843. The climate of South Australia. In: *Royal South Australian Almanack*. MacDougall, Adelaide, pp.80–98.

ROYAL AUSTRALIAN AIR FORCE, 1942. *Weather on the Australia Station*. Air Force Headquarters, Melbourne, Vic., I: 641 pp.; II: 470 pp.

RUSSELL, H. C., 1876. Meteorological periodicity. *J. Roy. Soc. New South Wales*, 10: 151–176.

RUSSELL, H. C., 1877. *Climate of New South Wales, Descriptive, Historical and Tabular*. Government Printer, Sydney, 252 pp.

RUSSELL, H. C., 1893. Moving anticyclones in the Southern Hemisphere. *Quart. J. Roy. Meteorol. Soc.*, 19: 23–34. (Reprinted in R. ABERCROMBY (Editor), 1896. *Three Essays on Australian Weather*. White, Sydney, 104 pp.)

SCOTT, R. H., 1876. Contributions to the meteorology of West Australia. *Quart. J. Roy. Meteorol. Soc.*, 3: 177–178.

TAYLOR, T. G., 1920. *Australian Meteorology*. Oxford University Press, Oxford, 312 pp.

TAYLOR, T. G., 1932. Climatology of Australia. In: W. KÖPPEN und R. GEIGER (Herausgeber), *Handbuch der Klimatologie*. Borntraeger, Berlin, 4(S): 108 pp.

TODD, C., 1862. *Meteorological Observations*. Government Printer, Adelaide, 43 pp.

VOEIKOV, A. I., 1884. *Klimati zemnogo shara*. St. Petersburg. (Reprinted 1948 by Akad. Nauk. S.S.S.R., Moscow; German edition: Woeikow, A., 1887. *Die Klimate der Erde*. Costenoble, Jena.)

WHITE, J., 1790. *Journal of a Voyage to New South Wales*. Debrett, London, 300 pp. (Reprinted 1962 by Angus and Robertson, Sydney, 288 pp.)

Chapter 3

The Australian Region and the General Circulation of the Southern Hemisphere

U. RADOK

Introduction

The general circulation of the southern "water hemisphere" has always excited special interest, as the closest approximation in nature to an idealized planetary state of large-scale flow. The first ideas concerning synoptic conditions in the Southern Hemisphere were dominated by observations from sailing ships in the region of the prevailing westerlies, and by the impressions of isolated observers on the Australian continent, the only southern land mass with an appreciable extent in longitude (just under one sixth of a full latitude circle). Here arose at the end of the 19th century the picture of the Southern Hemisphere circulation as a regular procession of highs and lows girdling the earth in 40 days (RUSSELL, 1893). This simple picture has long been abandoned, but no similarly complete model has yet taken its place. This is because the observational network of the Southern Hemisphere is far from adequate even today, and the network over the ocean areas has deteriorated, if anything, since the sailing ship days. This dismal state of affairs is compensated partially by the extension of observations upward into the free atmosphere. Above the continents and some of the islands of the Southern Hemisphere such "aerological" observations during the past 30 years have added greatly to our knowledge of the general circulation. Moreover, there exists the prospect of much fuller information yet from a planned hemisphere-wide observational system, employing constant-level drifting balloons and interrogating satellites. This system will revolutionise Southern Hemisphere meteorology in the next few years, but in the following discussion we shall be concerned exclusively with the traditional type of upper-air data.

Even before upper-air data for the Australian region began to occupy the thoughts of meteorologists, the original circulation model of Russell underwent substantial changes, which brought it into closer agreement with the polar-front theory then current in the Northern Hemisphere; J. Holmboe, L. Lammers, E. Kidson, and C. E. Palmer were the principal names associated with this development, and a summary of the resulting circulation model can be found in the Southern Hemisphere chapter of the *Handbook of Meteorology* (BERRY et al., 1945). Extensive controversies continued during the 1940's in Australia regarding the nature of the Southern Hemisphere fronts. Even without the classical cloud sequence, these are often marked over land by dust walls (Fig.1) but how should they be drawn over the blank spaces of the southern ocean? The answer is only now being provided by satellite cloud photos.[1] At the time the question had to be shelved,

---

[1] Continuous reception of satellite cloud photographs is carried out at Melbourne and Perth. (Editor)

Fig.1. Example of a dust wall, showing the presence of a front.

and attention turned to the new observations from the aerological networks of Australia, New Zealand, and South Africa, which for the first time revealed details of the upper-flow conditions over a considerable part of the Southern Hemisphere. In the late 1950's, the stations established for the International Geophysical Year in South America and on the Antarctic continent filled further gaps, and only the Southern ocean area continues to this day as an object of meteorological speculation. The density of the existing upper-air station network in the Southern Hemisphere is illustrated by Fig.2 against the outline of the Northern Hemisphere continents; familiarity with the upper-air network there will at once suggest a number of poignant comparisons between the two hemispheres.

In Australia, the first upper-air data go back to January 1931, when regular aircraft soundings up to 5 km started at the Laverton airfield near Melbourne. Similar soundings began near Sydney in June 1937, and near Perth in November 1938. During the war, aircraft soundings were initiated also at Darwin, Townsville and Brisbane, but apart from some postwar weather reconnaissance operations, all meteorological flying came to an end in the early 1940's with the introduction of the American-type radiosonde as the standard aerological tool in Australia.

Initially, the upper-air observations were used exclusively for the analysis of local stability conditions and no attempt was made to construct upper-air synoptic charts. Nevertheless, this possibility had not been overlooked, and provided the incentive for an ambitious enterprise in 1934. During the month of April of that year aircraft soundings were carried out, in addition to the regular flights at Laverton, also at Adelaide (South Australia), Hay (New South Wales), Daly Waters (Northern Territory), and Charleville and Cloncurry (Queensland). The results have never been analysed, but it is now clear that they could have given only a small part of the story. This is because in the free

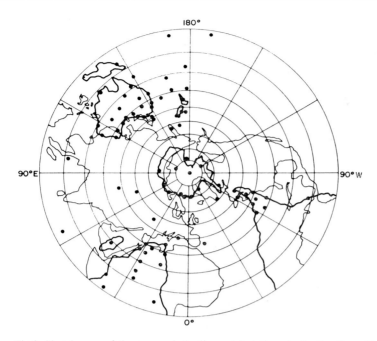

Fig.2. Sketch map of the upper-air (radiosonde) stations in the Southern Hemisphere (black dots). The southern lands are shown by the thick lines; the Northern Hemisphere lands are shown for the purpose of comparing latitudes, but are reversed so that western and eastern sides of land masses may remain homologous. The network of Australian stations, although deficient in the north, is the most satisfactory. (Cf. Fig.8 for rawin stations.)

atmosphere over the Australian region the crucial events occur in layers well beyond the reach of the aeroplanes available at the time.

Apart from being useful operationally, the regular soundings at Melbourne, Sydney and Perth represented a body of novel climatological information which was analysed in detail by LOEWE (1940, 1945) and RADOK (1946). The scope of these data can be demonstrated by the following comparison. In answer to an urgent war-time request, LOEWE (1942) estimated the mean thermal structure of the Australian atmosphere up to

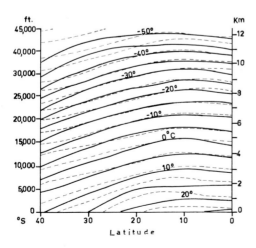

Fig.3. Observed (——) and estimated (– – –) annual mean isotherms over Australia. The only discrepancies are in the low-level 15°, 20°, 25° lines. (Estimated data from LOEWE, 1942.)

a height of 12 km, from the aircraft measurements made up to less than half that height, and from upper air data from stations in analogous climatic regions of the Northern Hemisphere. Fig.3 shows the inferred thermal field, together with that derived from the first five years of radiosonde data for nine stations ranging from Darwin to Melbourne, and a few observations from Macquarie Island (LOEWE and RADOK, 1950). Although there are minor discrepancies the estimated isotherms clearly establish the most important feature of the free atmosphere over the Australian region—its marked meridional temperature gradient. The "thermal wind" equation:

$$\frac{\partial V_g}{\partial z} = \frac{g}{fT}\frac{\partial T}{\partial y} \tag{1}$$

where V_g is the geostrophic wind velocity (identical for climatological purposes with that of the actual wind), g the acceleration of gravity, T = temperature °K, $f = 2\Omega \sin \varphi$ the Coriolis parameter (Ω = angular velocity of the earth, φ = latitude), and z and y are the vertical and meridional coordinates, demands that the strength of westerly winds in such a "baroclinic" region (with temperature changes along isobaric surfaces) must rapidly increase with height. A recent construction of the mean flow field for July over Australia (SPILLANE, 1965) illustrates this effect (Fig.4).

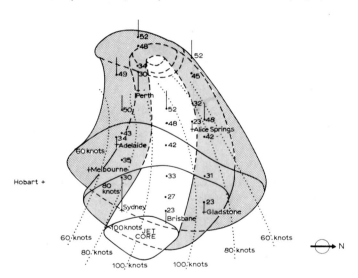

Fig.4. Perspective diagram of the July mean wind field over Australia (1956–1961). The vertical scale is greatly exaggerated, the jet core being more like a tape in cross-section. Day-to-day variations from the mean pattern may be considerable. The isotach surfaces of 60, 80 and 100 knot westerly wind are drawn, and projections on the earth's surface of their latitudinal extremes are shown as dotted lines. The values in lines are the heights in 1,000's of ft. of the isotach surfaces above selected stations. (After SPILLANE, 1965.)

The general significance of baroclinic zones in the free atmosphere was established during World War II by staff members of the University of Chicago (1947) who under the leadership of E. Palmén and C. G. Rossby discovered and discussed the associated narrow, meandering bands of high wind velocity in the upper troposphere, now known as "jet streams". Later work (for a survey, see RIEHL (1962) and for an exhaustive treatment of jet-stream meteorology, REITER (1963)) brought out basic differences between the transient jet streams associated with the polar front and the semipermanent jet streams of the subtropical and subpolar planetary baroclinic zones. A large part of recent meteorological research has been devoted to the dynamics of the jet stream, its role in creating and steering synoptic disturbances, and its contribution to the maintenance of the general circulation. From this point of view, many of the basic meteorological and climatological problems of the Australian region form part of the wider problem of the Southern Hemisphere subtropical jet stream.

The jet stream of the Australian region

The first study of the jet stream in the Southern Hemisphere was made by LOEWE and RADOK (1950) who used up to five years of radiosonde data for eastern Australia to construct mean meridional cross-sections of temperature, isobaric surface heights and zonal-wind velocity, for summer (December–February) and winter (June–August). Similar cross-sections were constructed almost simultaneously from more restricted data for a single meridian (170°E) in the Southwest Pacific by HUTCHINGS (1950) and for the South Atlantic by FLOHN (1950).

For the zonal wind calculations all these studies used the geostrophic equation:

$$V_g = \frac{g}{f} \frac{\partial z_p}{\partial y} \tag{2}$$

where z_p is the height of the surface with constant pressure p and the other symbols have the same meaning as in eq.1. Within the limitations of the implied assumption of straight unaccelerated flow, and those of the observational material (notably the scatter of stations around the nominal meridian of the Australian sections and the different periods of observation), there was complete agreement among these early studies concerning the main features of the mean zonal flow field. These will be illustrated here by cross-sections for the Australian region constructed especially for this survey from the complete upper-air data for Australia (unpublished) and New Zealand (ANONYMOUS, 1963), for the period 1956–1961. These new cross-sections are homogeneous in time and, moreover, refer to almost the same period as the latest published Australian radar wind observations (MAHER and MCRAE, 1964) and an upper-level synoptic chart study of the Australian jet stream (MUFFATTI, 1964).

The mean fields of temperature and zonal flow over the Australian region are shown for midsummer (January) and midwinter (July) in Fig.5 and 6. Here the use of pressure as the vertical coordinate clearly reveals the baroclinicity of the temperature distributions. In summer (Fig.5) the mean flow has a weak jet core near 35°S and a stronger jet is suggested south of 50°S; the precise location of the high-latitude jet requires the use of Antarctic observations which will not be considered here. It should be noted that over the Australian continent, in summer, the temperature field is far from zonal and, therefore, the cross-section approach is then much less adequate than it is in winter. In winter (Fig.6) the core of the jet stream has a speed of nearly 60 m/sec and is situated near the 12 km level in latitude 30°S. A pronounced minimum in the zonal flow intensity appears between 40°S and 45°S; another jet stream in Subantarctic latitudes can be surmised from Fig.6 and has been firmly established by the upper-air observations in Antarctica during and since the International Geophysical Year (PHILLPOT, 1962).

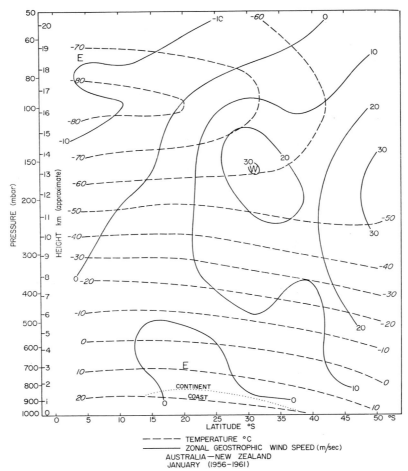

Fig.5. Vertical cross-section of temperature and zonal geostrophic wind in January. Note the position of the continent (thin dotted line) and the altitude reached by the easterlies above it.

The first upper-level mean charts for the Australian region (PHILLPOT, 1951) raised the possibility of significant deviations from the mean-flow patterns of Fig.5 and 6 occurring as a function of longitude in both summer and winter (GIBBS, 1952). This was confirmed by a study of individual 15-day 200 mbar mean charts for a three-year period (RADOK and GRANT, 1957), which showed moreover that time variations are equally important. Apart from the major seasonal change, some periods were found to be marked by strictly zonal flow with a single jet stream, while others had pronounced wave patterns or "split jet" configurations, with a single strong jet in the west and a double jet further east. The transition from the single to the double jet was found to occur at a preferred site for the formation of slow-moving intense "blocking" anticyclones and of mid-troposphere low-latitude depressions which play a large role in the rainfall regime of the Tasman Sea area.

This illustrates the intimate links between high-level flow and synoptic disturbances, which have come to be recognized in recent years and account for the many studies carried out on individual jet streams. In the Australian region such studies were delayed for a while by the lack of high-level wind observations, which are essential for a full analysis, especially in low latitudes. However, for two cases of relatively straightforward

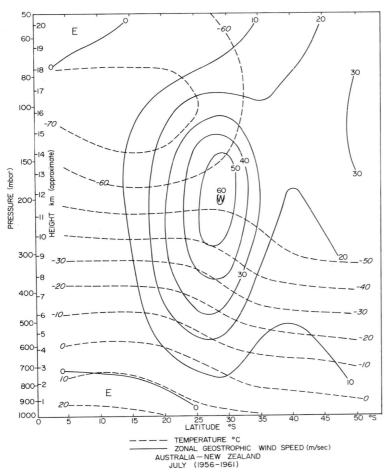

Fig.6. Vertical cross-section of temperature and zonal geostrophic wind in July. The westerly jet core is now much more pronounced than in January, and has moved slightly equatorwards. The easterlies have shifted much more, and only affect the northern half of the continent; they have descended from a top altitude of about 5,500 m to 3,000 m.

winter conditions, RADOK and CLARKE (1958) were able to establish the principal features of individual jet streams from temperature and pressure data alone. Fig.7 shows one of their cross-section series, together with the corresponding surface weather charts on which the positions of the jet core and other regions of dynamic interest have been marked. These show that the distribution of cloudiness is related more closely to the field of the large-scale Richardson number Ri (a parameter measuring the vertical static stability of the air and the rate of increase in wind velocity with height), than to the location of the jet core J or to that of the unstable conditions (I), which on theoretical grounds are suspected to exist on the equatorial side of strong jet streams.[1]

From an analysis of all jet stream occasions for an entire year, RADOK and CLARKE (1958) also established that the strongest jet streams over eastern Australia normally

[1] TRAJER (1967) has devised some useful nomograms, based on Australian experience, for the determination of the jet stream's strength and location relative to known winds to the north and south of it at the level of maximum wind, and for the determination of wind speeds at points in the region around the jet core. (Editor)

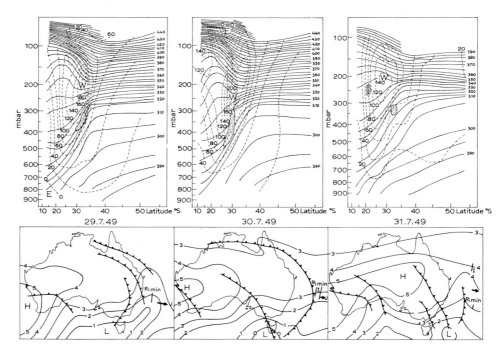

Fig.7. Cross-sections and 1,000 mbar charts for 29–31 July, 1949. Regions of dynamic instability and $Ri < 1$ are shown hatched and stippled in the sections and marked I and Ri_{min} in the charts. Broken lines are geostrophic wind isotachs (miles/h); framed values are observed zonal winds. Isentropes (full lines) are labelled in °K, contours in 10^2 ft. Where high clouds are shown figures give the amount of low cloud (octas). (After RADOK and CLARKE, 1958.)

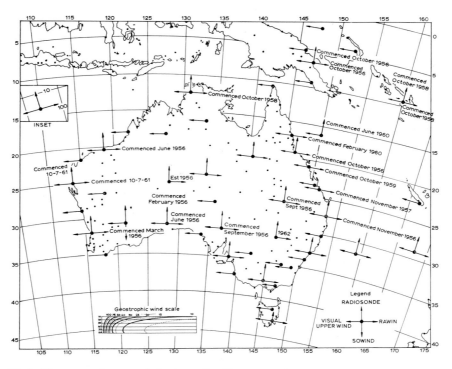

Fig.8. The upper-air network in Australia. The two major gaps correspond to desert regions. (After MUFFATTI, 1964.)

followed the appearance of relatively intense and sluggish anticyclones in the Great Australian Bight, which were subsequently accelerated eastward. By contrast, the weakest jet streams occurred in conjunction with stationary anticyclones over the Tasman Sea. Depressions in low latitudes were found to be associated mostly with strong jet streams; in particular, important cyclogenetic developments just off the Australian east coast followed almost exclusively upon the appearance of unusually intense jet streams over the Australian region.

The detailed study of jet stream situations has been facilitated by a recent increase in radar wind observations, which provide direct information on high-level flow conditions and an improved basis for the construction of upper-air synoptic charts. In New Zealand regular and frequent "rawin" soundings started in 1951. The expansion of the Australian network began a little later and is summarised by Fig.8. This has been taken from a study by MUFFATTI (1964), who used all the upper-air charts for the period 1956–1961 to prepare statistics of daily jet stream features over Australia. Some of Muffatti's results are reproduced in Fig.9–12 to illustrate the changes in jet latitude and speed with longitude and season, and also their variations from year to year.

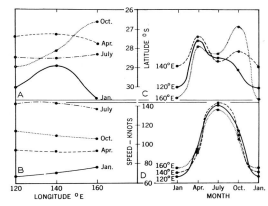

Fig.9. Mean latitude and speed of the lowest latitude 200 mbar jet axis in the westerlies (1956–1961). A. Position by latitude and longitude in January, April, July and October; notice the clustering of latitudinal positions along 140°E and their scattering by 160°E; B. mean speed for the same longitudes and months; C. mean latitude and longitude of the lowest latitude jet axis arranged by month; notice the April–July clustering and the October–January scatter; D. speed at each key meridian, by months. (After MUFFATTI, 1964.)

The mean jet latitude curves in Fig.9 can be interpreted as arising from a seasonal cycle in the length of the jet-stream undulations, from very long waves giving almost zonal flow in autumn and winter, through shorter waves with equatorward flow over Australia towards a trough east of the continent in spring, to a still shorter wave pattern with a mean trough over the eastern section of the continent in summer. The year-to-year variations, however, show that in some years the short mean trough can appear in winter (July, 1959) and the intermediate wave-length meridional flow in summer (January 1959); moreover, the direction of the meridional mean flow can on occasions be poleward (July, 1957). All this confirms the conclusions drawn by RADOK and GRANT (1957) from their 15-day mean 200 mbar charts.

The frequency distributions of jet latitudes in Fig.10 and 11 point to a change in the behaviour of the jet stream as it advances eastward across the continent. Whereas jet streams can cross the west coast with almost equal probability within a wide range of

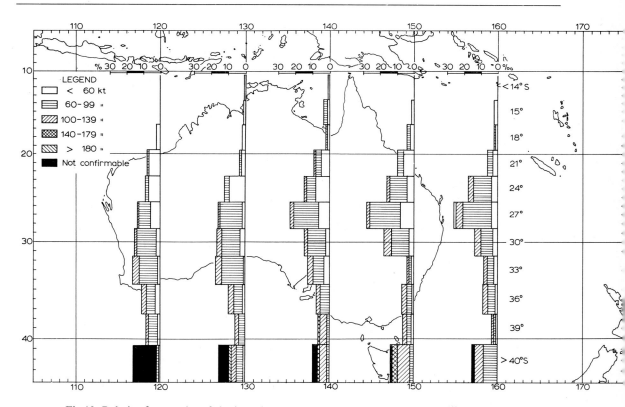

Fig.10. Relative frequencies of the jet axis at 200 mbar in January, 1956–1961. The bimodal nature of frequencies along each key meridian is clear; there is a progressive channelling effect as the jet passes over the mainland, the pattern becoming more peaked. (After MUFFATTI, 1964.)

latitudes, a majority of them are channelled into a narrow zone by the time they reach the Tasman Sea. This zone extends from 26° to 31°S in winter, while in summer the jet frequency has a sharp peak near 27°S.[1]

Both the seasonal features and the year-to-year variations of the jet stream are reflected in the behaviour of the depressions and anticyclones of the Australian region (cf. Chapter 5). For the Tasman Sea, HILL (1964) has traced the connection by means of careful analyses of jet streams associated with widespread steady rain in low latitudes. For the Australian region as a whole, Karelsky has analysed the length of time for which depressions and anticyclones occupied different squares of 5° latitude and longitude in a given month. Published results include monthly and seasonal mean values of these times (termed "cyclonicity" and "anticyclonicity" by KARELSKY (1961); cf. Chapter 5, Fig.7, 13) and of the numbers of systems together with their mean central pressures (KARELSKY, 1965; cf. Chapter 5, Fig.10, 38). Two preliminary papers (KARELSKY, 1954, 1956) included also information for individual months which would be suitable for collation with upper circulation anomalies.

The radar wind observations make it possible to apply an independent check to the mean wind fields derived from pressure and temperature data on the assumption of geostrophic flow. Although the most adequate picture evidently results from combining all available

---

[1] Detailed statistics of the jet stream in Australia giving frequencies, speeds and latitudes for every 10° of longitude have been published by WEINERT in 1968. (Editor)

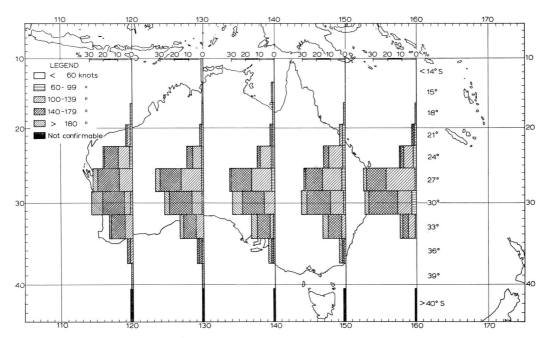

Fig.11. Relative frequencies of the jet axis at 200 mbar in July, 1956–1961. The frequencies are now unimodal, but noticeably skewed compared with the January pattern. (After MUFFATTI, 1964.)

information, the comparison of independently derived flow patterns provides a useful indication of the reliance that may be placed on such constructions. Comparisons will here be made between the zonal flow fields of Fig.5 and 6 and wind velocity isopleths for eastern Australia drawn by H. van Loon (personal communication, 1965) from the data of MAHER and MCRAE (1964). Since the periods of the radiosonde and radar obser-

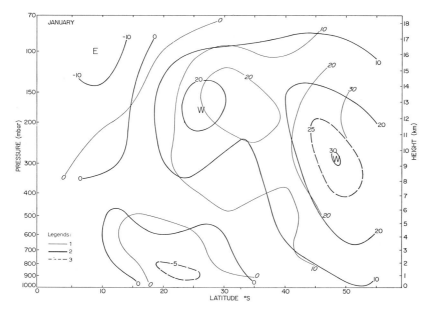

Fig.12. Mean zonal velocity isopleths (m/sec) in January. The thin lines (with italic numbers) show the geostrophic winds (1956–1961), the thick lines (with upright numbers) the radar winds (1957–1961). The geostrophic winds are also shown in Fig.5.

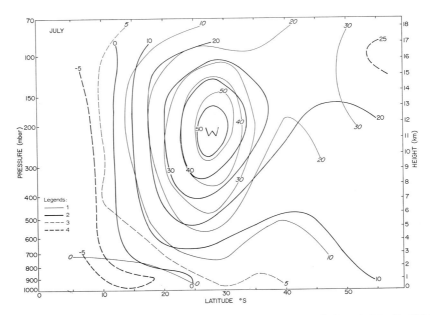

Fig.13. Mean zonal velocity isopleths (m/sec) in July. Same symbols as in Fig.12. (Cf. Fig.6.) The great strengthening of the jet core is very apparent; note its latitude.

vations are almost identical, any differences between the two sets of wind fields must arise from such causes as longitude variations, deviations from geostrophic balance, and subjectivity in fitting the isopleths to the data. Also a slight bias towards weaker winds may have been created in the radar data by range limitations.

Fig.13 shows the mean zonal wind field for July (previously shown in Fig.6) superimposed on the corresponding rawin velocity isopleths. The overall agreement is excellent, even though slight differences in the jet core location give rise to velocity discrepancies of the order of 10 m/sec in the region of strong winds. In low latitudes the differences probably must be attributed to the wider longitude range covered by the geostrophic cross-section; since they concern the direction of the wind, not its speed, Fig.13 provides no reason for doubting the validity of the geostrophic assumption in the lowest latitudes here considered.

The summer cross-sections (Fig.12) agree rather less well than those for winter. This reflects above all the fact that in summer the flow over the Australian region is far from zonal and requires cross-sections in several longitudes or mean charts for its adequate description. Such averaged charts are essential for placing the Australian jet stream into the framework of the general circulation.

The upper flow of the Australian region as part of the general circulation

The preceding discussion of the jet stream over the Australian region showed the existence of two major patterns, prevailing, broadly speaking, during the warm and cold halves of the year. Actually low-latitude (winter) jet streams are sufficiently common throughout the year to leave their mark also on the summer cross-section, and it is the presence or absence of the pronounced zonal-flow *minimum* near 40°S which introduces a major difference between the winter and summer mean flows. As regards other meteorological elements, a detailed climatological analysis of the early Australian radiosonde material (RADOK, 1952) showed that in winter all isopleths tend to follow the parallels, with little or no distortion by the continent being noticeable in the free atmosphere. In summer, by contrast, some effects of convection can be demonstrated over the continent to heights of the order of 5 km (500 mbar), and the majority of meteorological elements have closed isopleths which follow the outline of the coast (see Chapter 6).

Details of the low-level circulation created by the Australian continent in summer were investigated by MORIARTY (1955) who found one particular pressure and flow pattern to prevail, with minor variations, on almost 75% of all occasions. This pattern is reproduced in Fig.14 and shows stationary depressions near the northwestern coast and

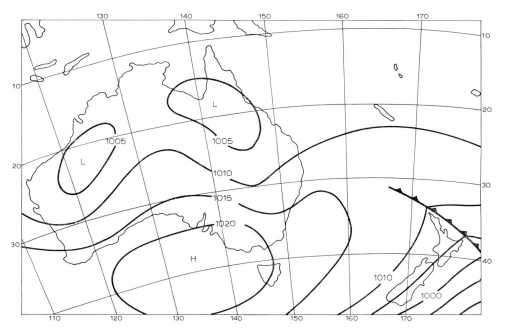

Fig.14. The mean sea-level synoptic charts for 12h00 G.M.T. January 20, 1960. The northwestern heat low (Taylor's Pilbara low) and the northeastern low (Cloncurry low) are well developed; the anticyclone is on a more southerly track than usual. (After DE LISLE and HARPER, 1961.)

over northern Queensland. The first of these depressions was identified by Moriarty as the main continental heat low (Chapter 5, Fig.5, 23) while he ascribed the second (the so-called "Cloncurry low") to orographic factors, i.e., disturbances of the low-level easterlies by the Australian Alps. Support for this interpretation has been provided by DE LISLE and HARPER (1961) who calculated large-scale flow patterns for a two-layer model atmosphere, with surface heating and equatorward rising boundary between surface easterlies and upper westerlies. Their results showed no sign of the Cloncurry low, but gave the observed position for the main heat low.

Hydrostatic considerations demand that a heat low be replaced by an anticyclone in the upper air. On the other hand, a dynamic disturbance in the easterlies might be expected to extend upward through a substantial part of the troposphere. These expectations can be tested by means of upper-level mean-flow charts for the Australian region, such as those constructed by FROST and STEPHENSON (1964). Their 700 mbar chart for

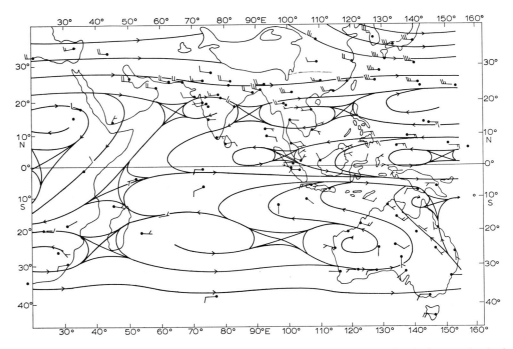

Fig.15. 700 mbar (3 km) streamlines for the Eastern Hemisphere, January. Notice the large anticyclonic flow over most of Australia, the slow speed of the westerlies, the southerly flow which follows the east coast and joins the monsoonal westerlies to form a cyclonic eddy over the Corral Sea. Note that the true monsoonal flow from across the equator does not reach Australia. (After FROST and STEPHENSON, 1964.)

January is reproduced in Fig.15; it shows both the anticyclones above the northwest permanent heat low, and the persistence of a cyclonic eddy at the 3-km level just north of where the Cloncurry low is found at the surface. It appears that this eddy marks the confluence of the southerlies which flow around the mean continental anticyclone with the low-latitude westerlies commonly in evidence north of Australia in summer (Chapter 5, Fig.25).

The origin and nature of these westerlies remains a subject of research. They form part of the complex, large-scale flow system known as the Asian monsoon which includes, as one of its relatively minor offshoots, the Australian monsoon (see Chapter 5, p.82). Studies of the momentum and energy budgets of the Australian monsoon (BERSON and TROUP, 1961; BERSON, 1961) have shown the low-level westerlies to be dynamically linked to the intense easterly stream which simultaneously exists in the upper troposphere over the low latitudes of the Australian region. Independently, FROST and STEPHENSON (1964) have suggested that the two streams form part of a zonal circulation cell over the Indian Ocean. But whatever their origin, the low-level westerlies are the main characteristics of monsoon conditions in northern Australia, where their onset brings the change from sporadic convectional showers to the prolonged rainy spells of "the Wet".

Even if the primary factor behind the Australian monsoon is the winter intensification and southward extension of the Northern Hemisphere Hadley circulation (BERSON, 1961), the strength of the upper westerlies in the Southern Hemisphere represents a second control on the monsoonal processes. KRAUS (1954) and TROUP (1961) have shown that the rainfall in tropical and subtropical Australia is negatively correlated with the

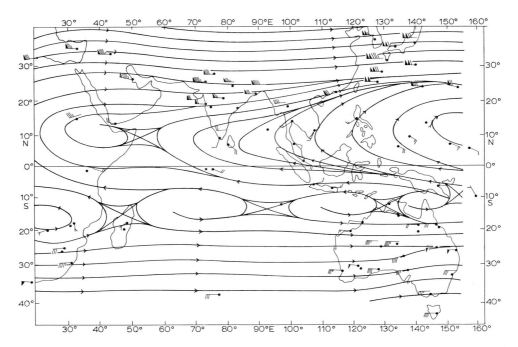

Fig.16. 200 mbar (12 km) streamlines from the Eastern Hemisphere, January. The westerly flow is in exclusive control from 20°S southwards; over tropical Australia there is a weak anticyclonic circulation and further north a moderate flow across the equator. (After FROST and STEPHENSON, 1964.)

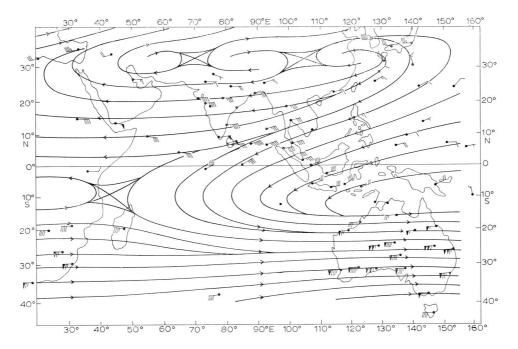

Fig.17. 200 mbar (12 km) streamlines for the Eastern Hemisphere, July. The westerly flow across Australia is greatly strengthened and reaches further north. In the tropical latitudes the westerly stream is part of an enormous transequatorial flow of east Asian and western Pacific origin. (After FROST and STEPHENSON, 1964.)

mean meridional pressure gradient at the 12 km (200 mbar) level. The physical background is provided by the 200 mbar mean flow chart of FROST and STEPHENSON (1964), reproduced in Fig.16. The strength of the upper westerlies is a measure of the intensity of the high-level anticyclonic belt and the associated divergence which controls the vertical extent of the convection in the lower troposphere. Thus the general circulation of both hemispheres plays a crucial part in the summer climate of the Australian region. In the southern winter the intense high-level Asian heat source produces very strong easterlies in low latitudes to the north of Australia and a substantial flow across the equator into the Southern Hemisphere. This is demonstrated by the mean 200 mbar chart for July (FROST and STEPHENSON, 1964) reproduced in Fig.17. Thus again the flow over the Australian region cannot be fully appreciated in isolation from the general circulation of both hemispheres.

In these circumstances it is not surprising to find evidence of a link between the transitions that occur from the winter to the summer pattern in the upper flow of both hemispheres. In the Northern Hemisphere the sudden shift of the jet stream from its winter latitudes south of the Himalayas to its summer latitudes north of the mountains has been established as one, or even the most important, factor controlling the onset of the Indian monsoon (YIN, 1946) and other circulation changes over Asia (ACADEMIA SINICA, 1957); a review of relevant studies has recently been given by HUTCHINGS (1964). For the Australian region, a similar sudden shift of the jet stream was first reported by RADOK and GRANT (1957) who showed that on the occasion in question it coincided with the reverse change in the Northern Hemisphere as a whole. This case and three others are illustrated in Fig.18 where the Southern Hemisphere jet stream latitudes represent 15-day averages and those for the Northern Hemisphere weekly averages. (Acknowledgement is made to the Extended Forecast Section of the U.S. Weather Bureau for these data.) It may be surmised from this diagram that in general there exists a lag between the major flow-pattern changes in the Northern Hemisphere and those which take place over the Australian region. However, the coincidence may well be closer for the Southern Hemisphere as a whole, and we are thus led back to the crucial question of how representative the circulation of the Australian region is of the general circulation of all the Southern Hemisphere.

A partial answer to this question has been provided by the stations established for the International Geophysical Year along the western coast of South America, which enabled SCHWERDTFEGER and MARTIN (1964) to construct reliable cross-sections of zonal flow for that sector of the Southern Hemisphere. With one exception the broad features resemble those of the corresponding Australian cross-sections (Fig.5, 6). The exception is the zonal velocity minimum south of the Australian continent in winter, which forms the main characteristic of the zonal flow field there, yet has no counterpart in the South American sector. The dynamic explanation of this minimum and its possible identification elsewhere on the hemisphere, therefore, must be regarded as the major problem in the present context.

BERSON and RADOK (1960) implicitly tackled this problem with an analysis of 200 mbar height and surface pressure differences for pairs of middle-latitude stations which happen to be available for a few longitudes of the Southern Hemisphere. As a working hypothesis they interpreted changes in the average zonal mean flow, for the latitude bands covered by these stations, as due to meridional displacements of a jet-stream profile

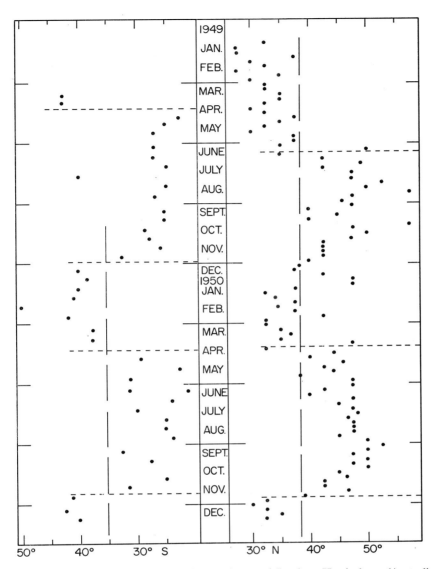

Fig.18. Mean jet stream latitudes in the Northern and Southern Hemispheres (Australian sector).

similar to that established for the Australian region, with due allowance for the general decrease in flow intensity in summer.

The most severe test of this hypothesis arises in the Indian Ocean sector, where cold Antarctic surface water extends into relatively low latitudes. From the upper temperature observations for Amsterdam (38°S) and Heard (53°S) Islands, BERSON and RADOK (1960) surmised a date for the winter to summer transition in the upper flow over the southern Indian Ocean somewhat earlier than that of the corresponding event over the Australian region. However, new wind observations for Kerguelen Island (49°S) reported by BARBÉ et al. (1963) have led VAN LOON (1964) to the conclusion that the zonal flow profile for the Indian Ocean sector resembles that of South America rather than that of the Australian region in its shape and, moreover, that it has peak velocities higher than those found anywhere else over the Southern Hemisphere. Over Kerguelen the upper westerlies remain in fact very strong throughout the year and the only distinguishing feature of the winter profile is that after a slight decrease in speed above the jet-core level the wind increases again with increasing height; this may represent a transition to the high-level Antarctic jet stream described by PHILLPOT (1962).

It is possible that the comprehensive upper wind and temperature surveys for the whole of the Southern Hemisphere now being engineered will add further unexpected features especially in the totally unknown Pacific region. However, these are not very likely to confute the view that the major circulation contrast of the Southern Hemisphere is that between the Indian Ocean sector, with a very strong and seasonally steady middle-latitude jet stream aloft and intense cyclogenetic activity in the lower and middle troposphere, on the one hand; and the Australian region with pronouncedly different winter and summer jet stream patterns aloft and preferred blocking tendencies of part of the lower tropospheric disturbances (VAN LOON 1956).

The reasons for this contrast, and in particular for the winter profile of the high-level zonal flow over the Australian region, with its intense low-latitude jet stream and pronounced velocity minimum near 40°S, remain problematical. VAN LOON (1961) has given a thermal explanation based on the temperature difference between the Australian continent and the relatively warm ocean waters south and southeast of it. However, the zonal character of conditions in the free atmosphere over Australia in winter is hardly compatible with the assumption of a cold source of sufficient intensity to account for the high-level flow features, as well as for the very marked blocking tendencies observed in the entire Australian and New Zealand region. As an alternative, the cause of the blocking may be sought in the upper zonal flow itself and the profile of the latter be derived from basic ideas concerning the formation of jet streams.

Soon after the discovery of jet streams, ROSSBY (1947) suggested that they are the result of large-scale lateral mixing in which the total rate of fluid rotation around the local vertical ("absolute vertical vorticity") is conserved. This hypothesis accounted for the general shape of at least some of the observed meridional profiles of high-level zonal flow. More recently H. Riehl (*World Survey of Climatology*, Vol. II) has shown that even closer agreement is obtained in such cases, by working in a coordinate system following the wave pattern of the mean zonal flow; this has proved a powerful tool for the clarification of the momentum and energy budgets of the jet stream both in nature and in laboratory experiments. However, in its original form Rossby's mathematical result cannot account for flow minima of the type observed in the zonal flow profiles for the Australian region.

The necessary extension of Rossby's argument was made in a little-noted paper by GRANT (1952), who showed theoretically that the mixing process envisaged by Rossby cannot create a jet stream on the equatorward boundary of a polar cap, but will do so when restricted to a zonal belt between two latitude circles. Changing the region of mixing has the effect of adding to Rossby's predicted jet-stream velocity expression:

$$(u/c_E)_{R_o} = \left[\frac{\eta}{(1 + \sin \varphi) \, \Omega} - 1 \right] \cos \varphi \tag{3}$$

where u is the zonal flow velocity, c_E the earth's velocity at the equator, η the (constant) absolute vertical vorticity, Ω the angular velocity of the earth, and φ the latitude, an extra term giving instead:

$$(u/c_E) = (u/c_E)_{R_o} + \frac{B - \eta}{\Omega \cos \varphi} \tag{4}$$

The constants B and η are determined by the initial velocity profile in the zone of mixing, through the requirement that the total vorticity and angular momentum must not change during the redistribution of the vorticity. For $B>\eta$ the velocity profile described by eq.4 has a minimum in middle latitudes and jet streams in high and low latitudes, provided that instability processes are assumed to produce a breakdown of the constant vorticity profile where its meridional gradients reach a critical magnitude. GRANT (1952) showed that the modified constant vorticity profile of eq.4 gives an improved fit to the zonal velocity profiles observed for the Australian region in winter.

The special shape of these profiles may then perhaps be explained by the fact that the Australian region is situated downstream from the Indian Ocean sector, where the most intense atmospheric activity of the entire hemisphere appears to be concentrated. This is probably due to the proximity of major heat and cold sources there; on the available information, it seems unlikely that the distant barrier of the Andes plays a significant part in this connection. The vorticity created in the Indian Ocean sector appears to be re-distributed to become almost uniform in the Australian region. On the short synoptic time scale, this condition takes the form of large-amplitude barotropic systems of slow eastward or even occasionally westward movement. It is relevant in this connection that such "blocking" systems can result directly from strong zonal flow, as a "hydraulic jump" phenomenon (ROSSBY, 1950).

In this way the Australian region in winter takes the character of a backwater region, with features unlike those prevailing over the rest of the Southern Hemisphere. In summer it has already been suggested that the circulation of the Australian region is dominated by the Asian monsoon and by the development of the region into a heat source in its own right. Altogether then the Australian region occupies a special position in the Southern Hemisphere and exhibits special climatic features which will be elaborated in the remainder of this volume.

References

ACADEMIA SINICA STAFF MEMBERS, 1957. On the general circulation over Eastern Asia. *Tellus*, 9: 432–446.

ANONYMOUS, 1963. Summaries of radiosonde data. 1956–1961. *New Zealand Meteorol. Serv., Misc. Publ.*, 119, 737 pp.

BARBÉ, G. D., DUMAS, P. and BALLAY, J., 1963. *Caractères Généraux de la Circulation Atmosphérique au-dessus des Iles Kerguelen*. Comité National Français des Recherches Antarctiques, 7: 21 pp.

BERRY, F. A., BOLLAY, E. and BEERS, N. R., 1945. *Handbook of Meteorology*. Mc-Graw-Hill, New York, N.Y., 1068 pp.

BERSON, F. A., 1961. Circulation and energy balance in a tropical monsoon. *Tellus*, 13: 472–485.

BERSON, F. A. and RADOK, U., 1960. Antarctic surges and the zonal circulation. In: *Antarctic Meteorology*. Pergamon Press, London, pp. 193–216.

BERSON, F. A. and TROUP. A. J., 1961. On the angular momentum balance in the equatorial trough zone of the eastern hemisphere. *Tellus*, 13: 66–78.

DE LISLE, J. F. and HARPER, J. F., 1961. A calculation of the effect of large-scale heat sources on Southern Hemisphere subtropical wind flow. *Tellus*, 13: 56–65.

FLOHN, H., 1950. Grundzüge der allgemeinen atmosphärischen Zirkulation auf der Südhalbkugel. *Arch. Meteorol. Geophys. Bioklimatol., Ser. A*, 2: 17–64.

FROST, R. and STEPHENSON, P. M., 1964. Mean streamlines for standard pressure levels over the Indian Ocean and adjacent land areas. *Proc. Symp. Trop. Meteorol., Rotorua*, pp.96–106.

GAUNTLETT, D. J., 1968. Objective analysis in the stratosphere and upper troposphere. *Australian Meteorol. Mag.*, 16:1–14.

GIBBS, W. J., 1952. Notes on the mean jet stream over Australia. *J. Meteorol.*, 9: 279–284.

GRANT, ALISON M., 1952. A re-examination of the zonal wind-profile under conditions of constant vorticity. *J. Meteorol.*, 9: 439–441.

HILL, H. W., 1964. The weather in lower latitudes of the southwest Pacific associated with the passage of disturbances in the middle-latitude westerlies. *Proc. Symp. Trop. Meteorol., Rotorua*, pp.352–365.

HUTCHINGS, J. W., 1950. A meridional atmospheric cross-section for an oceanic region. *J. Meteorol.*, 7: 94–100.

HUTCHINGS, J. W., 1964. Large scale perturbations in the tropical circulation. *Proc. Symp. Trop. Meteorol., Rotorua*, pp.123–143.

KARELSKY, S., 1954. Surface circulation in the Australian region. *Bur. Meteorol., Meteorol. Study*, 3: 74 pp.

KARELSKY, S., 1956. Classification of surface circulations in the Australian region. *Bur. Meteorol., Meteorol. Study*, 8: 36 pp.

KARELSKY, S., 1961. Monthly and seasonal anticyclonicity and cyclonicity in the Australian region—15 year (1946–1960) averages. *Bur. Meteorol., Meteorol. Study*, 13: 11 pp.

KARELSKY, S., 1965. Monthly geographical distribution of central pressures in surface highs and lows in the Australian region, 1952–1963. *Bur. Meteorol., Meteorol. Summary*, 39 pp.

KRAUS, E. B., 1954. Secular changes in the rainfall regime of southeast Australia. *Quart. J. Roy. Meteorol. Soc.*, 80: 591–601.

LOEWE, F., 1940. Discussion of eight years of aerological observations obtained by means of aeroplanes near Melbourne. *Bur. Meteorol. Bull.*, 27: 38 pp.

LOEWE, F., 1942. The distribution of temperature, pressure, humidity, and density, between sea level and 40,000 ft. in Australia and lower latitudes. *Meteorol. Serv. Res. Repts.*, Ser. 10, 5: 11 pp.

LOEWE, F., 1945. Discussion of seven years of aerological observations obtained by means of aeroplanes near Sydney. *Bur. Meteorol. Bull.*, 33: 47 pp.

LOEWE, F., and RADOK, U., 1950. A meridional aerological cross section in the southwest Pacific. *J. Meteorol.*, 7: 58–56 (Amendments, 305–306).

MAHER, J. V., and McRAE, J. N., 1964. Upper wind statistics, Australia: surface to 55,000 ft., January 1957–May 1961. *Bur. Meteorol., Meteorol. Summary*, 108 pp.

MAINE, R., 1966. Automatic numerical weather analysis for the Australian region. *Bur. Meteorol., Meteorol. Study*, 16: 156 pp.

MAINE, R. and SEAMAN, R. S., 1967. Developments for an operational automatic weather analysis system in the Australian region. *Australian Meteorol. Mag.*, 15: 13–31.

MORIARTY, W. W., 1955. Large-scale effects of heating over Australia, I. The synoptic behaviour of the summer low. *Div. Meteorol. Phys., C.S.I.R.O., Tech. Paper*, 7: 31 pp.

MUFFATTI, A. H. J., 1964. Aspects of the subtropical jet stream over Australia. *Proc. Symp. Trop. Meteorol., Rotorua*, pp.72–88.

PHILLPOT, H. R., 1951. Mean constant-pressure charts for Australia. *Bur. Meteorol. Res. Rept.*, unpublished.

PHILLPOT, H. R., 1962. Mean westerly jet streams in the Southern Hemisphere. In: H. WEXLER, M. J. RUBIN and J. E. CASKEY JR. (Editors), *Antarctic Research—Geophys. Monographs*, 7: 128–148.

PHILLPOT, H. R. and REID, D. G., 1953. Upper air temperatures for the Australian Region. *Bur. Meteorol. Bull.*, 42: 47 pp.

RADOK, U., 1946. Discussion of four years of aerological observations obtained by means of aeroplanes near Perth. *Bur. Meteorol. Bull.*, 37: 22 pp.

RADOK, U., 1952. *A Study of Meteorological Conditions in the Free Atmosphere over Australia*. Ph. D. Thesis, University of Melbourne, unpublished.

RADOK, U. and GRANT, ALISON M., 1957. Variations in the high tropospheric mean flow over Australia and New Zealand. *J. Meteorol.*, 14: 141–149.

RADOK, U. and CLARKE, R. H., 1958. Some features of the subtropical jet stream. *Beiträge zur Physik der Atmosphäre*, 31: 89–108.

REITER, E. R., 1963. *Jet Stream Meteorology*. University of Chicago Press, Chicago, Ill., 515 pp.

RIEHL, H., 1962. Jet streams of the atmosphere. *Dept. Atmosph. Sci., Colo. State Univ., Tech. Rept.*, 118 pp.

ROSSBY, C.-G., 1947. On the distribution of angular velocity in gaseous envelopes under the influence of large-scale horizontal mixing processes. *Bull. Am. Meteorol. Soc.*, 28: 53–68.

ROSSBY, C.-G., 1950. On the dynamics of certain types of blocking waves. *J. Chinese Geophys. Soc.*, 2: 1–13.

RUSSELL, H. C., 1893. Moving anticyclones in the Southern Hemisphere. *Quart. J. Roy. Meteorol. Soc.*, 19: 23–34.

SCHWERDTFEGER, W. and MARTIN, D. W., 1964. The zonal flow of the free atmosphere between 10°N and 80°S in the South American sector. *J. Appl. Meteorol.*, 3: 726–733.

SPILLANE, K. T., 1965. The winter jet stream of Australia and its turbulence. *Shell Aviation News*, 330: 15–19. (Reprinted in *Australian Meteorol. Mag.*, 16: 64–71.)

TRAJER, F. L., 1967. Subtropical jet stream nomograms. *Australian Meteorol. Mag.*, 15: 161–165.

TROUP, A. J., 1961. Variations in upper tropospheric flow associated with the onset of the Australian summer monsoon. *Indian J. Geophys.*, 12: 217–229.

UNIVERSITY OF CHICAGO STAFF MEMBERS, 1947. On the general circulation of the atmosphere in middle latitudes. *Bull. Am. Meteorol. Soc.*, 28: 255–280.

VAN LOON, H., 1956. Blocking action in the Southern Hemisphere. *Notos*, 5: 171–178.

VAN LOON, H., 1961. Charts of average 500 mbar absolute topography and sea level pressure in January, April, July, and October. *Notos*, 10: 105–112.

VAN LOON, H., 1964. Midseason average zonal winds at sea level and at 500 mbar south of 25 degrees south, and a brief comparison with the Northern Hemisphere. *J. Appl. Meteorol.*, 3: 554–563.

WEINERT, R. A., 1968. Statistics of the subtropical jet stream over the Australian region. *Australian Meteorol. Mag.*, 16: 137–148.

YIN, M. T., 1964. A synoptic-aerologic study of the onset of the summer monsoon over India and Burma. *J. Meteorol.*, 6: 393–400.

Chapter 4

Australian Climatic Factors

J. GENTILLI

Introduction

Climate is a complex result of many factors which have their origin outside the earth (astronomic) or in or on the earth (telluric). Telluric factors may have their origin in the air (atmosphere), on the surface of the ground (topographic), in the ocean (oceanic), or inside the earth itself (plutonic). Actual interrelationships defy such neat theoretical classifications, e.g., the fundamental factor of terrestrial climates is solar radiation (astronomic) which is intercepted, reflected and diffused by clouds (atmospheric), desert dust (topographic), and occasionally by volcanic dust (plutonic). But the desiccation which causes desert dust is usually due to both evaporation (caused by solar radiation, astronomic) and the advection of dry air (atmospheric).

Some climatic factors, or at least some aspects of them, are also part of the elements of climate. Thus, the duration and intensity of the sunshine is also an element of the climate of any locality and the direction, force, frequency and regularity of the wind also find a place among such elements. These interrelationships and dual or multiple roles of climatic factors and elements are part of the realities of climatological research and must be constantly tested and verified.

Climatic factors are basically the same everywhere, and mostly differ from place to place only in some quantitative aspect. For this reason, climatic factors will be mentioned in this chapter only insofar as their specifically Australian characteristics, manifestations, interrelations, or any other aspect make them unusual, peculiar or otherwise distinctive. The small latitudinal span of Australia allows only a moderate variety in the seasonal rhythm of solar and thermal climates, hence the relatively modest importance of day length as a factor. On the other hand, the greater zonal (longitudinal) span causes considerable modifications in the air masses that travel over the surface of the land, so that the degree of continentality is greater than might be expected. As BARTON (1895) aptly put it, "Australia occupies, in the drier latitudes of its hemisphere, a longer extent than either of the two other continents. Between the rising of the summer sun from the Pacific horizon to his setting behind that of the Indian Ocean, he has poured down his mostly unclouded rays at an angle but little removed from the vertical, over a solid and unbroken land surface 2,000 miles long by at least 1,000 miles wide, nearly as much as the corresponding regions of Africa and South America put together."

In the winter, the cooling of the land mass causes an increase in the density of the overlying air, and a consequent alteration in the barometric pattern (Chapter 6, Fig.2).

Solar radiation

Latitude and solar angles

The low mean latitude of Australia results in a very intense insolation with a high angle of incidence, low photo-periodism and frequent high temperatures. South of Cape York, at 10°30′S, nearly half the continent is in the infratropical zone. The extratropical part reaches only as far south as 29°8′S on Wilson's Promontory on the mainland, and 43°40′S on Maatsuyker Island which is the southernmost exposed part of the continental shelf. The high angle of incidence of the sunshine over so great a part of Australia is widely known. It is less known that this phenomenon had a noticeable effect on early Australian architecture, causing the addition of wide verandahs to most one-story buildings. The usefulness of the verandah in the infratropical latitudes is undoubted, provided, of course, that free air circulation is allowed between the roof and the ceilings. On the other hand, in extratropical latitudes, little or no provision was made for roof-space ventilation, and the absorption of solar heat through the roof and ceiling is very high. This retention of the verandah in extratropical latitudes, for which other reasons were at times adduced, does not seem justified (MARSHALL, 1955).

The angle of elevation of the midday sun at typical Australian localities is so great that the exclusion of sunshine from the windows can easily be achieved for infratropical locations and for the high-sun period at extratropical locations by shading the walls or windows affected (COMMONWEALTH HOUSING COMMISSION, 1944). The COMMONWEALTH EXPERIMENTAL BUILDING STATION (1952) designed an apparatus, the solarscope, which permits the exact reproduction of the angle of incidence of the sunshine for any locality of known latitude at any time of the year. PHILLIPS (1951) designed solar charts and a shadow-angle protractor which enable the user to compute the direction of the sun's rays at any hour of any day for latitudes 5°, 12°30′, 20°, 27°30′, 32°30′, 35°, 37°30′, 42°30′S. These charts and the protractor make it easy to determine the extent of the interception of solar rays and projection of shadows by eaves and screens, as well as to calculate the times of sunlight penetration through windows and other openings. In an appendix to this work he also describes the solarscope. Detailed data derived for Perth were published by GENTILLI (1954). MARSHALL (1955) noticed in South Australia, that only in very recent years did a few architects and householders take advantage of solar planning. The recent advent of air conditioning for large office buildings has had a deleterious effect because it has made solar planning less necessary. It may be said that in Australian cities, air conditioning now covers (at great cost) a multitude of architectural sins, among which lack of solar planning is usually predominant.

The C.S.I.R.O. Divisions of Building Research and Mechanical Engineering have now developed a digital computer programme which produces drawings of any building or any group of buildings for any solar angle, so that shadow projection and screening may be obtained for any latitude and time of the year.

Duration of sunshine

The duration of daylight does not vary much in Australian latitudes: at 40°S, the longest day is just over 15 h, and the shortest, 9 h 20 min (see Table I). The duration of sunshine

TABLE I

DURATION OF DAYLIGHT AT SELECTED LATITUDES[1]

Latitude	Length of astronomical day on the 21st day of each month																							
	Jan.		Feb.		March		April		May		June		July		Aug.		Sep.		Oct.		Nov.		Dec.	
	h	min	h	min	h	min	h	min	h	min	h	min	h	min	h	min	h	min	h	min	h	min	h	min
10°S	12	37	12	23	12	08	11	51	11	38	11	33	11	38	11	50	12	07	12	23	12	37	12	42
20°S	13	09	12	41	12	08	11	35	11	07	10	55	11	06	11	34	12	07	12	41	13	09	13	20
30°S	13	45	13	03	12	10	11	17	10	33	10	13	10	30	11	14	12	08	13	01	13	45	14	04
35°S	14	07	13	15	12	11	11	07	10	12	9	48	10	10	11	03	12	09	13	13	14	07	14	30
40°S	14	32	13	28	12	12	10	55	9	49	9	20	9	46	10	51	12	09	13	27	14	31	15	01

[1] Data from various issues of *Astronomical Ephemeris*, H. M. Stationery Office, London.

TABLE II

DURATION OF SUNSHINE AS A PERCENTAGE OF DURATION OF DAYLIGHT[1]

Locality	Jan.	Feb.	Mar.	Apr.	May	June	July	Aug.	Sept.	Oct.	Nov.	Dec.
Brisbane	*51*	57	57	63	**63**	**62**	**65**	**70**	**68**	**65**	60	59
Perth	**75**	**75**	**73**	**68**	55	48	51	54	60	62	**69**	**74**
Sydney	54	53	*53*	55	56	53	60	63	60	57	53	53
Adelaide	68	70	65	54	47	43	42	49	52	55	61	66
Canberra	59	55	59	60	50	43	47	58	59	59	58	59
Melbourne	55	56	*53*	45	*42*	*35*	*38*	*42*	*45*	*44*	*44*	*48*
Hobart	53	*52*	*53*	47	45	43	46	49	50	46	50	*48*

[1] Bold-face figures indicate maxima, italics indicate minima.

is considerably less because of the curvature of the earth's surface and interception by obstacles at sunrise and sunset, and by clouds at any time (see Table II). Perth and Sydney, at similar latitudes, have about 10 h of daylight at the winter solstice, and 14 h at the summer solstice. At the winter solstice, Perth usually has less than 5 h of sunshine while Sydney has more than 5 h. At the summer solstice, Perth has over 10 h and Sydney over 7 h, as a result of frequent cloudiness. Perth is the sunniest capital city from November to April, Melbourne the least sunny one from March to December. Fig.1 and 2 show the hours of sunshine experienced in January and July, respectively. Fig.3 shows

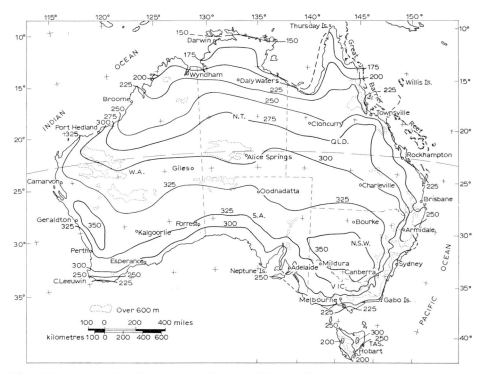

Fig.1. Mean hours of sunshine, January. Because of the cloudiness caused by the monsoonal situations affecting northern Australia, the greatest duration of sunshine occurs further south, in the two continental lobes. (Simplified from *Climatological Atlas*.)

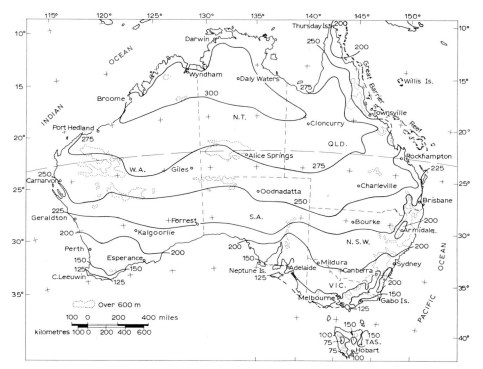

Fig.2. Mean hours of sunshine, July. This is the time of greatest cloudiness in the south of the mainland and in Tasmania, while the trade-wind flow takes clear continental air to the northwestern area. Hence, north of the tropic, the west coast is sunnier than the east coast. (Simplified from *Climatological Atlas*.)

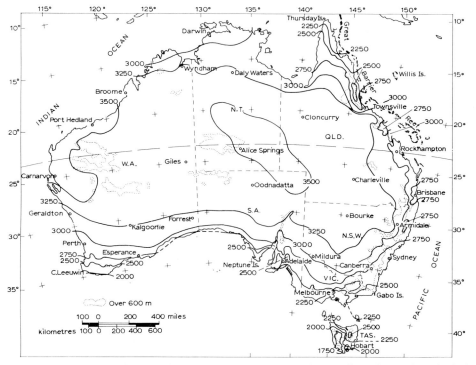

Fig.3. Mean number of hours of sunshine per year. The combined effect of the mid-latitude fronts in the south, the inter-anticyclonic front in the east and the monsoonal developments in the north increases cloudiness and reduces the duration of sunshine. (Simplified from *Climatological Atlas*.)

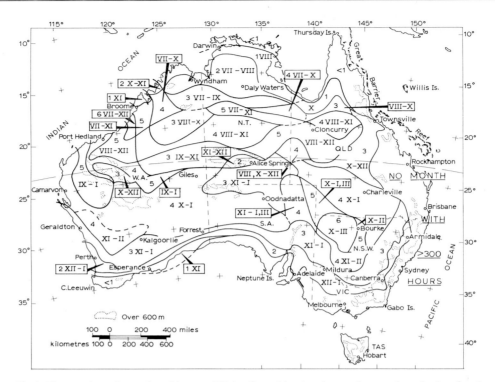

Fig.4. The number of months with over 300 h of sunshine has been taken as the criterion for the delimitation of Australia's sunniest regions, which are the northwest of Western Australia, and inland New South Wales. The roman numerals show the actual months.

the yearly duration of sunshine, also in hours. Fig.4 shows the mean frequency of sunny months. (See Table III.)

More than 80 Campbell-Stokes sunshine recorders are now in use.

A series of maps showing the duration of sunshine for each month and for the year was published in 1953 (COMMONWEALTH METEOROLOGY, 1953).

TABLE III

FREQUENCY OF DAYS WITH GIVEN SUNSHINE
(After KEOUGH, 1951)

Locality	Number of days with hours of sunshine					
	of 90 summer days			of 92 winter days		
	>9	>6	>3	>9	>6	>3
Brisbane	43	60	73	48	70	80
Perth	70	82	88	13	37	62
Sydney	39	57	70	28	59	73
Adelaide	52	68	78	8	30	57
Melbourne	41	61	76	3	23	49
Sale (Vic.)	45	64	78	5	39	66
Hobart	39	59	76	3	33	62

Intensity of sunshine

The measurement of the caloric intensity of insolation at Sydney was the object of experiments by Russell in the last century (RUSSELL, 1885), but the research was not pursued any further at the time.

Since 1927, data on the intensity of solar radiation had been obtained at the Mount Stromlo observatory near Canberra (RIMMER and ALLEN, 1950). PRESCOTT (1940) computed an equation adapted from ÅNGSTRÖM's (1924) formula, for the estimate of the intensity of solar radiation from the actual duration of sunshine received. The equation postulated was:

$$Q = Q_A \left(a + b \, \frac{n}{N} \right) \quad \text{or:} \quad \frac{Q}{Q_A} = a + b \, \frac{n}{N}$$

where Q = radiation actually received; Q_A = total radiation that would be received if the atmosphere were perfectly transparent (Angot's values, "airless earth"); n = actual duration of sunshine; N = maximum possible duration of sunshine; a and b are empirically determined coefficients.

BLACK et al. (1954) applied the formula to additional data from Mount Stromlo and Dry Creek (see Table IV). The values found for Q/Q_A and n/N are, among others, a

TABLE IV

RELATIONSHIP OF SOLAR RADIATION AND SUNSHINE
(After BLACK et al., 1954)

Locality	Period	Mean values		Constants		Correl.
		Q/Q_A	n/N	a	b	(r)
Mt. Stromlo	1928–1939	0.593	0.631	0.25	0.54	0.89
Dry Creek	1947–1950	0.600	0.591	0.30	0.50	0.95

function of latitude and cloudiness. In fact, ALBRECHT (1955) pointed out that values of sunshine directly obtained from sunshine recording apparatus are much more suitable for this kind of estimate than those obtained from observations of cloudiness, usually exaggerated by the inclusion of thin cirri in the cloud recordings. The "optical thickness" of clouds was studied by ALBRECHT (1957) for the purpose of assessing its effect on the incoming solar radiation.

The effect of different types and amounts of cloud on insolation at Adelaide was studied by MAINE (1958). Low clouds reduce insolation by about 5% for each okta covered, up to 5 oktas; greater cloudiness has a more pronounced effect, to a maximum of about 8% per okta for total cloud cover. The opacity of high clouds (for low cloud being 1) is about 0.9 for As, 0.7 for Ac, and 0.3 for Cs.

Measurements of the intensity of radiation were begun in 1947 by I.C.I. Alkali (Australia) Pty. Ltd., at Dry Creek (Osborne, South Australia). In 1950, F. Albrecht, working for the Bureau of Meteorology and the Department of Meteorology of the University of Melbourne, established the main station of the Australian radiation network at Box Hill (13 km southeast of Melbourne), and other stations in 1952 at the airports of Darwin, Alice Springs, Perth, Garbutt (Townsville, Queensland) and Williamstown, New South Wales (ALBRECHT, 1954). In 1955, observations began at Rabaul, New Guinea. In April 1957, the Box Hill instruments were moved to East Kew, Melbourne. The measurements were made by Robitzsch bimetallic actinographs (LOEWE, 1956).

Values of the net gain of radiant energy by a horizontal surface, in cal./cm² h, were calculated by BONYTHON et al. (1955) from the duration of sunshine.

LOEWE (1956) noted that Darwin has no day with more than 700 cal./cm². Garbutt (Townsville) has 30, Alice Springs 56, Guildford (Perth) 64, Williamtown 29, Box Hill (Melbourne) 27. As to days with more than 750 cal./cm², Guildford has 35, Alice Springs 21, Box Hill 12, Williamtown 6, Garbutt 2, Darwin 0. The cloudiness and high moisture content of the air on the eastern and northern coasts reduce the incoming radiation.

At Darwin more than half of all days receive from 451–550 cal./cm². At Alice Springs and Garbutt the frequency distribution is bimodal. At Alice Springs the two peaks are at 351–450 (winter insolation) and at 601–700 cal./cm². At Garbutt they are closer together, at 401–450 and 551–650 cal./cm². At Guildford, the distribution is trimodal, with peaks at 251–300, 451–550, and 701–800 cal./cm², corresponding to the typical weather of winter, early spring (September–October) plus early autumn (March), and summer, respectively. At Box Hill, the frequency is more diffuse; there is a faint trimodal pattern but the lower amounts of insolation (below 300 cal./cm²) are more frequent, and the higher ones (above 600) much less so.

The absolute maxima reach 825–830 cal./cm² and have been recorded at Guildford (Perth), Box Hill (Melbourne) and Garbutt (Townsville) (LOEWE, 1956).

HOUNAM (1963) used the data from the same series of observations, and after statistica tests developed the equation:

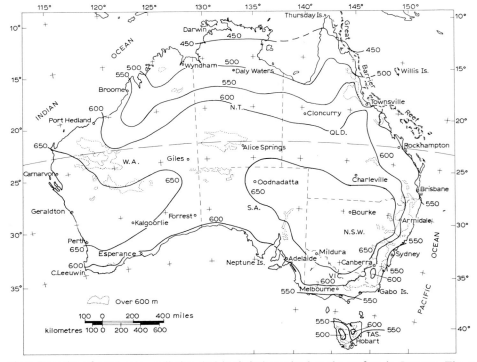

Fig.5. Intensity of total solar radiation (cal./cm² day) received at the surface in January. The map is based on a few actinometric stations and on cloudiness records. (After HOUNAM, 1963.)

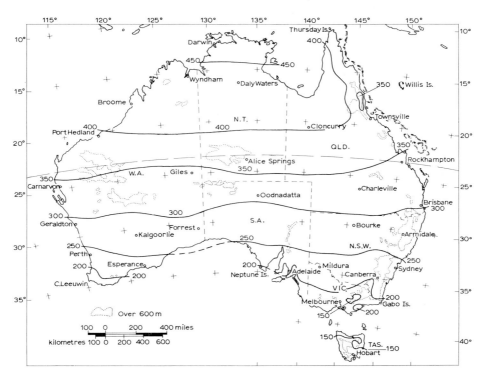

Fig.6. Intensity of total solar radiation (cal./cm² day) received at the surface in July. The pattern is essentially zonal, with some depletion along the trade-wind coast north of the tropic, and towards the southwest where the mid-latitude fronts are more frequent at this time of the year. (After HOUNAM, 1963.)

$$\frac{Q}{Q_A} = 0.26 + 0.50\,\frac{n}{N}$$

His maps, slightly simplified, are shown in Fig.5 (January) and Fig.6 (July). They are the best maps of estimated incoming radiation available so far for Australia, but it should be noted that the coefficients in the equation are based upon the average of 458 monthly values obtained at 6 stations, with observations ranging from 3 to 10 years. Coefficient a (mean value 0.26) varied between -0.22 and $+0.72$, and coefficient b (mean value 0.50) between -0.07 and $+1.28$. The statistical variability was reduced by taking yearly values instead of monthly ones.

The maps show the effect of monsoonal clouds in January in the north, while the dry areas in the south receive abundant insolation, and in July by contrast, the simple latitudinal arrangement, from the dry and sunny north to a cloudier south, with a much lower angle of incidence of the sunshine.

The great intensity of the sunshine available on the mainland during most of the year makes it economic to install *solar heaters*, of which several models are available, mostly based on specifications by the Commonwealth Scientific and Industrial Research Organisation (C.S.I.R.O.). The installation is mostly on the northern slope of the roof (usually inclined at 30°). In tests conducted at Perth, where June is one the of cloudiest months, out of 30 June days, there were only 8 during which the temperature of solarly heated water had to be boosted, for a total period of 24 h and 30 min. The remaining 22 days had a mean air temperature of 17.5°C; the average temperature of the solarly heated water was 45.9°C at noon, and by 17h45 it had only fallen to 44°C. It may be added that of these 22 days, 8 had showers of rain and another 6 were partly or temporarily cloudy. Another practical application of solar heating devised by C.S.I.R.O. (1963) is the *desalination of water*, achieved by evaporating the salt water under a sheet of transparent plastic. The vapour condenses on the sloping plastic and the drops run down the side to a suitable container.[1]

By the end of 1968 there were 5 solar still plants in operation, with glass replacing the plastic. The largest one, at Coober Pedy in South Australia (29°00′S 134°40′E), consists of 76 still units for a total surface of 3,530 m²; it treats water drawn from a bore, with a salinity of 24,000 p.p.m., at a rate of over 20,500 l per day in fine hot weather, decreasing to under 8,000 in cool cloudy weather. Three solar still batteries, of less than 630 m² each, are in Western Australia, and one in new South Wales (AUSTRALIAN WATER RESOURCES COUNCIL, 1968).

Impurities in the air

The intensity and quality of insolation is affected by dust and other impurities in the atmosphere. In the dry conditions which abound in Australia, the unusually high transparency of the air may be greatly reduced by dust storms; dust from Australia, carried by the westerlies, has even been known to settle as far away as New Zealand. Dust haze is not rare over some considerable distance off the northwestern coast, in the belt of tropical easterlies. Watery mist, on the contrary, is restricted to the littoral and to topographic depressions. Air pollution from industries has not been a serious problem so far, except over very small areas (see KIDSON, 1925, for early Melbourne data); pollution by automobile exhaust fumes is noticeable in the cities. The fact that such pollution has not reached dangerous proportions is probably due to the frequently low moisture content of the air and the rarity of low-level inversions. Smoke haze is very frequent in summer because of bushfires, which are usually localized in the more closely settled regions, but may spread to hundreds of square kilometres in the outback (LUKE, 1961), as can be seen in the air photographs. BIRCH (1968) observed that smoke trails may be dense enough to appear clearly on the radar screen.

Extraordinary red sunsets were observed in 1883 after the great Krakatoa eruption, but the then Government Astronomer of Victoria failed to see their significance (ELLERY, 1884).

The 1963 eruptions of the Agung volcano in Bali, ejected large quantities of very fine dust which by early April had reached the northern Australian skies, and was over the whole mainland by May, affecting insolation during several months. The colour of the sky, usually a very clear blue, altered to a distinctly milky blue. A solar "Bishop's ring" of 42° diameter was observed at Albany, Western Australia (ROSS, 1963) and a lunar halo 30° wide at Canberra; the sun's "meteorological corona" extended some 30°–40° from the disc (HOGG, 1963). Vividly red sunsets and sunrises were exceptionally frequent over some months, and at times the reddish glow extended well past the zenith. The

---

[1] In 1963, allowing for reasonable interest and depreciation, water produced by this method cost about $A 0.44 (approx. U.S.$ 0.50) per 1,000 l.

TABLE V

EXTINCTION OF VERTICAL LIGHT BY SCATTERING AND ABSORPTION; DATA
FOR MOUNT STROMLO
(After Hogg, 1963)

Wavelength (Å)	Normal extinction	Extinction June–July 1963	1963 as % of normal
3,600	0.54	0.91	168
4,400	0.26	0.64	246
5,400	0.18	0.48	267
6,700	0.12	0.32	267
8,000	0.08	0.27	337

red colour appeared well before sunset and its brightness was, therefore, far greater than usual. The extent of diffusion of the light was very great, and twilight was prolonged. Hogg and Westerlund (Hogg, 1963) measured the abnormally high extinction coefficient of the light received at Mount Stromlo in June–July, 1963, for typical wavelengths (Table V). The dust content of the air continued to increase until about the end of August, but from early August to October there were no twilight displays because the dust-haze had caused "the almost complete extinction of the sun long before it had set" (Bigg and Miles, 1964).

A complete survey of Australian observations of volcanic dust from the first eruption was carried out by Weinert (1967), who also gives an isochronic map of the spread of related optical phenomena (exactly one month from the first sighting at Broome to the last one at Hobart).

The nature of the surface

The nature of the ground has a dual effect as a factor of climate, because it reduces and modifies solar radiation through its albedo and it deflects and modifies air streams.

Albedo and the heat balance

Estimates of albedo in Australia must, for the time being, be based on the application of European values (e.g., Gayevskii, 1953; Landsberg, 1960) to the vegetation, soil and crop regions. There are observations by De Vries (1956) at Deniliquin, New South Wales, which would give the same albedo (0.23) to dry land and irrigated pasture, but more thorough tests are required in the different environments. Satellite photographs will give exact data.

The most important difference between the European and the Australian environments is due to the much sparser vegetation of Australia which leaves far more of the soil exposed and contributes far less humus. Most soils, in consequence, are paler than European soils. In the more open vegetation, the sunshine can penetrate more readily because of its higher angle of incidence. European values have, therefore, been revised upwards. Another factor of outstanding significance in Australia is the seasonality of

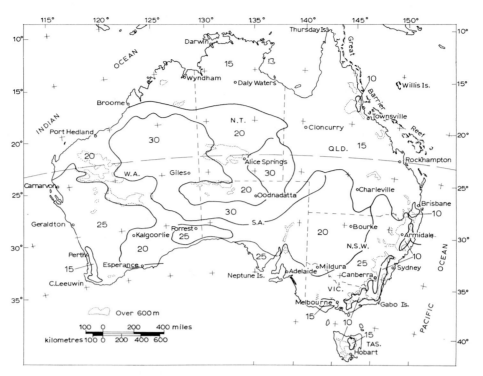

Fig.7. Estimated albedo of the surface, in percentage of the light received, in summer.

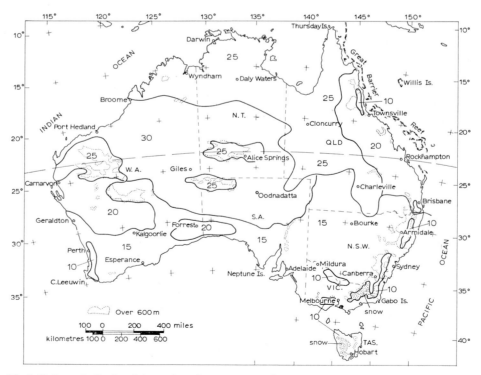

Fig.8. Estimated albedo of the surface, in percentage of the light received, in winter.

the rains, leading to rapid and drastic changes in the plant cover, except where there are trees, which in Australia are very nearly all evergreen. Fig.7 and 8 show the estimated albedo in summer and winter, respectively.

The northern savannas have a sparse tree cover, the trees are small, the grasses are only green in the summer, and much bare ground is exposed in the winter. The estimated summer albedo of 15 is, therefore, increased to 25 in the winter. South of the savanna are various types of desert, for which an average albedo of 30 has been estimated, in view of the large areas of bare sand or stones, to some extent reflecting a little less than higher-latitude deserts because of the predominant red colour.

The southward limit of summer-growing grasses varies from year at year, but in the main it has been placed just south of the tropic in the west, and at nearly 30°S in the east. The summer albedo is reduced accordingly. The wheat-belt further south undergoes the opposite change and probably has a summer albedo of 25, partly because of the pale red-brown earth, partly because of the bleached stubble. The Nullarbor Plain, with silvery Chenopodiaceae, on calcareous red soil on a perfectly flat surface, has a slightly higher albedo.

The southern forest may have an average albedo of 15, probably 10 on the very humid southeastern highlands. The northeastern forests, maintained by orographic mist, are hidden by clouds which greatly increase the albedo.

In the winter, the northern half of Australia is dry, its grasses have all but disappeared, and the albedo there is much higher. The southern half is green with grasses, crops and young leaves, and the average albedo would not be much over 15, except on the Nullarbor Plain and in the mulga areas, where bare soil is always conspicuous.

Only a very small area of the Australian Alps and Tasmanian Plateau is blanketed with snow in the winter months, and its surface is far from uniform and smooth.

The black-earth region of Queensland and the volcanic plains of Victoria, because of the darker soil surface, have a lower albedo than the surrounding land.

HOUNAM (1963), using BRUNT'S (1939) equation:

$$Q_b = 1440 \, \sigma \, Ta^4 \, (0.47 - 0.067 \, \sqrt{e_d}) \quad \left(0.1 = 0.9 \, \frac{n}{N}\right)$$

where Q_b is the outgoing radiation in cal./cm² day; e_d, the mean surface vapour pressure in mbar; Ta, the mean air temperature in °K; σ, Stefan's constant, calculated the outgoing radiation over Australia. The hottest and least cloudy areas have the greatest heat loss, as is to be expected.

In the same paper, Hounam, assuming an albedo value of 0.23 for the whole of Australia, computed the net radiation received by multiplying the total radiation by the albedo and then subtracting the outgoing radiation.

The effect of topography

Lack of high mountains is one of the outstanding characteristics; most of the immense Western Australian Plateau only rises to 300 m and most of the eastern highlands hardly reach 1,000 m. The name "Great Dividing Range" is now discounted because there is no major "range", but rather a multitude of tablelands, cuestas and ridges of the most

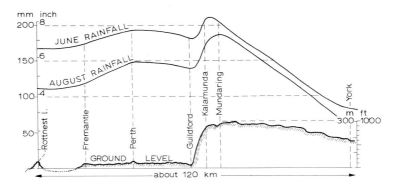

Fig.9. Effect of the Darling Scarp on the rainfall in June (rains coming more north of west) and August (rains coming more south of west). The section is taken from Rottnest Island in the west through Fremantle and Perth to York in the east, a distance of some 100 km. The ground, shown by the thick line (scale on the right) rises abruptly east of Guildford, which lies some 8 m below the mean level of the coastal plain. The rainfall increases even between Rottnest and Fremantle. There is a slight depletion at Guildford, and a sharp increase just above the scarp (which is only 300 m high) followed by steady and marked depletion further east.

Fig.10. Cumulus mammatus, east of the Mount Lofty Range, South Australia. (Photo B. Rofe.)

varied geological origin and so low that they hardly "divide" anything, except where a steep scarp rises near the coast, as is the case with the Blue Mountains and the New England Tableland. Certainly there are topographically induced local climates, and the altitude is sufficient to give rise to interesting ecoclimates, but such occurrences are on a very small scale.

The general meridional alignment of relief along the eastern and southwestern margins causes an orographic uplift of the air streams which is clearly reflected in the amount of rain. The amount is sharply increased on the windward side, and gradually decreased on the leeward side of the obstacle. Genuine orographic clouds are restricted to the upper seaward slopes of the Atherton Tableland in northeastern Queensland, and of the Tamborine Mountains, New England Tableland, Blue Mountains, Australian Alps, etc., further south. The increase in rainfall, however, occurs even where the altitude of the obstacle is quite inadequate to cause condensation, e.g., on Mount Lofty, South Australia and on the Darling Scarp, Western Australia, the orographic uplift causes a wave in the oncoming stream, and precipitation develops some 1,500 or 2,000 m above the obstacle, and already 20 or 30 km in front of it (Fig.9). Conversely, the descent of humid air streams coming from either side of Mount Lofty or off the Darling Scarp may cause local instances of *cumulus mammatus* (Fig.10). The vertical displacement is too small to produce widespread katabatic winds, but local examples occur in some valleys, as for instance along the Darling Scarp, Western Australia, where many trees lean towards the west as a result of the frequent easterlies which descend from the plateau.

Australia has another peculiarity—being the driest continental mass, it only requires a small amount of energy to evaporate any surface waters that may occur in the interior, after the sporadic falls of rain. This is why, notwithstanding the high albedo, there still remains a substantial amount of energy, as heat, which is transported by air currents (see Chapter 5, Fig.5, 23).

Ocean temperature

The small size of the Australian continental land-mass makes it more sensitive to oceanic influences where such influences are conveyed by a favourable air flow. The physical characteristics of the waters around Australia have been mapped repeatedly and from different sources (SCHOTT, 1935; MCDONALD, 1938; K.N.M.I., 1949; U.S. NAVY, 1957, 1959) but the information is not equally plentiful for every area; that for the northern waters and for the Great Australian Bight is particularly scanty. Another factor that may modify the effect of oceanic influences is the slow change in ocean temperatures, which has been going on throughout the high and middle latitudes of the world for many decades; in February, 1855, icebergs were seen at 46°S, 134°E (TOWSON, 1859): less than 1,000 km from Tasmania and 1,200 km from the mainland, but this century such sightings took place much further south.

The absence of a definite cool current along the western coast is an outstanding characteristic of the Australian continent (unfortunately the mythical "West Australian Current", supposed to be "cold", still continues to appear even in reputable atlases). The presence of warm waters near the coast in winter (Fig.11) results in the extension of precipitation much farther towards the equator (GENTILLI, 1952) and makes south-

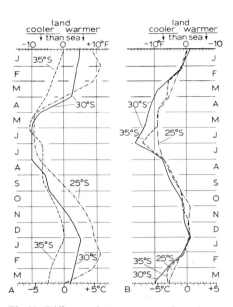

Fig.11. Difference between mean air and sea-surface temperatures for each month: A. on the west coast; and B. on the east coast of Australia. On the west side the land is cooler than the sea during the winter, and hotter for the rest of the year. On the east coast the land is cooler than the sea throughout the year. (Modified from KARELSKY, 1954.)

western Australia far wetter than any correspondingly situated land. Its rainfall is quite reliable. The differences in rainfall amounts between the various continents are shown in Table II of Chapter 6.

Unique to the western coast of Australia, among the mid-latitude regions, is the reversal of its surface ocean flow. During most of the year, from October to April, there is a northward drift which is primarily due to the general anticyclonic (anticlockwise) circulation of the air above the Indian Ocean. This anticyclonic circulation is only noticeable as a long-term seasonal average (reflected in the ocean surface circulation) and not in the day-to-day succession of moving anticyclones. No cool subantarctic water is drawn or driven along the western coast; the flow is from the south, but the surface water is drawn around Cape Leeuwin from the Great Australian Bight, which thus fulfils, in a much more modest way, a role similar to that of the Gulf of Mexico for the Gulf Stream. Hence, this south-to-north flow is warm, not cool (Fig.11). Slightly cooler water does flow northwards, but further offshore; even there, the rarity of fogs is the best proof of the absence of truly cool surface waters. The mean January air temperature at Fremantle, 22.5°C, is identical with that at Newcastle, on the eastern coast, where the ocean surface is warmer.

In the winter, the surface flow along the western coast is reversed, under the prevailing anticyclonic circulation over the continent, and warm water from the Timor Sea drifts southwards, to round Cape Leeuwin, enters the Bight, and eventually drifts eastwards through Bass Strait. The southwestern coast has a milder winter because of this warm coastal drift (Fig.11) The mean July temperature at Fremantle is 13.6°C, at Newcastle, 12.5°C. Even so, there is a definite correlation between the temperatures of the sea surface and of the coastal stations along the eastern coast.

The field of oceanic influences over coastal climates deserves much further study, with special regard to the northern and southern coasts, and to the factors which make the eastern coast rainfall so variable (Moss, 1958). This last field is being investigated by Priestley (1964, and earlier papers) especially from the standpoint of the correlation between the temperature of the sea surface and the amount of rain. The role of the ocean at the onset of the sea breeze will be mentioned later.

Index to existing literature

Land surfaces: C.S.I.R.O. (1960), Department of National Development of Australia (1952).
Ocean surfaces: Schott (1935), McDonald (1938), K.N.M.I. (1949), Meteorological Office (1949), U.S. Navy (1957, 1959).
Effect on rainfall: Gentilli (1952), Karelsky (1954), Moss (1958), Priestley (1964).

References

Albrecht, F. H. W., 1951. Intensität und Spektralverteilung der Globalstrahlung bei klarem Himmel. *Arch. Meterol. Geophys. Bioklimatol., Ser. B*; 3: 220–243.

Albrecht, F. H. W., 1953. Das Impulsaktinometer. *Geofis. Pura Appl.*, 26: 153–171.

Albrecht, F. H. W., 1954. Establishment of a radiation network connected to ordinary weather stations. *Geofis. Pura Appl.*, 29: 115–154.

Albrecht, F. H. W., 1955. Methods of computing global radiation. *Geofis. Pura Appl.*, 32: 131–138.

Albrecht, F. H. W., 1957. Untersuchungen über die "optische Wolkendichte" und die Strahlungsabsorption in Wolken. *Geofis. Pura Appl.*, 37: 205–219.

Ångström, A., 1924. Solar and terrestrial radiation. *Quart. J. Roy. Meteorol. Soc.*, 50: 121–125.

Australian Water Resources Council, 1968. Solar distillation in Australia. *Water Resources Newsletter*, 11: 49–50.

Barton, C. H., 1895. *Outlines of Australian Physiography*. Alston, Maryborough, 180 pp.

Bigg, E. K. and Miles, G. T., 1964. The results of large-scale measurements of natural ice nuclei. *J. Atmospheric Sci.*, 21: 396–403.

Birch, R. L., 1968. Radar observations of smoke. *Australian Meteorol. Mag.*, 16: 30–31.

Black, J. N., Bonython, C. W. and Prescott. J. A., 1954. Solar radiation and the duration of sunshine. *Quart. J. Roy. Meteorol. Soc.*, 80: 231–235.

Bonython, C. W., Collins, J. A. and Prescott, J. A., 1955. The evaporation pattern over Australia for the months of January and July. *Trans. Roy. Soc. S. Australia.*, 78: 99–109.

Brunt, D., 1939. *Physical and Dynamical Meteorology*. Cambridge Univ. Press, Cambridge, 428 pp.

Bureau of Meteorology, 1953–1966. *Australian Radiation Records*. Bur. Meteorol., Melbourne, Vic.

Commonwealth Experimental Building Station, 1952. The solarscope. *Commonwealth Exptl. Building Sta., Tech. Mem.*, 186: 43 pp.

Commonwealth Housing Commission, 1944. *Final Report. Min. Post-War Reconstruction*, Canberra, A.C.T., 328 pp.

Commonwealth Meteorology, 1953. *Sunshine*. Bureau of Meteorology, Melbourne, Vic., 13 maps.

C.S.I.R.O. (Commonwealth Scientific and Industrial Research Organization), 1963. *Solar distillation. Water Resources Newsletter*, 1: 20–21.

C.S.I.R.O. (Commonwealth Scientific and Industrial Research Organization), 1960. *The Australian Environment*. Melbourne Univ. Press, Melbourne, Vic., 151 pp.

Department of National Development of Australia, 1952. *Atlas of Australian Resources*. Dept. Natl. Develop., Canberra, A.C.T., 40 pp.

De Vries, D. A., 1956. The thermal behaviour of soils. In: *Australia-UNESCO Symposium on Arid Zone Climatology, Canberra 1956—Papers from Australia and New Zealand*. Government Printer, Canberra, pp.11a–11k and 12a–12e.

Drysdale, J. W., 1951. Climate and design of buildings. *Commonwealth Exptl. Building Sta., Tech. Studies*, 35: 15 pp.

Drysdale, J. W., 1959. Designing houses for Australian climates. *Commonwealth Exptl. Building Sta., Bull.*, 6: 47 pp.

Ellery, R. L. J., 1884. The recent red sunsets. *Trans. Roy. Soc. Vic.*, 20: 124–125.

FUNK, J. P., 1959. Improved polythene-shielded net radiometer. *J. Sci. Instr.*, 36: 267–270.

GAYEVSKII, V. L., 1953. K voprosu o roli albedo v formirovanii radiatsionnogo rezhima povyerkhnost. *Tr. Gl. Geofiz. Observ.*, 39 (101).

GENTILLI, J., 1952. *A Geography of Climate*. Univ. of Western Australia Press, Nedlands, pp.102–104.

GENTILLI, J., 1954. Climate and comfort at home. *Suppl. Wesfarmers News*, September 2, 1954, pp.1–20.

HOGG, A. R., 1963. The Mount Agung eruption and atmospheric turbidity. *Australian J. Sci.*, 26: 119–120.

HOUNAM, C. E., 1963. Estimates of solar radiation over Australia. *Australian Meteorol. Mag.*, 43: 1–14.

I.C.I. ALKALI (AUSTRALIA) PROPRIETARY LIMITED, 1947 (continuing). Summary of insolation and hours of bright sunshine. (Daily data, released for each financial year), Dry Creek, S.A.

KARELSKY, S., 1954. Surface circulation in the Australian region. *Australia, Bur. Meteorol., Meteorol. Studies*, 3: 45 pp.

KEOUGH, J. J., 1951. Selected Australian climatic data. *Commonwealth Exptl. Building Sta., Tech. Studies*, 36: 95 pp.

KIDSON, E., 1925. Atmospheric pollution: observations with the Owen's dust-counter during 1923–1924. *Australia, Bur. Meteorol., Bull.*, 17: 4 pp.

K.N.M.I. (Koninklijk Nederlands Meteorologisch Instituut), 1949. *Sea Areas round Australia*. Staats-drukkerij- en Uitgeverijbedrijf, 's-Gravenhage, 79 pp.

LANDSBERG, H. E., 1960. *Physical Climatology*. Gray, DuBois, Pa., 446 pp.

LOEWE, F., 1956. Notes on global radiation in Australia. *Australian Meteorol. Mag.*, 15: 31–41.

LOEWE, F., 1962. The intake of solar radiation by slopes with cloudless sky. *Australia, Bur. Meteorol., Bull.*, 45: 57 pp.

LUKE, R. H., 1961. *Bushfire Control in Australia*. Hodder and Stoughton, Melbourne, Vic., 136 pp.

MAINE, R., 1958. Maximum temperature prediction. *Australian Meteorol. Mag.*, 22: 50–63.

MARSHALL, A., 1955. Climate and housing. *Proc. Roy. Geograph. Soc., S. Australian Branch*, pp.11–13.

MCDONALD, W. F., 1938. *Atlas of Climatic Charts of the Oceans*. Govt. Printing Office, Washington, D.C., 130 pp.

METEOROLOGICAL OFFICE, 1949. *Monthly Sea Surface Temperatures of Australian and New Zealand Waters*. H.M. Stationery Office, London, 12 charts.

MOSS, J. M., 1958. A preliminary investigation of ocean temperatures and maximum dew points in the New South Wales region. *Bur. Meteorol., Conf. Estimation Extreme Precipitation, Melbourne, Vic.*, pp.127–147.

PHILLIPS, R. O., 1951. Sunshine and shade in Australia. *Commonwealth Exptl. Building Sta., Tech. Studies*, 23, 43 pp.

PRESCOTT, J. A., 1940. Evaporation from a water surface in relation to solar radiation. *Trans. Roy. Soc. S. Australia*, 64: 114–118.

PRIESTLEY, C. H. B., 1964. Rainfall—sea surface temperature associations on the New South Wales coast. *Australian Meteorol. Mag.*, 47: 15–25.

RIMMER, W. B. and ALLEN, C. W., 1950. Solar radiation observations at Mount Stromlo, 1927–1939. *Obs. Mt. Stromlo, Mem. Commun.*, 11: 41 pp.

ROSS, A. D., 1963. Sky effects from volcanic eruptions. *The West Australian*, April 25th, 1963.

RUSSELL, H. C., 1885. On a new form of actinometer. *J. Roy. Soc. New South Wales*, 18: 73–74.

SAPSFORD, C. M., 1957. An estimation of solar energy radiation for Australia. *Australian J. Sci.*, 20: 99–105.

SCHOTT, G., 1935. *Geographie des Indischen und Stillen Ozeans*. Boysen, Hamburg, 413 pp.

SELLERS, W. D., 1965. *Physical Climatology*. Univ. of Chicago Press, Chicago, Ill., 272 pp.

TOWSON, J. T., 1859. *Icebergs in the Southern Ocean*. Brakell, Liverpool, 16 pp.

U.S. NAVY, 1957. *Marine Climatic Atlas of the World, 3. Indian Ocean*. Superintendent Documents, Washington, D.C., 267 charts.

U.S. NAVY, 1959. *Marine Climatic Atlas of the World, 4. South Pacific Ocean*. Superintendent Documents, Washington, D.C., 267 charts.

VINACCIA, G., 1939. *Il Corso del Sole in Urbanistica ed Edilizia*. Hoepli, Milano, 384 pp.

WEINERT, R. A., 1967. The movement and dispersion of volcanic dust from the eruption of Mt. Agung, Bali, 17 March, 1963. *Australian Meteorol. Mag.*, 15: 225–229.

Chapter 5

Dynamics of the Australian Troposphere

J. GENTILLI

Introduction

Situated in the lower middle latitudes, Australia is mostly affected by the descending air which forms the belt of subtropical high pressure and divergence. During the summer, this belt moves southwards by some 5–8° of latitude, and at times allows the equatorial belt of low pressure (intertropical convergence) to reach the northern shore (Australian monsoon). During the winter, the high-pressure belt moves northwards, and allows the mid-latitude jet stream to meander over the land. All told, the summer circulation is rather more *meridional*, the winter circulation rather more *zonal*. In the summer, northern Australia has an alternation of monsoonal northwesterlies and tropical easterlies; central and southern Australia, the dry variable winds of the high-pressure belt (anticyclonic belt); and Tasmania, the mid-latitude westerlies. In the winter, tropical easterlies dominate northern Australia, the high-pressure belt is narrower and broken into anticyclonic cells, and westerlies and the jet stream control the climate of southern Australia and Tasmania.

Air masses of Australia

Frontal analysis was introduced into Australia relatively late. KIDSON (1935) wisely proposed not to apply "the names used in Europe (tropical, polar, maritime, continental, etc.) to the various (Australian) air masses. Some of them are inappropriate to this region and they are likely to clog rather than aid the discussion..." Within 10 years, air-mass analysis was found unsatisfactory for practical purposes, and was abandoned, especially after the introduction of streamline analysis for tropical regions.

Australian air masses may still offer an interesting field for research because of their unusually limited range of characteristics. There is no polar continental air anywhere near Australia, and it is doubtful whether any Australian polar maritime air would be recognized by Northern Hemisphere climatologists, having been modified by hundreds of kilometres of travel over mid-latitude waters. However, any really large depression well to the south of Australia sends a strong southerly flow, which may be as cold as 2–3°C when it reaches the southern shore in winter, and 15–16°C in summer. Relative humidity (RH) may be as high as 80 in the early morning, especially on cloudy days, but a few hours of sunshine will lower it to about 70 in winter, and 20–30 in summer. The source areas of Australian air masses are shown in Fig.1.

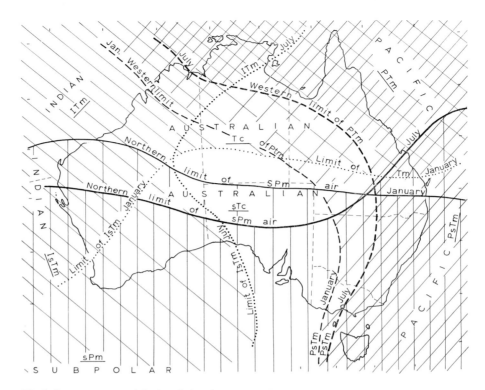

Fig.1. Source areas and limits of the air masses affecting Australia. There are three primary (typical) types: Australian tropical continental *(Tc)*, Indian tropical maritime *(ITm)*, and Pacific tropical maritime *(PTm)*. There are also four secondary (varietal) types: Australian subtropical continental *(sTc)*, Indian and Pacific subtropical maritime *(IsTm* and *PsTm)*, and subpolar maritime *(sPm)*. The limits shown are based on weather charts of 5 years and must be taken as indicative only. Subpolar air is already modified when it reaches as far as shown on the map. Anticyclonic circulation drives Indian and Pacific air towards the limits shown, and continental air well beyond the mainland. In July, the colder land surface strengthens the anticyclonic regime, so that the peripheral air masses (Indian, Pacific and subpolar) are held back and penetrate the interior less frequently and less deeply than in summer.

Tropical maritime air may be from the west (Indian *Tm*) or from the east (Pacific *Tm*); the continent is the source of tropical continental air. These tropical air masses dominate the climates of Australia; in the Northern Hemisphere, corresponding air masses control the climates of India and Mexico. These countries lie under the belt of tropical sinking and divergence (in summer, largely nullified in India by monsoonal developments) and are thus affected by the relative position of the *airshed*, if we may here propound the use of this term, analogous to watershed, for the line of highest pressure. Tropical air-masses might, therefore, be reclassified into subtropical (originating polewards from the airshed ridge) and tropical proper (originating equatorwards from the airshed ridge). What most treatises consider "tropical" is actually subtropical air; we shall show it by the prefix *s*.

Our notation therefore differs from that of other authors. TWEEDIE (1966) distinguishes the following air masses: tropical Indian (our *ITm* and *IsTm*), tropical Tasman (our *PTm* and *PsTm*), equatorial *(Eq)*, tropical continental (our *Tc* and *sTc*), and southern maritime (our *sPm*). As may be expected, soundings show that actual temperature and dew point of equatorial, tropical Indian and tropical Tasman air masses are very similar, but for the fact that "tropical" Tasman air is a little cooler—which is why we prefer to distinguish it as Pacific *subtropical* maritime.

IsTm or Indian subtropical maritime air affects the western and southern shores of Australia. It is the air of the cool change between two anticyclones in summer, the air of the westerly streams in winter. In summer it may be at 20–22°C and with a RH of 70–75. In winter it may be at 10–12° and its RH may be 80–85, so that it soon deposits some dew on cold surfaces, and it brings substantial frontal rains. Fig.1 shows its extreme limits in a five year period. In summer it may spread very far eastwards and reach the Pacific (Fig.2).

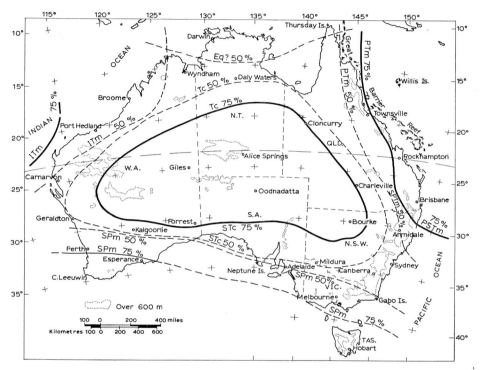

Fig.2. Air-mass frequency, January. The thick lines show the limits of the 75%-frequencies, the broken lines the 50% ones. The boundary between *sTc* and *sPm* masses is well defined because no other air masses reach that area; the 50–50 line runs through Kalgoorlie and Adelaide. On the northern side the boundaries are blurred because four distinct air masses may occur; Daly Waters is at what might be termed the aerological crossroads. For further explanation see Fig.1.

PsTm or Pacific subtropical maritime air affects the southeastern shore (including eastern Tasmania) and is generally very humid and unstable. In summer its 22–24°C and near-saturation humidity make it a source of thunderstorms and torrential rains, especially against steep orographic or frontal barriers. In winter its 14–16°C and near-saturation humidity make it easily lifted by any southwesterly stream of no matter how thoroughly modified polar air. Hence, a belt of cyclogenesis may appear over the western Tasman Sea. In summer it may reach the Indian Ocean across northern Australia; in winter it does not spread so far westwards (Fig.1).

Subtropical continental air is generally dry, usually between 30–50 RH, and its temperature is extremely variable, from 4–5°C on wintry mornings to 20–22°C in the afternoon of the same days, if the sky stays clear. During a summer night its temperature may also

TABLE I

STABILITY FOUND BY MEAN AEROLOGICAL SOUNDINGS. TEMPERATURE OF A SURFACE PARTICLE OF MEAN
MONTHLY TEMPERATURE AND DEW POINT, WHEN LIFTED TO 500 MBAR, MINUS THE MEAN MONTHLY
500-MBAR TEMPERATURE (°C)[1]
(After McRae, 1956)

Station	Jan.	Feb.	Mar.	Apr.	May	June	July	Aug.	Sept.	Oct.	Nov.	Dec.
Darwin	**4.0**	**4.0**	**3.5**	**0.7**	−1.4	−4.5	−4.3	−3.4	−1.3	**2.6**	**4.1**	**4.2**
Townsville	**2.0**	**2.0**	**0.7**	−0.8	−4.0	−6.7	−6.7	−7.8	−3.7	−1.2	*0.7*	**1.8**
Port Hedland	**1.7**	**3.0**	**2.7**	−1.1	−7.3	−12.5	−12.5	−13.7	−5.3	−3.7	−0.5	**2.4**
Cloncurry	−2.7	−2.7	−2.5	−7.7	−10.0	−13.0	−13.5	−13.8	−8.7	−7.0	−3.4	−1.0
Amberley (Brisb.)	−2.8	*0.4*	−4.6	−3.0	−6.2	−6.1	−7.3	−6.6	−3.0	−1.4	−1.4	−1.3
Norfolk I	−3.5	−3.2	−5.0	−3.5	−1.7	−2.7	−5.5	−4.2	−4.0	−3.6	−4.0	−4.0
Guildford (Perth)	−2.4	−3.0	−3.0	−2.4	−2.7	−3.7	−4.4	−4.4	−4.0	−4.3	−4.0	−4.4

[1] Values in bold face type = real latent instability; values in italics = pseudo-latent instability; negative values = conditional stability.

be 20–22°C, and the RH may rise to 60–70, but a clear morning will make temperatures soar to 36–38°C and RH fall to 30–35. It is very stable (Table I).

Needless to say this *sT* (subtropical) air in Australia is largely derived from superior air, now seldom recognized by meteorologists. It can be traced back to the 10,000 m level of the subtropical high-pressure belt, and it changes but little in its descent, apart from the effects of its obvious adiabatic warming and consequent drying. Over the oceans the air is soon modified, but over the land it remains unchanged. It dominates the Australian troposphere, being by far the most frequent and widespread air mass (Fig.2). *Tm* air is warmer and more humid than *sTm* air; there is no appreciable difference between air from either the Indian Ocean or the Pacific in this respect, mainly because of the orientation of the Australian coastline north of the tropic, which provides two major reservoirs of *Tm* air in the Timor and Coral seas and a minor reservoir in the Gulf of Carpentaria. *Tm* air is very often unstable (Table I). It is a northwesterly stream of *Tm* air that is dragged by the jet stream in winter, and brings the heavy winter rains to the southwest. It is *Tm* air that forms the tropical cyclones. It is *Tm* air that brings the torrential summer rains to the northeastern slopes. It is also mostly *Tm* air that brings the summer thunderstorms during the monsoon season, because it is only occasionally that transequatorial air enters Australia. Generally, *Tm* air will be at 22–24°C in winter and 26–28°C in summer, with a very high RH throughout; it brings the very heavy dews that characterize the tropical coast.

Equatorial air *Eq*, does not differ substantially from very humid and unstable *Tm* air; the distinction may be convenient in the summer months, when *Eq* air of transequatorial origin may enter Australia as a northwesterly stream. There is no doubt that the immense northwesterly stream that overran Australia in February, 1955, with a dew point of nearly 24°C, was equatorial air (see Chapter 6).

Tc air shows the greatest extremes of temperature, from 15–16°C on a clear winter night to 25–26°C in the respective afternoon, with RH 60–65 falling to 30–35. On a summer night radiation cooling may lower the temperature of the air to 18–20°C, but by the afternoon the temperature has soared to 38–40°C and RH may be at 20. This is the air of the great "heat waves" which can occur anywhere in Australia, even in Tasmania. During his explorations, STURT (1833, 1849) recorded 38–45°C in the shade, on one occasion, 57.2°C: "the men had the upper leather of their shoes burnt as if by fire".

As a minor point of interest, it should be known that MARTIN (1853) had implicitly the concept of air mass in mind when he wrote that the hot winds of Australia "bear some affinity to the hot winds experienced in the Mediterranean, in Egypt, Arabia, Persia, Bombay, and Mexico; but whether these all belong to a common system of atmospheric circulation, or are caused in the several countries by local circumstances, it is not easy to decide authoritatively; and my own impression is, that the form, extent, and latitude of the regions where they prevail—the characteristics of the soil and the quantity and nature of the vegetation, all exercise a powerful influence in the production of hot winds during summer."

Modification of air masses is frequent, especially when maritime air flows for some time over the land (e.g., *Tm* from the Coral Sea may travel inland for some distance before reaching the subtropical shore from the land side, or *sTm* air from the Great Australian Bight, may flow westwards over the land for some distance, and still retain its coolness and some of its humidity). *Tc* or *sTc* air may flow over the sea and reach the land again without great modification, as happens around the heat low that forms over the northwest. The "coastal front" so well developed in Western Australia (CASSIDY, 1945b) is the

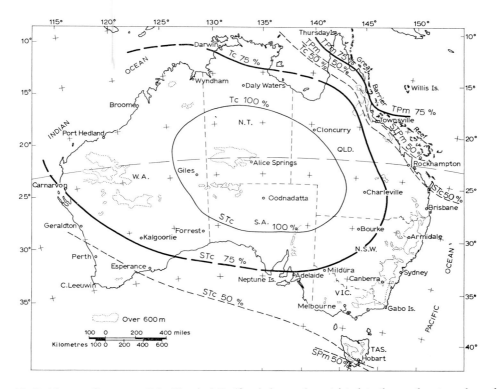

Fig.3. Air-mass frequency, July. Tropical Pacific air is mostly restricted to the northeastern shore, but occasionally sweeps south in deep troughs along the coast. Indian air is even less plentiful, having been pushed away from the northwestern shore by the further extension of anticyclones and trade winds; it reaches the southwestern coast with every deep trough associated with midlatitude fronts, but not often enough to attain a 50% frequency. Notice the 100% continental core. For further explanation see Fig.1,

result of diurnal modification and is therefore, not very effective; it is a sea-breeze front of continental proportions.

In Fig.2 and 3 are shown the mean frequencies of air masses in the months of January and July. Thus, it can be seen that at Brisbane, the frequency of *PsTm* air is over 75% in January and below 50% in July. On the other hand, at Perth, *sPm* air reaches a frequency of over 50% in January, being brought in by every oncoming anticyclone; but it is absolutely marginal *sPm* air, just cool and only moderately humid; typical *sTc* air dominates the climate during most of the year.

The travelling anticyclones

The descending air which causes the belt of high pressure and tropical divergence is subdivided, as a result of the Coriolis effect, into a series of travelling anticyclones. The air, falling from about 10,000 m, rotates anticlockwise and warms up adiabatically in its descent. Because of its relative dryness the rate of warming up is rapid, and high temperatures are common, especially when summer insolation heats the underlying land surface. The loss of outgoing radiation is great, and both night and winter temperatures are relatively low. Over most of Australia, where anticyclones are prevalent, the mean daily amplitude of surface temperature varies between 14–17°C, whereas it is only 5–8°C in the I.T.C. belt and in the belt of westerlies.

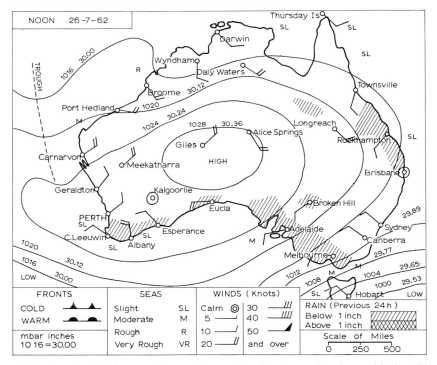

Fig.4. Normal anticyclone on the most frequent track. Pressure in mbar and inches. The anticyclone is normal in its size and proportions, and follows a track just south of the tropic. More northerly and especially more southerly tracks occur very often (cf. Fig.7). There is a trough to the west, and rain from the inter-anticyclonic front along the eastern shore. Some rain from the peri-anticyclonic front has fallen northwest of Longreach. At higher altitudes the depressions must extend above the south coast, because there have been widespread rains well within the anticyclonic surface field. (West Australian weather map.)

The first scientific discussion of moving anticyclones in Australia, made possible by the increased availability of observations from Western Australia since 1887, and the consequent compilation of 1,400 weather maps in about 5 years, is due to RUSSELL (1893). He noticed that these anticyclones "follow one another with remarkable regularity and are the great controlling force in determining local weather. The fixed anticyclone over the Indian Ocean in these latitudes, which is found in books of reference, must give place to a moving series".

It is true that maps of mean pressure or of resultant winds give a most misleading picture of pressure and wind conditions in the anticyclonic belt, where pressure rises and falls and winds regularly change with the passage of each anticyclone. On the north side of the anticyclones, the winds gradually steady onto a northwestward course to become the trade winds. Within each anticyclone the successive directions of surface winds are regular and well-known. Between anticyclones they are conflicting—neatly so in summer, with a quasi-stationary front and light clouds, more irregularly in winter, with lower pressure, multiple cold fronts and stormy weather. On the south side of the anticyclones, the winds are westerlies, irregular and often violent.

All in all, these travelling anticyclones follow definite patterns in shape, size and rate of movement. Their nearest counterpart in the Northern Hemisphere is the Azores anticyclone in its eastward advances over western Europe and North Africa, but the Azores anticyclone is soon modified by continental influences. The variations in the Australian

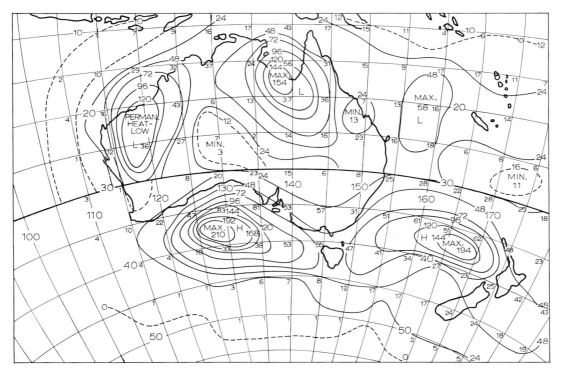

Fig.5. Summer anticyclonicity and the permanent heat low. This is a composite map: south of latitude 30°S it shows the frequency of anticyclones in summer, and north of that latitude the frequency of low-pressure centres in the same season. The northwestern permanent heat low (Pilbara low) is the result of heat transported by the trade winds issuing from the frequent anticyclones further to the southeast. The northeastern low (Cloncurry low) is not due to heat transport but to dynamic causes (see Chapter 5). The Tasman Sea anticyclone at times may exert a blocking effect. (Adapted from KARELSKY, 1954.)

anticyclone have been studied by KIDSON (1925) and KARELSKY (1954, 1956, 1961, 1965). The most remarkable climatic effect of anticyclonic circulation is the transport of heat from the heart of the continent towards the northwest, where a *stationary heat-low* (Fig.5) arises during the summer months (November–March). On most days the maximum temperature of this large area rises above 40°C; the small mining town of Marble Bar, lying on solid rock on the lee side of the plateau usually reaches the highest maximum temperatures in Australia.

The pressure trough between successive anticyclones, associated with an equatorward meander of the subtropical jet stream and its subsequent acceleration, is a most important feature in the regional atmospheric circulation. Its main effect in winter is the formation of upper-air depressions which often bring copious rains (our "inter-anticyclonic front"), while in the summer it causes the advection of hot northerly air southwards.

The peripheral anticyclonic stream in its southeastern quadrant is maritime on the west coast and continental on the east coast. In its northwestern quadrant it is the opposite, continental on the west coast and maritime on the east coast. Hence, other things being equal, the summer anticyclones bring northeasterly tropical continental air to the southwest and northeasterly tropical maritime air to the southeast (cf. Fig.1). Northeasterly winds in the Southern Hemisphere come from lower to higher latitudes and, therefore, increase their relative humidity, but the high temperatures of the dry interior are such that saturation point is very seldom reached (Fig.6). On the other hand, saturation point is more easily reached on the eastern coast, after the winds have travelled over broad expanses of ocean. Southwesterly winds travel equatorwards and their relative humidity

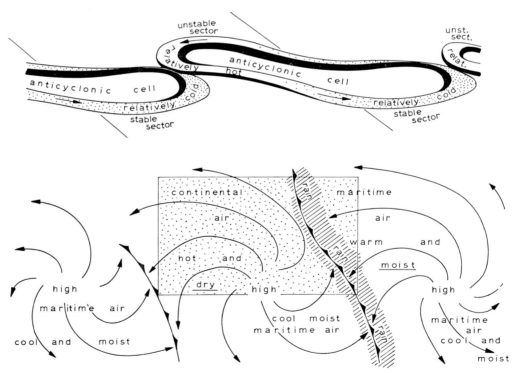

Fig.6. Diagram of stability and instability within the anticyclonic cells (above) and of the different efficiency of the inter-anticyclonic front as a rain-bearing factor (below); the important difference is in the moisture content of the northeasterly stream, respectively *Tc* and *PTm* air.

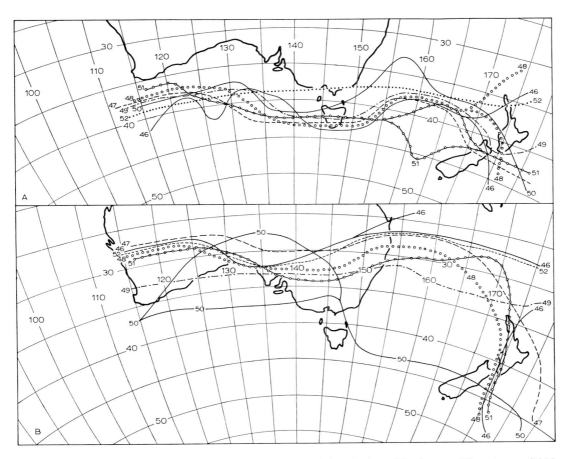

Fig.7. Belts of maximum anticyclonicity in summer (A) and winter (B) of seven different years (1946–1952). In the main, the point of entry at about 115°E shows great regularity from one year to another, but especially in winter there may be great variations in the latitude of the point of exit. Notice the alternative belts for the summer of 1951 and the winter of 1950. (After KARELSKY, 1954.)

becomes lower, and when they reach the hot Australian coast they fail to yield any rain, although there may be a stationary front and some clouds. Southwestern Australia hardly ever gets any rain from peripheral anticyclonic winds in summer, while the southeast gets some good rains from the northeasterlies, but not from the southwesterlies.

In winter, the whole system is situated further north (Fig.7) and any maritime air, on reaching the colder continent, may release rain. Substantial variations in the size and latitude of anticyclones may occur even within a few weeks (Fig.8, 9).

Thus, any description of annual and diurnal variations in pressure fails to convey a true image of the real pressure changes. These are due to the succession of these more or less distinct anticyclonic cells, each one with higher pressure (1,020–1,030 mbar, rarely to 1,035 mbar) at the centre, and cols or troughs of 1,000–1,010 mbar to the west and to the east (Fig.10, 11). The winds blow along the isobars in altitude, but near the ground, friction causes them to be deflected by the Coriolis effect some 30° or more, so that they issue from the high-pressure core at an angle. Because of its general eastward progress, each anticyclone brings its own 5–6 day periodicity in the physical characteristics of the air and in the direction of the wind, until the next anticyclone arrives and repeats

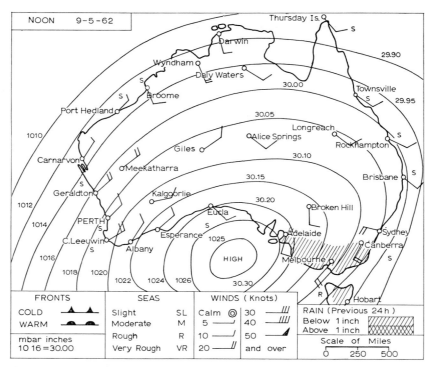

Fig.8. Large anticyclone on a southerly track. The pressure of 1028 mbar around the centre is not particularly high, but the absence of depressions in the vicinity has allowed this anticyclone to reach an unusual extent: its major axis (SW–NE) is about 6,000 km, its minor axis 4,500. The wind direction at Sydney shows that there is a trough or a depression to the southeast; winds at Broome, Wyndham and Darwin are also deflected. (West Australian weather map.)

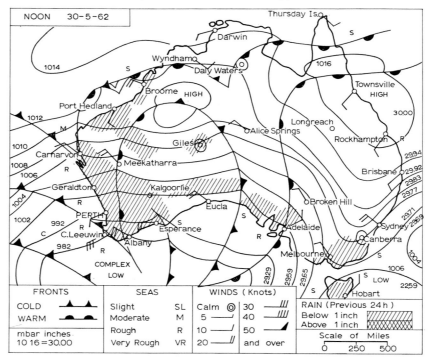

Fig.9. Very small (residual) anticyclones on a northerly track. This situation occurred only three weeks after the preceding one; very large and complex depressions to the south have eroded an anticyclone, until only two small parts of the core, the northernmost ones, are left. They may perhaps coalesce again over the Coral Sea if conditions are favourable. This drastic anticyclolysis is not very common over the mainland. (West Australian weather map.)

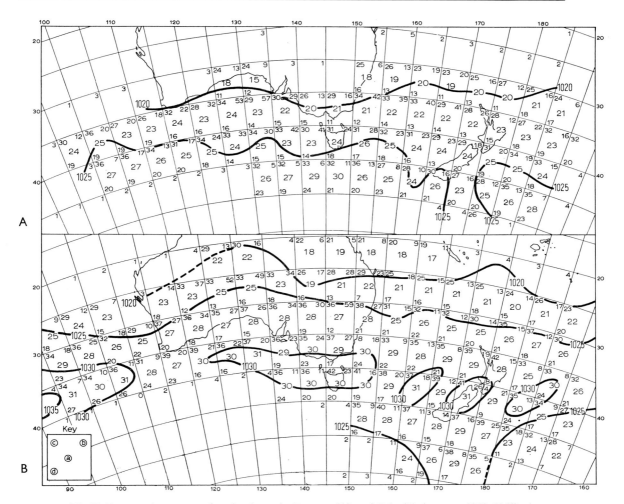

Fig.10. Pressure in centres of anticyclones in January (A) and July (B) (average 1952–1963). A mean pressure of 1,020 mbar which just skirts the south shore in January is found even north of the tropic by July. In January a mean pressure of 1,025 mbar is normal to the south of Tasmania, while in July, pressures above 1,030 mbar are normal there and over Bass Strait. a = average pressure in mbar: e.g., a = 27, i.e., 1027 mbar; a = 82, i.e., 982 mbar; b = total number of anticyclonic centres in 5° squares in 12 years; c = maximum pressure in anticyclonic centre; d = minimum pressure in anticyclonic centre. (After KARELSKY, 1965.)

the cycle. Occasionally the denser air may be subdivided into two or even three anticyclonic cells (Fig.12).

Although they may be strengthened by outbursts of Antarctic air over the Indian Ocean, Australian travelling anticyclones owe their prime origin, as mentioned above, to the descending air masses of the tropical latitudes of the Indian Ocean region. They are typical warm-core anticyclones, with high pressure already noticeable at heights of 10,000 m or more. The anticyclonic high-pressure pattern extends longitudinally over some 2,000–3,000 km, and latitudinally over 1,000–2,000 km (cf. Fig.8 and 9).

"The moving anticyclones are about five times as large in area as the low pressure V between them, and they are also moving to the east..." (RUSSELL, 1893).

On the average, some 40 anticyclones pass over Australia during a year; they are slightly more numerous in spring–summer than in autumn–winter, when the denser air is more

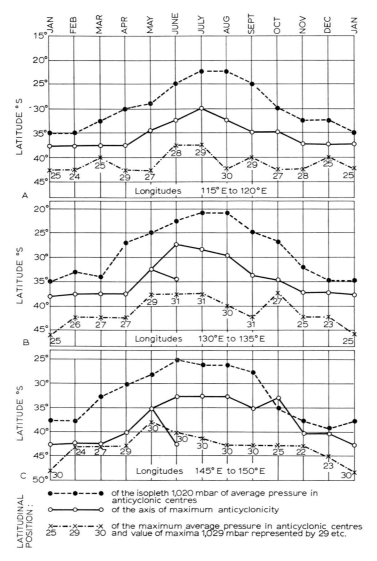

Fig.11. Monthly latitudinal position of anticyclones across the meridional belts at longitudes: A. 115–120°E; B. 130–135°E; C. 145–150°E. The position of the anticyclones has been assessed from the position of the isopleth 1020 mbar in anticyclonic centres, the axis of maximum anticyclonicity, the line of maximum average pressure in the anticyclonic centres. The various graphs should be enlarged and combined in various ways to obtain the best view of the seasonal and geographical changes (After KARELSKY, 1965.)

sluggish, and a "cold pool" may develop west of the highlands, producing a "blocking" effect.

Russell's observation that "the latitude of anticyclone tracks varies with the season, being in latitude 37–38°S in summer and 29–32°S in winter" is confirmed by the detailed analyses carried out by KIDSON (1925) and KARELSKY (1965), as shown by Fig.10.

Over most of Australia, the mean anticyclonic track is found at its farthest north in June–July, but near the eastern shore and over the Tasman Sea, the greater lag in water temperatures has the effect of retarding the northward deflection of the tracks until August–September (KIDSON, 1925).

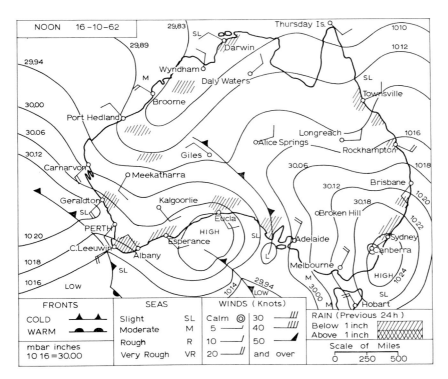

Fig.12. Subdivision of anticyclonic cells, by the development of peripheral and high-level depressions and fronts. (West Australian weather map.)

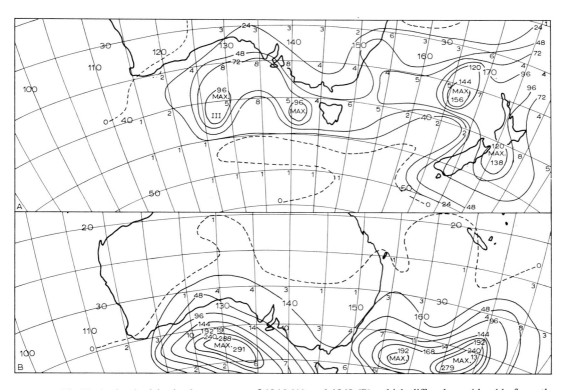

Fig.13. Anticyclonicity in the summers of 1946 (A) and 1948 (B), which differed considerably from the normal as shown in Fig.5, the former because of very low frequency of anticyclones, the latter because of more anticyclones than usual. (After KARELSKY, 1954.)

Russell correctly stated that "when an anticyclone track is far from the mean, the weather is also far from the mean"; Fig.13, from KARELSKY (1954), shows the pattern of anti-cyclonicity in two very different weather years, 1946 and 1948.

The magnitude and latitude of an anticyclone are closely related to the speed and latitude of the subtropical jet stream (discussed in Chapter 3). A large anticyclone travelling south of latitude 30 shows that the jet stream, where it occurs, forms ample meridional bends. Easterly winds are then prevalent within much of the region. A relatively elongated anticyclone travelling along latitude 25 or 30 reveals a strong zonal circulation further south, with a powerful and more direct jet stream. Towards the south, a westerly regime dominates the weather. Some of the climatic implications of these different regimes are mentioned in Chapter 6.

RUSSELL (1893) believed that the anticyclones passed over Natal on their way to Austra-lia, and proposed the setting up of a forecasting network based on the assumption that such anticyclones could be followed around the world. HARDING (1893) (in discussion of RUSSELL's 1893 paper) stated that careful observation of weather charts from Nata showed that anticyclones never moved across the whole width of the Indian Ocean as postulated by Russell; it is now known that Harding was correct and that these anticy-clones begin to appear over the central Indian Ocean.

Russell stated that "the average time of passage over any place is 8.7 days", but did not allow for the trough between anticyclones. The modal transit time of an anticyclone is 5 days, the mean 6.4 days (KIDSON, 1925); the whole spiral of stable anticyclonic air travels eastwards at a rate of 500–600 km/day. VEITCH (1965) analyzed the principal components of the pressure variation with interesting results; he found that "approxi-mately one-half of the variation present in the first principal component, about 18% of the total variation present in the Australian region, is due to some process beyond the normal procession of cyclonic troughs and anticyclones through the eastern Australian area. This process gives rise to relatively slow quasi-periodic oscillations in the pressure with "periods" of 10–40 days".

The shape of the anticyclone over the western plateau or the lowlands is an ellipse with a 3/2 axis ratio. At times, on reaching the eastern highlands, the lower part of the anticyclone is delayed or even blocked ("cold pool" effect) and its north–south axis may become the major one; this often happens when deep lows "erode" the anticyclone (Fig.14).

With the eastward advance of the anticyclone, the wind gradually veers to southerly and loses force until the anticyclonic core has gone past. West of the core, it veers to south-easterly, then to easterly, gradually becoming stronger again.

Near the outgoing (western) edge of the anticyclone the wind is northeasterly to norther-ly, gusty, at 10–15 m/sec, and the air, coming from lower to higher latitudes generally unstable (Fig.6). As RUSSELL (1893) puts it, "the two winds, northerly and southerly, pass each other as if struggling to get through between two obstacles, i.e., the preceding and following anticyclones"—an anticipation of the concept of meteorological fronts or a re-statement of Dove's theories?

The intensity of weather is in proportion to the difference in pressure between the anti-cyclone and the trough or col, but "the relation of the pressures varies frequently *before* the wind responds, and it seems as if the pressure were controlled from above by the more or less rapid descent of air, which feeds the anticyclone."

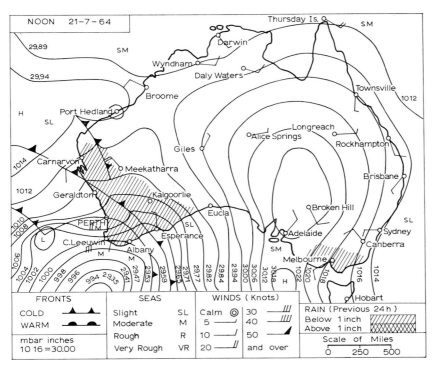

Fig.14. Changing shape of an eroded anticyclone, blocked by the Eastern Highlands and thus gradually reduced in its west–east axis. This "cold pool", which is common in July, is the nearest to a stationary anticyclone that may ever arise in Australia, but the effect only lasts a few days, enough to give a higher frequency of dry northeasterlies to the region around Adelaide. (West Australian weather map.)

Along the north (equatorward) edge of the anticyclonic field, air always issues from the southeast or east, and constitutes the tropical easterlies or *trade winds* (see p. 74). Along the south (poleward) edge of the spiral, the air becomes unstable (Fig.6) and the outflow is much more variable.

In summer, air from the western anticyclonic edge, very dry and over-heated by the continental interior, may reach temperatures above 38°C, especially when it is made more persistent (e.g., over 3–4 days) by a delay in the progress of the anticyclone, or by the addition of inland air actioned by a northern depression. In the old days these hot winds were noticed more because people were less equipped for withstanding the discomfort they brought. In coastal districts, they may occur 3–4 to 7–8 times during a summer. Then temperatures at Sydney or Perth rise by 14–16°C in a day and reach over 38°C, very rarely 42–43°C. On the south coast these winds are known as *northers*. Adelaide, drier, reaches slightly higher extremes than Sydney; Melbourne, further south, is usually cooler, but such hot winds have reached even Hobart and, greatly attenuated, New Zealand. DE STRZELECKI (1845) experienced such a wind at the top of Ben Lomond (northeastern Tasmania), at an elevation of 1,500 m, but did not feel it below, at 900 m, to the windward, where he evidently found himself in a different air-stream.

In the interior of Australia, where the temperature may rise 20°C in a day, "the intense heat of these winds raises the thermometer, in the shade, to 47°C, or even 49°C; the grass becomes dry, like hay..." As to introduced crop plants, "the red and blue grape lose their colour and watery elements; green leaves lose their colour, turn yellow, and wither..." (MARTIN, 1853). "Vegetation droops and dies, the leaves shrivel like half-burnt paper and the ripening fruit is scorched and half-roasted on the trees. Luxuriant crops of wheat or barley, with the grain still tender in the ear, are shrivelled up by the heat and scarcely fit to cut for hay" (BARTON, 1895). The effect on human beings is not usually great because these winds are dry, and the body is cooled by the evaporation of body moisture from the skin, but danger arises when the lost water cannot be replaced. There have been, and still are, every year, deaths through dehydration of adults lost in the outback, and of babies in isolated localities. Most tragic and recent, the death of a family of four, about Christmas, 1963, on the Birdsville track, north of Marree, South Australia, where the maximum temperature in the sun was over 60°C and in the shade 43.9°, 43.9°, 47.8° and 43.9°C on four consecutive days.

In the southern interior, especially in Victoria, these hot winds laden with dust, are known as *brickfielders*. (The term was originally used in Sydney to denote the "southerly bursters"—see discussion of cool change below—and was brought to the Victorian goldfields by prospectors from Sydney.)

When they blow, "the fine particles of dust penetrate everything; there is no chest so tightly locked, no drawer so secret, but will show signs of its intrusion; it gets into the eyes, ears, hair, nostrils—everything; imparts to every mouthful an unsought grittiness" (BARTON, 1895). The terms "Cobar shower", "Darling shower", etc. were used ironically in New South Wales for such winds.

In the winter, the anticyclonic situation is intensified by the cooling of the land. The slightly higher latitude of southeastern Australia causes a greater loss of heat, and the slight concavity of the land favours the stagnation of cold air. Especially in July, anticyclonic situations tend to linger on over southeastern Australia (KIDSON, 1925), gradually changing shape with the eastward advance of low-pressure troughs, so that at times the anticyclone has a meridional elongation instead of the usual zonal one (Fig.14). Then chilly north winds blow across the Riverina and over South Australia and northern Victoria, bringing a dry spell. Cold air flows over the eastern highlands and onto the eastern coast, and is accentuated by a land-breeze effect. Over southwestern Australia the cool, dry anticyclonic air descends the plateau as a chilly and rather gusty easterly which tends to be channelled by gravity along the many gullies and valleys, where westward-sloping trees bear witness to its lasting influence.

The aerology and dynamics of the transition from one anticyclone to the next are more complex than was imagined even a few years ago. Summer conditions are discussed in the section that follows, under a separate heading, because the cool change they bring is a most distinctive feature of Australian climatology. TROUP (1956) investigated conditions between May and September, at Perth and Kalgoorlie, in order to find the so called "meridional front" unaffected by any coastal front, sea-breeze or similar phenomenon. In order to determine the structure of the atmosphere where the transition from one anticyclone to the next takes place, the investigation was limited to 57 situations which showed an approximately meridional disturbance line. In about half of these there was a clearly recognizable frontal discontinuity (far less pronounced than in the classical cases of the Northern Hemisphere) with or without a marked inversion, usually at about 2,500 m. This study confirmed that neither warm fronts nor occlusions appear in these situations. The so-called "meridional front" is as often as not non-frontal in character, the disturbance at the surface having in such cases many of the features of a cold front but being in fact associated with a rise or disruption of the subsidence inversion.

Cool changes and southerly busters

In summer, the contrast between the converging airstreams from two consecutive anti-cyclones is very pronounced, because the northerly flow from the first anticyclone is hot, and the southerly flow from the second anticyclone is cool (Fig.6). The change from the hot air of the first anticyclone to the cool air of the second one is locally known as the *cool change*, and has been the object of thorough analysis by LOEWE (1945) and BERSON et al. (1957, 1959).

This is one of the various types of "dry cold front", and must be distinguished from true "coastal fronts" which "in summer... tend to form in certain preferred areas, owing probably to the large low level heat source provided by the continent. A frequent location of frontogenesis in summer is roughly southeast—northwest through the Cape Leeuwin—Kalgoorlie area, with a stationary or slow-moving col to the south. Another preferred position lies about the coastal areas of western Tasmania and the Coorong of South Australia. These latter... once formed, are probably not distinguishable from those which form in other regions without the presence of a parallel coast" (CLARKE, 1960). On the other hand, the differences between these larger scale fronts and the sea-breeze front are clearly shown by their mesostructure, as revealed by the streamlines and isentropes obtained by CLARKE (1961). They have a life of several days, sweep over large areas, and their deformation field is hyperbolic (CLARKE, 1960).

These cool changes during the summer (December–March) "take a unique place among the weather changes in any part of the world. Rain... is rare and sporadic. Yet, the changes in wind, relative humidity, and temperature (Fig.15) may be regarded as regular and beneficial breaks in a regime typical of semi-arid tracts. Thus, in the inhabited coastal regions, spells of extreme heat are very rarely protracted and it is usually not difficult to predict in a general way, a day or so ahead, that a (hot) spell is imminent or

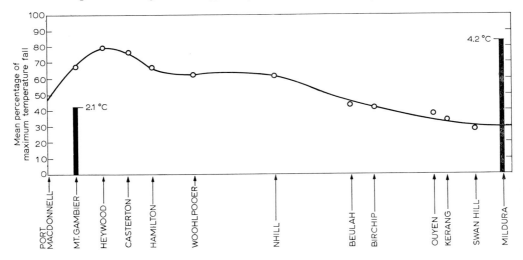

Fig.15. Mean fall in temperature at the time of a cool change at Mount Gambier, S.A., and Mildura, Vic. (large figures and thick vertical bars) and mean percentage of the largest temperature fall (scale on left) on each occasion for 15 dry cool changes, at each locality shown by the arrows. The mean percentage of the largest fall decreases slowly towards the interior, but on the other hand the actual largest fall is much greater there than near the sea, because of the much higher maximum temperature. (Adapted from BERSON et al., 1957.)

about to come to an end... Cool changes, especially well-marked and abrupt afternoon or evening changes, have a tendency to arrive earlier in southern Victoria than indicated in the forecasts based on the synoptic map" (BERSON et al., 1959).

To the west of the Australian land mass the sequence of oncoming anticyclones is a planetary phenomenon relatively undisturbed by continental influences. The synoptic chart shows that the usual alignment of the front occurring between two successive anticyclones is from northwest to southeast. In such a situation the overheating of the land surface causes a pressure trough to form along the west coast. When this trough is well developed, the trough line begins in the tropical latitudes and extends south along the west coast. At Perth it usually precedes the front by some 500 km and 12 h. The air that flows into the trough comes first from the northeast and then from the north. Between the trough line and the front the surface flow has a westerly component and may be humid *Tm* or dry *Tc*, depending on the area of origin and the trajectory. The trough itself is a pericontinental summer phenomenon which entails a northerly or northwesterly flow in the upper troposphere. There may be cloudiness and perhaps thunderstorms if the tropical air is very unstable. A kinematic chart can show the successive positions of fronts and trough lines as they appear on a sequence of synoptic charts. The interval between the trough line and the front that follows it becomes narrower as the whole system moves eastwards, and often disappears within a day or two.

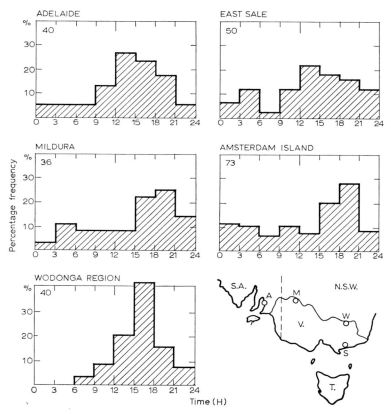

Fig.16. Time frequencies of cool-change passages, at 3-hourly intervals, in southeastern Australia and at Amsterdam Island in the Subantarctic. At Adelaide and East Sale, near the shore, the change arrives very early, aided by the usual sea-breeze. In the Wodonga region (inland Victoria) and at Mildura, which is even further from the sea, the greatest frequency of cool changes is after 15h00. (Adapted from BERSON et al., 1957.)

If the inter-anticyclonic trough is narrow, there may or may not be a weak front marking the line of change. The pre-change stream is a northeasterly, very hot and dry. The contrast in temperature and humidity between pre-change and post-change streams is quite pronounced, but the dryness of the northeasterly allows little or no cloud formation. If the front comes near the shore just before dawn, its progress is delayed for a few hours by the shallow land-breeze. If it arrives around noon, its progress is speeded up by the strong development of the sea-breeze. These combinations account for most of the frontal "preference" in this coastal belt for the forenoon and afternoon hours. Thus the sea-breeze to some extent overshadows the true frontal cool change (GENTILLI, 1969). On the south coast, from South Australia eastwards, there are interrupted highlands which have some local effect on the advancing front.

If it arrives in the afternoon, between 13h00 and 19h00, the cool change is accelerated or strengthened and at times deflected by the sea-breeze, which adds its local effect (Fig. 16). In the early morning, on the contrary, the cool change is weakened or retarded and somewhat deflected by the offshore land-breeze, at least to a depth of a few hundred metres above the ground. Above this level, the cool change may be effective to a height of 4,000–5,000 m near the southern shore, 2,500–3,000 m towards the interior.

The cool change is often double, the leading change preceding the trailing change by nearly 200 km and about 3 h (Fig.17). BERSON et al. (1959) made a detailed analysis of the air in the vicinity of a cool change complex and discussed various theories on the subject.

There is a most dramatic description of the effects of a sudden cool change which hit Adelaide on February 1, 1964. Wind gusts of up to 103 km/h followed a 3-day heat wave. Several buildings were unroofed, plateglass windows were shattered, fences were flattened, and trees snapped like matchsticks in the southeasterly blast. Two tall radio masts crashed in the city and a 20 m steel chimney toppled onto a factory. A workshop built of stone collapsed. Widespread areas were blacked out when branches ripped from trees were flung against power lines. At varying times from midnight, eleven suburbs were without electricity. Fruit trees in suburban gardens were stripped, and the persistent wind made tatters of canvas blinds.

Usually the cool changes are conditioned more by the heat and dryness of the pre-frontal air than by any other factor. The southerly component of the cool air is weak when compared with the westerly component, which at times may be remarkably strong. LOEWE (1945) and BERSON et al. (1957) found a much greater frequency of cool changes in the afternoon, and because of the general trend of the coastline from Adelaide to Melbourne, it is suggested that this is to a great extent again due to the sea-breeze effect. This is supported by the fact that BERSON et al. (1957) found that the highest mean wind-speeds associated with a large number of cool changes occurred in the afternoon between Mount Gambier (37°48′S 140°40′E) and Essendon, Victoria (37°39′S 144°28′E) while between Essendon and Gabo Island (37°32′S 149°55′E) the highest mean speeds occurred in the morning. In the former area the coastline has a general northwest–southeast trend and both cool change and sea breeze travel in the same direction. In the latter area the trend is towards the cool change, or at most slightly against it. In general, these cool-change fronts increase their speed as they progress eastwards; this was thought by HUNT (1894) to be a significant factor in making the cool changes so remarkable along the New South Wales coast.

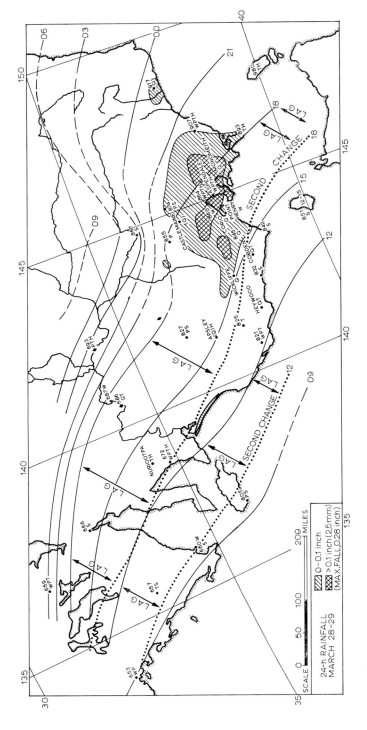

Fig.17. Isochrones of cool changes, showing the 3-hourly progress northeastwards of the first cool change (thin lines, small figures) and two 6-hourly positions of the second cool change (dotted lines, large figures) with short arrows showing the lag between first and second changes at 12h00 and 18h00 respectively. Precipitation shown by shading. (Adapted from BERSON et al., 1957.)

On the coast of New South Wales, the sudden change from one summer anticyclone to the next may take the striking form of a *southerly buster* (or *burster*), heralded by a typical roll cloud "as if a thin sheet of cloud were being rolled up before the advancing wind... The change is sometimes very sudden. There may be a fresh northeastern breeze, and in ten minutes a violent gale from the south. These 'bursters' usually end with a thunderstorm and rain" (WALLACE, 1893). The drop in temperature may vary between 10° and 20–22°C in 24 h, being very pronounced at first, and more gradual later on. The wind then reaches an average speed of 30 knots, with gusts up to 70 knots, and often with thunderstorm activity (HUNT, 1894).

Southerly busters occur about 30 times per year on the average; in some years their frequency may be down to one-half, in other years it may be even doubled. It should be noted that until the early 1830's (e.g., BRETON, 1833), the term "brickfielder" was synonymous with "southerly buster", because the southerly busters used to blow across Sydney's brickfields, sending heat and dust over the settlement. Then the town grew, the brickfields were moved out and the usage changed.

The southerly buster differs from the south-coast cool change because: (*a*) it is more definitely a southerly wind; (*b*) it is stronger and comes with a squall; and (*c*) it very often brings the characteristic cloud roll. These characteristics will now be discussed in turn.

The alignment of the southerly buster's front is definitely different from that of the usual cool-change front. The characteristic change to a post-frontal airstream with a southerly direction along the coast of New South Wales is partly due to the combined effect of the Eastern Highlands and Bass Strait. As was pointed out by HUNT (1894), the highlands exert a slight braking effect on the advancing anticyclone. TAYLOR (1920) also stated that the special features of the southerly buster are largely due to topography. A most significant factor is the contrast between the rough surface of eastern Australia, and the smooth water expanses of Bass Strait and the Tasman Sea. As MARES (1930) put it "due to an irregularity in the trough line resulting from a more rapid easting being made over Bass Strait than over the mainland, the southerly bursts initially on the extreme south coast... As a rule its rate of movement northward is about 20 miles per hour, so that after reaching Jervis Bay it should be experienced at Sydney about four hours later." BERSON et al. (1957) were able to trace the "change" lines over the mountains of south-eastern New South Wales. They found that most changes experience a strong distortion, "penetrating far along the coast before their effects were felt in the mountains..." In the eastern part of the mountains the post-frontal wind shift "is to the southeast, instead of southwest as over Victoria." There was also evidence of a similar frictional braking over the highlands of Tasmania.

The speed of progress of the southerly buster is not great; it moves along the coast of New South Wales at a speed which is little more than half its former speed of progress along the south coast. The force of the wind, however, is much greater. The writer estimates as 14 knots the speed of the post-frontal wind in a small number of west-coast cool changes (less than the 17-knot speed of the sea-breeze for the corresponding period), while BERSON et al. (1957) found a mean wind speed of 19 knots in 312 cool changes in South Australia and Victoria, and HUNT (1894) gave a mean speed of 28 knots for 991 southerly busters at Sydney. Indeed, the southerly buster comes in very strong gusts: on January 30, 1966, it capsized hundreds of small boats between Port Kembla and Newcastle, damaging many.

Cloud rolls normally do not accompany cool changes on the west and south coasts because the pre-frontal northeasterly is hot and dry, but higher-level clouds may be present near the front and travel with it. By contrast, along the coast of New South Wales cloud-rolls frequently accompany the southerly busters.

Such combination of peculiarities is unique to the southeast coast; its main cause may be sought in a combination of local features. The cloud roll and the strong initial squalls of the post-frontal stream are conspicuous along the coast between 29° or 30° and 38° to 40°S where there is a steep scarp close to the shore, with a ridge of high country (above 1,000 m) immediately landwards. At times the hot continental air spills over the scarp and gives the coastal stations their highest temperatures. Often, however, it is contained up to an altitude of at least 1,000 m by a layer of maritime air, while the pressure trough still moves eastwards over the Tasman Sea. Air which flows into it from the northeast is of tropical origin, warm, moist and unstable. The characteristic conditions of the southerly buster only occur in this narrow coastal band, 45–90 km wide. Further landwards and seawards the southerly buster loses its cloud roll and most of its squalls, and becomes a common cool change again.

It is possible that this combination of topographic, oceanographic and aerologic conditions is necessary to give the southerly buster its typical characteristics, with the result that it is localized to the central and southern coast of New South Wales. It fails to occur on the Queensland coast, north of 29° or 30°S. During the warmer part of the southern year, September to April, the anticyclonic cells mostly travel far enough to the south to send the post-frontal airstream along the coast of New South Wales as a southerly. Further north this airstream becomes deflected to the westward by the quasi-stationary low that occurs over central Queensland, and a near-monsoonal situation arises. In the cooler months between April and September, when there are practically no southerly busters, the anticyclonic cells have tracks further to the north and are usually separated by wider troughs. Any southerly airstream north of latitude 30°S then gradually diverges from the coast, which follows a different alignment.

The trade winds

The winds which leave the anticyclonic spirals on the equatorward side are the southeast trade winds (tropical easterlies). There is no clear-cut definition of these winds, except that they are more or less southeasterly and very constant. This constancy may be taken over a short period (e.g., the wind to blow from the same direction 80% of the time or more during a given month) or over a long period (e.g., to blow at least 50% of the time during 4–6 months). Only in the winter months do southeasterly winds exceed a 50% frequency and this only in northeastern Australia.

As a rule, the direction varies only slightly from southeasterly (just north of each anticyclone) to easterly (north of the pressure col between two anticyclones). Because of the succession of anticyclones further south, the direction of flow of the trade winds is thus subject to a 5–6 day cycle (Fig.18).

The air of the trade winds is unstable, but it is held down by the *trade wind inversion*, very pronounced over the sea. Over the continent, in winter, trade-wind air is cooled by the land surface and can travel a long way towards the equator and beyond, always

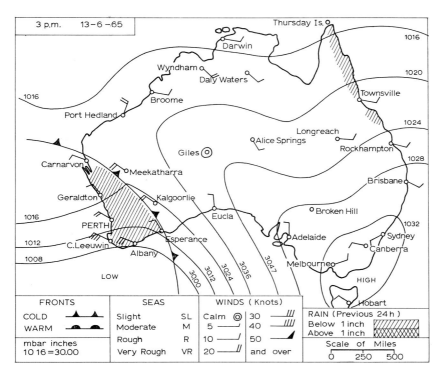

Fig.18. The trade winds over Australia in winter—a typical situation from Wyndham to Townsville, with rain along the narrow coastal belt north of Rockhampton (the "Trade Wind Coast"). The slight deflection of the wind at Broome is due to the col between the anticyclone shown on this map, and the one following from the Indian Ocean; this deflection will travel eastwards with the anticyclonic system. (West Australian weather map.)

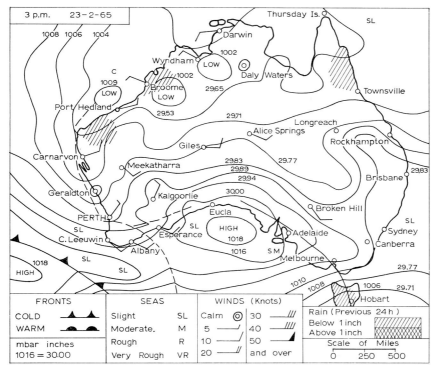

Fig.19. The contraction of the trade-wind belt in summer, caused by the monsoonal lows and thunderstorms in the north. (West Australian weather map.)

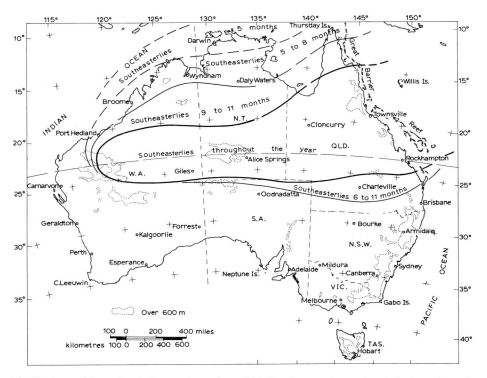

Fig.20. Australia's trade-wind core, shown by a thick line, has southeasterly winds throughout the year, with only slight variations in their direction caused by the passage of troughs (cf. Fig.18). To the south is the belt of anticyclonic centres, with rhythmically variable winds, and to the north the summer monsoonal belt with its northwesterlies.

keeping near the ground. In summer, the overheated land surface makes the air much more unstable, and numerous thunderstorms result over northern Australia, especially when the trade wind *Tc* air meets the even more unstable monsoonal *Tm* air (Fig.19). Only a short tract of the eastern coast is subject to the trade winds the whole year round, or nearly so. This is the stretch from Cooktown to Bundaberg (Fig.20). Farther south, the travelling anticyclones bring variable winds, among which southeasterlies still abound, but are by no means dominant, except in summer as far south as Brisbane Between December and February, the heating of the interior causes some deflection of the winds, which assume a more easterly component, but may still be called modified trade winds, because the modification does not warrant the name of "monsoon". At Cooktown, for instance, easterly-component winds occur in 45% of the mornings and 73% of the afternoons in January, when westerly-component winds occur in 41% of the mornings and 20% of the afternoons. Since land-breezes here come approximately from the west and sea-breezes from the east, it may be said that in January a trade-wind element is noticeable in at least 45% of the observations, and a monsoon element in at least 20% (Fig.21, 22).

Most of the southeasterly winds recorded at Sydney in the afternoon are mere sea-breezes, as is shown by the almost complete absence of similar winds from the morning records. At Lord Howe Island, only a little farther north than Sydney, but away from continental influences, southeasterly winds have a frequency of more than 20% between November–March, and, all in all, are the most frequent winds, although by no means

dominant. It may be validly argued that, since they occur *within* each anticyclonic system, they alternate with winds from every other direction as the anticyclone advances, so that their constancy, as distinct from frequency, is very low.

At Brisbane, unmistakable trade winds occur in the warm semester, when several anti-cyclones follow a more southerly path, but only in October–December are they more frequent than any other wind—and very slightly so (Fig.21, 22).

Inland and further north, at Charleville, Queensland, the trade winds are more frequent in winter than in summer, and at the tropic their dominant role is evident, with a 50% frequency in March and April, and nearly 40% in October, January and February.

At Cloncurry, Queensland, in the interior, the winter predominance of the trade winds is well established, with about 50% frequencies in May and June, and more so at Daly Waters, Northern Territory, where the June–July frequencies near 70%. Thursday Island has the greatest frequency of southeasterlies, over 80% of all winds from May to September (Fig.21, 22).

Towards the western coast, the problem is complicated by oceanic factors in the winter and by heat transport in the summer, so that it is only from Broome northwards that the trade winds are clearly recognisable in the winter.

Trade wind air is still recognisable in the cyclonic spiral that occurs almost permanently during the summer around the Pilbara–Hamersley area (Fig.23; cf. Fig.5). However, the change in the direction of flow of this trade-wind air is so pronounced that it must be discussed together with the monsoon (see following section).

On the eastern coast, north of Brisbane, a similar deflection of the trade winds takes

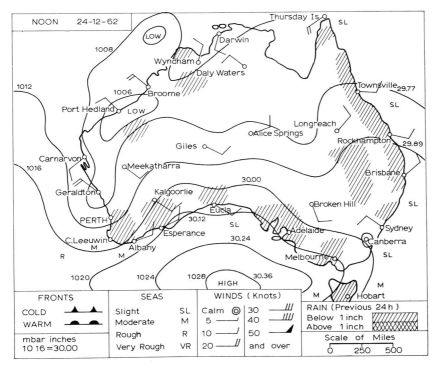

Fig.23. The heat low and the deflection of the trade winds. The heat low owes its existence to the transport of heat by the trade winds, but on the other hand it has a strong effect on their direction in the northwestern region. A noticeable deflection occurs also around the low, not shown, over northern Queensland, cf. Fig.32. (West Australian weather map.)

place around the Queensland low, and the wind often comes as an easterly or even northeasterly, instead of southeasterly (Fig.23); this effect may be felt hundreds of miles offshore.

The monsoon

By definition, a monsoon is a wind which reverses its direction with the seasons. The occurrence of such seasonal reversal may be verified from surface data for oceanic and for inland stations, but at coastal stations the seasonal wind is usually overshadowed below 600–700 m by the land-breeze, in the morning, and by the sea-breeze, in the afternoon. The *seasonal* wind near the surface, as distinct from the diurnal breeze, is, therefore, better revealed by the morning records if onshore, and by the afternoon records if offshore.

An analysis will be made of the Australian flow patterns which in some aspects at least may be considered monsoonal according to these criteria. Early works must be studied with caution because of the tendency (e.g., HUNT, 1908; QUAYLE, 1918) to term "monsoonal" any barometric trough or depression which originated or passed over the interior. In the winter, most of northern Australia is dominated by a flow of southeasterly *Tc* or *sTc* air (trade winds, Fig.18) which observers in the last century used to call the "winter monsoon." This pattern is only slightly disturbed by the variations in pressure that occur further south, due to the passage of successive anticyclones separated by cols or troughs.

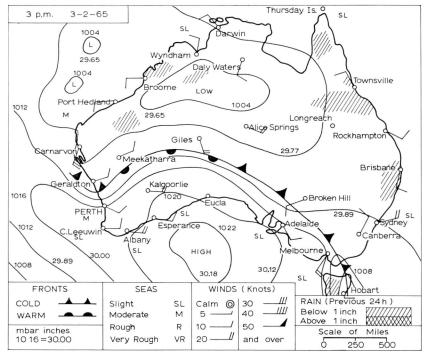

Fig.24. The summer westerly and northwesterly flow over northern Australia. This is a typical monsoonal situation, with multiple loci of low pressure and scattered convectional rains. It is essentially variable even within a few hours, and is likely to be replaced by a southeasterly flow (trade winds) after a few days. (West Australian weather map.)

This steady winter pattern may be contrasted to the much more variable summer pattern.

In the summer, because of the overheating of northern Australia, which causes a noticeable drop in pressure, the northern edge of the trade-wind belt is truncated, and *westerly* or *northwesterly Tm* or *Eq* winds take its place (Fig.23, 24).

The onset of the monsoon has been studied by TROUP (1961), who considers monsoonal bursts "as due to the movement of cyclonic vortices, often small in horizontal extent and weak in pressure gradients, along preferred tracks which tend to be displaced further poleward as summer progresses". Early thunderstorms and isolated falls of rain are not enough to characterize the onset of the monsoon, for which Troup considers necessary a fall of rain over an area, or at 4 or more stations out of 6 in a sample. This prerequisite led to the exclusion of 56% of the November rainfall and 12% of the January rainfall.

TROUP (1961) also found that, in four years, the monsoon brought 1–6 spells of heavy rain per year; a spell of heavy rain is a period of N days in which the total rainfall exceeds 19 ($N + 1$) mm, provided that no day is totally dry or no two consecutive days total less than 19 mm. These spells of heavy rain indicate the presence of the I.T.C.Z.; monsoonal rains are always intermittent. Westerly winds are present with the onset rains, but not always. There is a marked negative correlation between monsoonal rainfall and strength of the upper westerlies in low latitudes, "attributed to subsidence on the equatorward side of the jet stream, and to restriction of convective cloud development by strong vertical shear". Usually the onset of the monsoon corresponds to an almost sudden shift southwards of the subtropical jet, late in December or early in January, by some 10 degrees of latitude.

It has already been shown that the low-pressure pattern is extremely variable; the limits of the conflicting streams of southeasterly and northwesterly air accordingly change from day-to-day, and "average positions" become somewhat mythical. The *southeasterlies* (trade winds) blow over the interior and usually reach diagonally as far north as 25°S in the northwest (where they end into the heat low which they contributed to build up), 20°S near the Kimberleys, and 18°S near the Gulf of Carpentaria and Cape York Peninsula (Fig.24).

The *west-northwesterly* streams, which include the monsoon proper, are better studied by longitudinal regions, according to their characteristics.

From about 113–124°E and south of about 15°S latitude, the prevailing surface winds are from the west-southwest, very shallow and dry. In January, Broome has northwest winds at 500 m in 8% of the observations, at 1,000 and 3,000 m in only 5%. Easterly-component winds are 40% of the total at 1,000 m, 66% at 2,000 m, and 85% at 3,500 m. This shows that the *primary trade wind* still dominates the upper air at this time of the year. The dominant surface westsouthwest winds are streams of modified southerly *sTm* or *sTc* air, deflected around and into the northwestern heat-low (Fig.24).

From about 124°–130°E and south of about 12°S latitude, the prevailing winds blow from west-northwest and are mostly of southerly origin, but at times may include transequatorial streams.

From about 130°E onwards, and north of 18°S, northwesterly (monsoonal) winds are important, but even in Darwin they are not as frequent as the southeasterlies (Fig.21 and 22). At Darwin in January, westerly-component winds occur 70% of the time at 500 m, 65% of the time at 1,000 m, 50% at 2,000 m, 30% at 3,500 m, and never at 5,500 m (where, on the other hand, easterlies occur most of the time). If a distinction is made between northwesterly and westerly surface wind, we find that in December and January the monsoonal northwesterly is the most frequent surface wind by a very narrow margin indeed. Westerly winds are almost as frequent in December–January and more frequent in February; they may be a continuation of the summer winds of the Kimberleys. Calms or local thunderstorms are almost as frequent as these winds, as may be expected when the intertropical convergence is over the area.

At Groote Eylandt, summer winds are 20–30% from the northwest and 20% from the north. Calms also number some 20% (Fig.21, 22).

At Thursday Island, in January–February, more than 40% of the wind comes from the northwest, as against less than 15% from the southeast. From April to November, more than 50% of the wind comes from the southeast. These figures show that the trade winds, blowing unhampered over the Pacific, are dominant by far, but that on many summer days, though by no means on all, the monsoon takes over. Calms are rare.

At Townsville, Queensland (19°10′S) and Cloncurry, Queensland (20°45′S) northwesterly winds are infrequent, and further south they are rare (Fig.21, 22).

Summing up, low-level westerly winds are frequent, but very shallow, during November and December, when the northerly component is also prominent. In January (Fig.25) and February, westerly winds become much more prominent at low levels and up to 3,000 m, but the northerly (transequatorial) component is relatively uncommon. Dep-permann (1943) shows these westerlies as part of the flow of "South Indian westerlies", whereas his "Northers" (monsoonal streams) end at about 9°S.

McDonald (1938) shows rains in less than 10% of the December–February observations all along the northern Australian coasts, with the exception of the Kimberley coast, where the direct onshore advection of maritime air, combined with orographic factors, trigger off some 80–120 thunderstorms a year, and where observations of rain increase to 10–15%. Obviously the zone of convergence, and not a very active one because of the almost parallel streams which cause it, is usually outside the Australian continent.

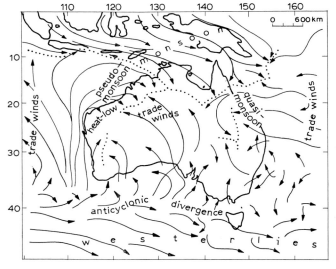

Fig.25. The mean flow pattern in January. The words show the terms used in the text; the dotted lines are only indicative and show very labile boundaries, apt to vary with the weather situation of each day.

The main Australian monsoon, some 3,000 m thick and at times coming from beyond the equator, is intermittent, and mostly limited to the far north, east of 130°E, and to January or early February. Occasionally it occurs earlier or later, rarely does it spread further south. In February 1955, a unique northwesterly monsoonal stream, over 2,000 km wide, overrode the easterly and anticyclonic air for some 2,500 km over most of Western Australia, reaching beyond latitude 32° south and bringing saturation point and torrential rains to over 2,500,000 km² of land.

West of 124°E, there is a distinct monsoonal flow, localized and shallow, and in reality only a shallow stream of southeasterly (trade wind) air deflected by the Pilbara–Hamersley heat-low (Fig.25); above this shallow northwesterly, the southeasterly flow continues unhampered. Between 124–130°E, either thick northwesterly or shallow westerly air may occur.

In the southeast, south of about 18°S, monsoonal winds are very uncommon, but a *monsoonal tendency* may be detected in the deflected trade winds, which become easterlies or even northeasterlies (Fig.23, 25).

The evidence therefore shows that northern Australia has a monsoon from November to March, but this monsoon is usually shallow (about 1,000 m) in November–December and in March. It is only a stream of eastern Indian Ocean air (trade wind air) deflected by the northwestern Australian heat-low and, therefore, not connected with the Asian winter monsoon. The relative humidity of the Indian Ocean air-stream is not very high by the time it reaches so far north, and so the amount of summer precipitation is moderate, and more variable than world average. It is true that during January and February the air streams issuing from Asia often push southwards far enough to reach the Australian coast, but they are only intermittent. Most of northern Australia's rainfall, which occurs within the westerly monsoonal stream, is due to local instability thunderstorms of convectional origin (Fig.19).

Even when a true monsoonal stream reaches northern Australia, it seldom penetrates far inland, mostly remaining north of Cambridge Gulf and the southern shore of Carpentaria. Occasionally it reaches further south, and very seldom indeed beyond the tropic. This was observed and reported, e.g., by WICKHAM (1846).

BERSON (1961) calculated the energy balance of the Australian monsoon; he found that "the lower branch of the circulation is self-maintaining and in addition provides more than the energy necessary to sustain the circulation in the overlying troposphere where the net loss of sensible heat is incurred. The remaining excess energy is transferred... to the north- and southward with rather indefinite net zonal transfer. Excess energy is there made available to be consumed in reducing radiative cooling or by release of potential energy during descent... The effect is much smaller than on the winter side of the equatorial trough zone at large where it represents a substantial item in the heat budget."

The expanse of water (Timor, Arafura, Coral seas) to the north of Australia, remaining cooler than the land, keeps summer temperatures down and thus weakens the intertropical convergence, which is so effective over the large land masses of equatorial Africa and South America. Northern Australia has much more pronounced continental characteristics than the much bigger South America. The "Australian monsoon" has been grossly overrated; monsoonless continents have a much better tropical rainfall than Australia, which in the corresponding latitudes is not only drier, but also hotter.

Tropical cyclones (hurricanes)

Because of its shape, broadly trapeze-like north of the tropic, Australia is the only continent which experiences almost the same incidence of tropical cyclones or hurricanes on both its western and eastern coasts.

Tropical cyclones are not common events. Queensland and the Northern Territory combined have 3.3 per year, on the average; from 1870–1955 inclusive, the western coast recorded 178, or an average of 2.1 per year. It should be noted, however, that earlier records were generally incomplete, especially over wide expanses of ocean and before the use of weather satellites, and that the definition of tropical cyclone has varied from time to time and place to place. Many low-pressure systems in tropical Australian waters never reach hurricane force; they are classified as "tropical depressions" (or "tropical disturbances") if their wind speed does not exceed 33 knots (Fig.26). If the wind speed is greater, they are called "tropical cyclones" and are classified of major entity if the winds exceeding 33 knots extend beyond 100 miles (about 160 km) from the centre.

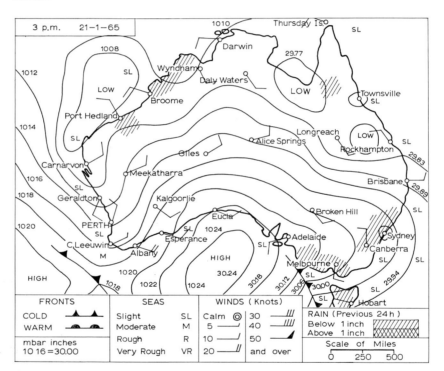

Fig.26. Two tropical depressions over northern Australia: the northwestern one could well develop into a tropical cyclone while still over the sea. Such depressions may bring copious rains without violent winds. (West Australian weather map.)

The highest wind speeds recorded in Australia occurred in gusts during tropical cyclones, 109 knots at Willis Island, Queensland, on 8 February 1957, and 125 knots at Onslow, Western Australia, in February 1963.

A number of tropical cyclones recorded in Western Australia pass over the desert and would have gone unnoticed but for the thoroughness of meteorological analysis; also, some tropical cyclones, formed in the Timor, Arafura or Coral seas, travel for thousands of miles over the ocean and may never be recorded, unless detected by a satellite or reported by a ship or an aircraft. Today, no cyclone may go undetected.

In 1917, the western coast recorded five tropical cyclones, and sequences of three per year are not rare, while on the other hand, there have been periods of two years without any (Table II). On the eastern coast, there were six tropical cyclones in 1940, and on several occasions there were 3 or 4 in one year.

TABLE II

MONTHLY FREQUENCY OF TROPICAL CYCLONES IN WESTERN AUSTRALIA

Decade	Aug.	Sept.	Oct.	Nov.	Dec.	Jan.	Feb.	Mar.	Apr.	May	June	July	Total
1870–1879	–	–	–	1	4	3	7	4	2	2	–	–	23
1880–1889	–	–	–	–	1	7	5	3	1	–	–	–	17
1890–1899	–	–	–	1	3	4	–	1	1	–	–	–	10
1900–1909	–	–	–	–	1	5	2	3	3	–	–	–	14
1910–1919	–	1	1	1	1	5	4	6	–	–	–	1	20
1920–1929	–	–	–	–	4	3	5	10	1	–	–	–	23
1930–1939	–	–	–	–	1	5	4	6	–	–	–	–	16
1940–1949	–	–	–	2	2	8	7	12	2	1	–	–	34
1950–1959	1	1	1	1	5	7	11	7	6	–	–	–	40
1960–1969	–	–	–	–	2	12	12	9	2	–	–	–	37
1870–1969	1	2	2	6	24	59	57	61	18	3	–	1	234

The calendar year is a most unsuitable period for the recording of tropical cyclones, which are most frequent from November to March (Fig.27); for instance, the official list of Western Australian tropical cyclones for the year 1917 shows five occurrences, one in January, two in March, one in October and one in December, whereas the actual cyclone season, October 1917 to March 1918, recorded only three in October and De-

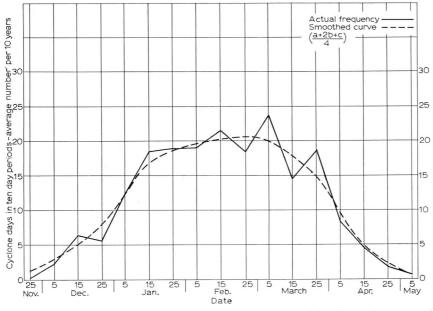

Fig.27. Frequency of cyclone days (in 10-day periods, averaged for 10 years), over northern Australia generally. (After BRUNT and HOGAN, 1956.)

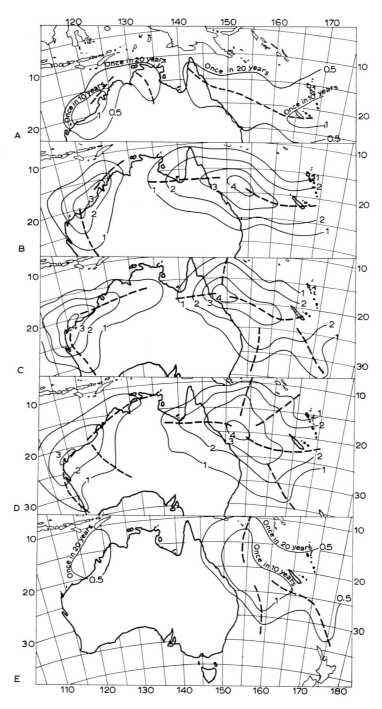

Fig.28. Mean 10-year frequency of tropical cyclones crossing 5° tesserae, shown by means of isopleths (thin lines) and small figures. The heavier dashed lines show the axes of maximum frequency. Notice the very low frequencies in December (A) and April (E), and the changed pattern in February (C) compared with January (B) and March (D), because of the more frequent monsoonal situations in that month. (Adapted from BRUNT and HOGAN, 1956.)

cember 1917 and in February 1918. In 1923, there was a violent cyclone which hit Carnarvon, Western Australia, in January, then on March 8, two cyclones were recorded, one of them off Wallal, Western Australia, the second one between the northwestern coast and Timor. Between March 18 and March 24, the fourth cyclone of the season occurred from Condon, Western Australia southwards. Had it occurred only a few years earlier, the third cyclone of this season would have gone unnoticed, or at least unrecorded. The season with the greatest number of recorded cyclones in Western Australia is 1950–1951, with a cyclone in December 1950, five cyclones in February 1951, and one in March 1951, a total of seven. And yet only the December 1950 cyclone passed over the mainland—two of the February 1951 cyclones skirted the northwestern coast without crossing it and the remaining cyclones of the season were far out at sea.

The track of any tropical cyclone is a function of several variables, namely the rotation, intensity and size of the cyclone itself, the sphericity of the earth and its angular velocity at the points concerned, and the surrounding meteorological conditions, with special regard to pressure and moisture.

Tropical cyclones do not originate within 4° south and north of the equator, and probably not beyond 20° north or south (Fig.28). The tracks which appear to originate

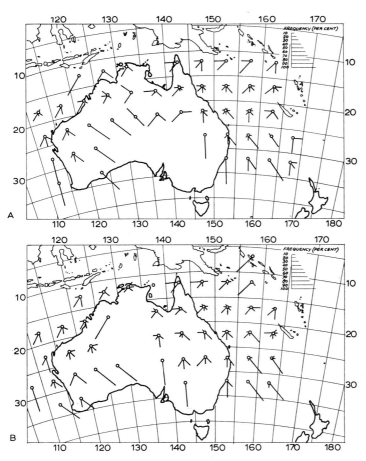

Fig.29. Direction of movement of tropical cyclones through 5° slabs in February (A) and March (B). The maps show quite clearly that the critical latitudes for tropical cyclones are those between 15 and 20°S in February and 15 and 25°S in March; changes of direction are quite frequent there, while they are most uncommon once the cyclones have progressed further south. (After BRUNT and HOGAN, 1956.)

at 22–24° north or south, almost certainly failed to be detected while they already existed in lower latitudes.

On the poleward side, tropical cyclones may travel so far as to merge with a middle-latitude depression, which they intensify considerably. COOKE (1905) was the first to verify that some tropical cyclones from the northwestern coast "came overland to the Great Bight." His forecasts of the progress of the cyclone of February, 1902 proved remarkably accurate.

Tropical cyclones can only travel around the travelling anticyclones. Should a tropical cyclone meet an anticyclone head-on, it would be destroyed by the inflow of dry air at higher pressure. This cyclolysis is not rare in Australia between 15–25°S (Fig.29), where cyclones are likely to meet large anticyclones, especially towards the end of the cyclonic season (late March–April). On the other hand, a cyclone may slide along the col of lower pressure between two anticyclones, and reach the middle-latitudes without losing its identity, even though some of its characteristics may change. For example, no thermal fronts occur in tropical cyclones, which consist entirely of *Tm* air, but while a tropical cyclone travels further south between two anticyclones, it is fed by different air masses (*sTm* and *sTc*). Whereas these two air masses were similar enough to form, at most, only a stationary front between the two anticyclones, the large amounts of humid *Tm* air and the additional dynamic impulse supplied by the cyclone's rotation are enough to form an active front. If the cyclone works its way to middle high latitudes, it becomes indistinguishable from any frontal cyclonic depression. There are no observations of cyclones of this origin to the south of Western Australia, but there are several records of Western Australian tropical cyclones which have ultimately reached Tasmania or New Zealand as ordinary depressions (GENTILLI, 1956).

Information on all tropical cyclones which occur in the Australian region, with details of their tracks for every day on record, is now published annually by the Bureau of Meteorology.

The area affected by tropical cyclones varies considerably, not so much because of variations in the size of the cyclones themselves, but because many of them cover a short track only, and die out without ever reaching the mid-latitudes. Some travel further, but lose most of their identity. A few (e.g., only 7 out of 72 recorded in Western Australia between 1924 and 1955) reach the southern shore of the continent without any apparent loss of intensity. Most of these cyclones, whether they cross the continent or whether they die out in the tropics, follow fairly regular tracks, except in the critical cyclolytic latitudes (Fig.29). They originate in the Timor Sea, usually between 10°–15°S, and travel southwestwards at a speed of about 8–24 km/h. Between 20°–25°S most of them gradually recurve southwards and then southeastwards (Fig.29), so that some four out of ten cross the coast between Onslow, Western Australia, and Broome. Anomalous tracks are not rare, an outstanding example being provided by the cyclone of February–March, 1956 (Fig.30), which first travelled almost due east from La Grange, Western Australia, to a point northeast of Alice Springs, then almost reversed its course, crossing the shore in a westward direction near Cape Lévêque, gradually recurving parallel to the coast and passing to the west of Fremantle, crossing inland between Mandurah and Bunbury, and passing out to sea again about 80 km west of Albany, Western Australia. The exceptional length of the track, its coastal location, and the almost unabated force of the wind throughout make it quite outstanding (GENTILLI, 1956).

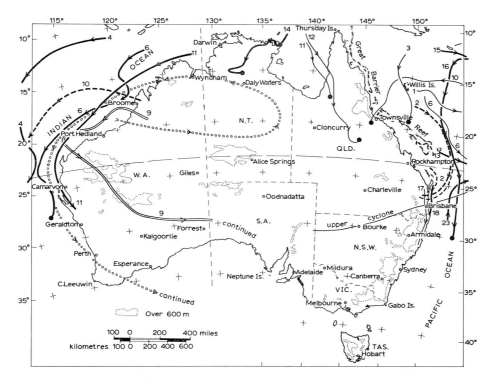

Fig.30. Tropical cyclones of the season 1962–1963. The meridian shown by a dashed line is the boundary between western and eastern regions; numeration is independent on either side. The thick lines (continuous or dashed or binary merely for graphic convenience) show tropical cyclones, the thin lines tropical depressions that did not attain high wind speeds. The final point of filling is shown by a dot. Only western cyclone No.9 (shown by a binary line) proceeded far enough to join a mid-latitude front. The eastern cyclone No.2 at one stage became triplets, as shown by the short dashed lines. An upper cyclone track is also shown. The beaded line shows the exceptional track of the cyclone of February–March 1956, which travelled well inland along an anticyclonic loop before coming out to sea again and describing a full trajectory. (Data from Bureau of Meteorology.)

The eastern coast cyclones usually originate between 8–15°S and 155°–165°E, near the Solomon Islands. They follow tracks similar to those of the western coast cyclones (Fig.28 and 29), and their characteristics are essentially the same. Most of them cross the coast between Cooktown and Mackay, Queensland, and soon lose intensity. The cyclone of January, 1918 gave the lowest Australian pressure reading of 933 mbar and brought 627 mm of rain in one day to Mackay, Queensland, where it caused considerable damage; the rainfall amounted to 1,411 mm in 3 days. A new lowest record was set in 1961, when the barograph trace at Onslow, Western Australia, descended well below the printed graph to 922 mbar; the exact time was 01h30 on January 25.

If the track that it followed after it had become frontal is added to its earlier track as a tropical cyclone proper, the cyclone of February 18, 1923 is quite remarkable because after having travelled around Australia, it went very nearly as far as Tasmania, covering some 7,500 km in 14 days. The track of the tropical cyclone of February–March, 1956 (Fig.30) extended for 5,950 km from its unknown inception off King Sound, Western Australia, to its exit west of Albany, Western Australia, 16 days later. It was still clearly recognisable as a frontal cyclone on the weather charts for March 5 and 6, thus adding some 1,600 more km to its original track, and reaching the exceptional track length of

7,550 km. It covered the longest distance over Western Australian territory of any cyclone on record. An extraordinary anomaly occurred in the case of this cyclone because it described a large loop *to the left*, i.e., anticyclonically (Fig.30).

Cyclone Audrey of January, 1964 had different peculiarities (WHIGHT, 1964). It continued for 8 days, from January 7 to January 15, and in that time traversed some 4,800 km along a path curving from Thursday Island, Torrens Creek, Northern Territory, Goondiwindi, Queensland and Newcastle, New South Wales, to the North Island of New Zealand. From Thursday Island to Mitchell River, it covered 480 km down the coast in 87 h, at an average speed of 6–8 km/h, but from Mitchell River to Burketown, a distance of 320 km, it took only 21 h, at an average of 13 km/h. By January 13, the speed had increased to 18 km/h, on the 14th to 50 km/h, and on the final day to nearly 100 km/h.

While it is more common for tropical depressions to develop into tropical cyclones, the reverse also happens. Cyclone Judy, of January 1955, which slowed down to become a rain depression, brought deluges of up to 350 mm to northern Queensland (Fig.31).

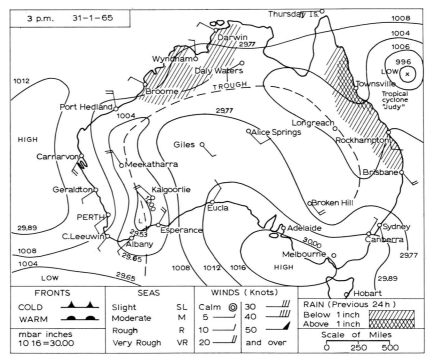

Fig.31. Torrential rain brought to the northeast coast by a tropical cyclone. (West Australian weather map.)

TABLE III

INITIALS OF AUSTRALIAN CYCLONES' NAMES

Warning centre	Initials of names to be used from 1963–1964 onwards												
Brisbane	A	D	G	J	M	P	T	W	C	F	I	L	P	V
Perth	B	E	H	K	N	R	U	A	D	G	J	M	R	W
Darwin	C	F	I	L	O	S	V	B	E	H	K	N	S	A

In general, the pastoralists agree in praising the cyclones' rains for their beneficial effect; they do promote the growth of excellent feed for the sheep. In the northwest, the rains may be heavy enough to make the dry Gascoyne River flow and reach Carnarvon, where banana and vegetable crops are irrigated, and to replenish wells and soaks (Table IV). On the other hand, the damage caused by the wind and floods of a single cyclone may run into many thousands of dollars.

TABLE IV

RAINFALL BROUGHT BY CYCLONES IN THE 1947–1948 SEASON,
WESTERN AUSTRALIA

Rainfall (mm)	Dec. 28–31, 1947 (km²)	Feb. 19–23, 1948 (km²)	Apr. 9–20, 1948 (km²)
< 25	278,000	470,000	375,000
25–50	131,000	250,000	207,000
50–75	101,000	240,000	82,000
75–100	49,000	230,000	16,800
100–125	37,500	172,000	5,400
125–150	28,500	111,000	4,400
150–175	17,200	67,500	3,100
175–200	7,700	59,000	2,100
200–225	1,500	44,000	1,500
225–250	–	19,500	1,000
250–275	–	15,500	–
> 275	–	7,700	–
Total area affected (km²)	651,400	1,686,200	698,300
Average (mm)	51.25	75.75	32.00

Since the first European settlement of northwestern Australia, in the 1860's, tropical cyclones have caused over 750 deaths and sunk or wrecked some 250 pearling luggers, a coastal steamer and two small freighters. The total damage to mainland property during this period amounted to several million dollars. The coastal steamer *Koombana*, only 2 years old and built for the northwestern run, disappeared on March 21, 1912, between Port Hedland and Broome, with the loss of 150 lives. Other disastrous cyclones occurred in 1875, off Exmouth Gulf, causing the total loss of a small pearling fleet with 59 men; in 1887, off Wallal, causing the loss of 18 pearling luggers from Broome, with 140 men aboard; in 1894, off Cossack and Onslow, sinking several luggers and drowning 40 men; twice in 1908, off Broome, almost wrecking the pearling fleet and drowning 100 men; in 1909, also off Broome, sinking 4 luggers with the loss of 24 men; and again in 1910, sinking 26 luggers and drowning 40 men. As late as 1935, a cyclone off Broome sank 20 luggers and killed 140 men. Only improved forecasting and communications, and to some extent, a reduction in pearling activities, have brought an end to the toll of human lives. The small town of Cossack never recovered from the 1897 cyclone. Onslow lost most of its jetty in the two 1958 cyclones, and there was some talk of abandoning the site. In the Kalgoorlie–Boulder area of Western Australia, houses have been flooded and railway lines washed away. Sections of the Yalgoo–Wiluna railway have been under 2 m of water. On many lines, including the transcontinental railway over the

Nullarbor Plain, trains have been stalled by floods for days, and the Kalgoorlie airport has been unserviceable for many hours after cyclonic rain has fallen.

A final, and minor, point: the term "willy willy" to denote tropical cyclones (and whirlwinds), is not from any aboriginal language and seems to have been coined some time during the last century by some colonial settler (BAKER, 1945).

Subtropical troughs and depressions

The subtropical jet stream, which plays such an important rôle in the dynamics of Australian climates, gives rise to most interesting and characteristic developments near the east coast. As CAMPBELL (1968) recalls, theory requires that there should be areas of divergence in the upper troposphere, at the right front and left rear quadrants of a jet velocity maximum in the Southern Hemisphere. The mean speed of the subtropical jet at 200 mbar is 77 knots at longitude 140°E in March, 92 knots in April, 116 knots in May, 129 knots in June (WEINERT, 1968). High maxima are therefore most likely to occur in May–June in the subtropical and near-tropical latitudes, where the preferred jet tracks lie at this time of the year, and in July further south. Jet maxima are slightly ahead of the inter-anticyclonic trough; the upper-air divergence, the upward motion, the dynamically induced instability at lower levels, the pressure trough at the surface, are all links in the one dynamic process. Cyclonic vorticity may result (CLARKE, 1956) but this is not always the case. CAMPBELL (1968) found a close correlation between rainfall episodes around Brisbane and Newcastle and jet speed maxima during the autumn. The correlation continues around Brisbane but much less near Newcastle during the winter, possibly because of the warm and humid air; cyclonic vorticity in altitude is always present during the major rainfall episodes. From the frequencies of the episodes studied by Campbell and the average frequencies of inter-anticyclonic troughs it may be seen that only one autumn trough in 7 or 8 shows cyclonic vorticity resulting in major rainfall. About 1 in 4 shows cyclogenesis resulting in any significant rainfall episodes around Brisbane and 1 in 5 around Newcastle. Very heavy rainfall in May–June over tropical Western Australia (e.g., Appendix, Tables I, XVII) is also associated with high jet maxima.

Mid-latitude depressions

Mid-latitude depressions usually occur well to the south of Australia, so much so that they used to be called "antarctic" (QUAYLE, 1915), but the winds and fronts associated with them have a deep influence on Australian weather and climate.

In the summer, in general, mid-latitude depressions are too far south and often too weak to affect Australia. The high frequency of depressions in the lower latitudes (Fig.32), is very largely due to the persistent heat-low which varies only slightly in its position according to the transit of successive anticyclones (see p.59); the remaining and very small share of that frequency goes to the tropical cyclones (see p.86). Only the insignificant mean frequency of 1–1.5 cyclonic centres in January, beyond 30°S, may be attributed to mid-latitude depressions. It should be noted that frontogenesis is slightly more frequent to the southeast of the continent than to the southwest, for the reasons outlined in Fig.6 and as is shown by the sequence in Fig.34, where at the end of December,

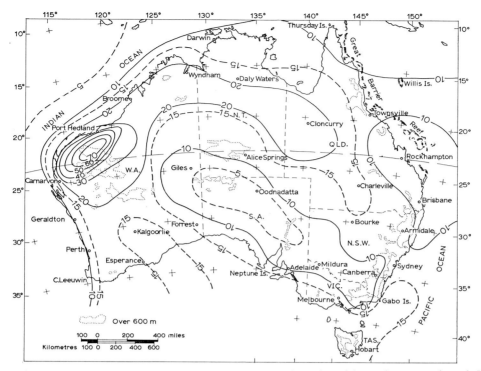

Fig.32. Number of depressions (× 10) in January. The total number of depression centres in each 5° slab is shown, averaged for a 10-year period. The northwest heat low (Pilbara low) is far more frequent than the Cloncurry low. The elongated area with fewer than 10 depression centres per slab in 10 years is the axis of anticyclonal transits during this month. Mid-latitude cyclones are well to the south of the mainland. (Compiled from data in KARELSKY, 1954, 1965.)

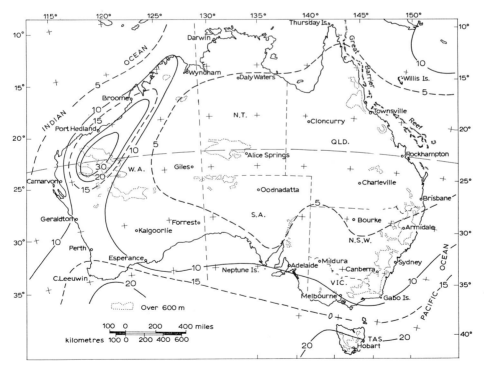

Fig.33. Number of depression centres in each 5° slab in April, averaged for a 10-year period. The northwest heat-low is now uncommon. Transits of mid-latitude depressions are a little more frequent and slightly closer to the mainland; they already control Tasmanian weather. (Compiled from data in KARELSKY, 1954, 1965.)

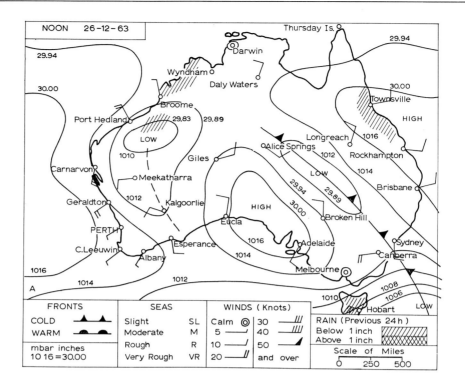

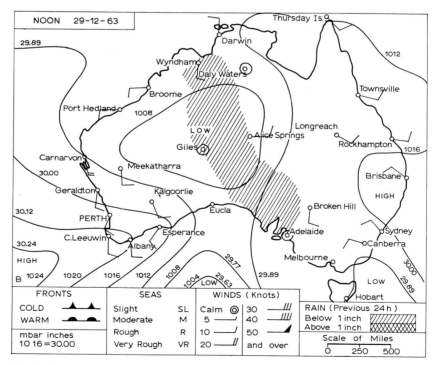

Fig.34. Summer sequence (26–31 December, 1963) from western trough to eastern inter-anticyclonic front. This sequence is unusual in its early development into a rain-bearing situation already over central Australia, probably because of an upper transgression of a strong monsoonal stream. (West Australian weather maps.)

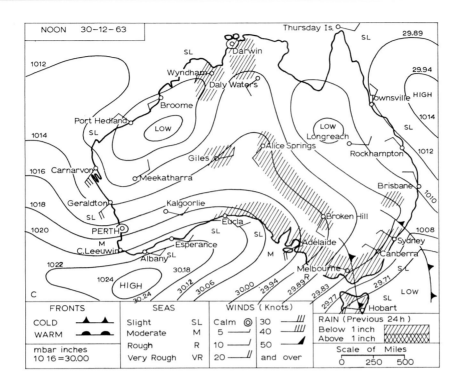

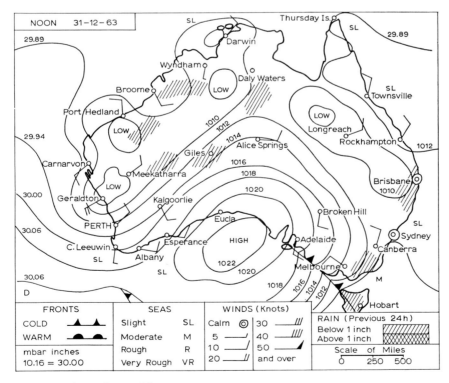

Fig.34C, D (Legend see p.96).

a trough on the western side of the continent (December 26, 1963) gradually gives way to a frontal situation by the time it has crossed the continent, and the westerly flow at the surface (December 30, 1963) uplifts *Tm* or *sTm* air instead of *Tc* or *sTc* as it was doing previously. No doubt a mid-latitude depression was travelling to the south of the continent throughout this period, but the northward extension of the fronts on December 30, 1963 only occurred with the change of the northeasterly flow from continental to maritime characteristics. It should also be noted from Fig.34 that the previous sequence, ended on December 26, 1963, produced an even more extensive frontal incursion over eastern Australia.

By autumn time, the frequency of western heat-lows and troughs has decreased considerably, and mid-latitude depressions travel slightly further north than in the summer, so that an average of two cyclonic centres is recorded by April (Fig.33). The northward extension of frontal situations occurs with increasing frequency (Fig.35) and does so already west of the mainland.

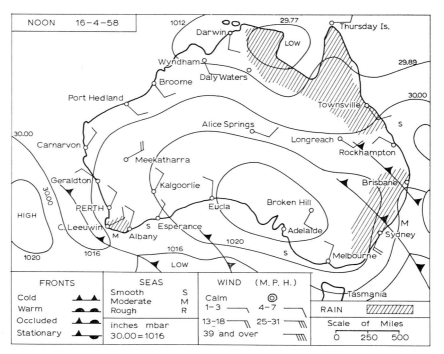

Fig.35. Day with greater frontal activity, April. The fronts are barely affecting the mainland except on the east coast, where they extend northwards into the inter-anticyclonic situation. (West Australian weather map.)

By the end of April or the beginning of May, the zonal circulation becomes more intense, the jet stream flows faster and mostly enters Australia at a lower latitude (cf. Chapter 3, Fig.8) and mid-latitude depressions go past much nearer to the southern coast. Well developed fronts do appear and bring the first widespread rains to the southwestern regions (Fig.36).

Satellite photographs show very clearly the arched bands of clouds which mark each front. The prevailing northwest–southwest alignment and the continuing parallelism of these bands as they travel eastwards across Australia are quite characteristic.

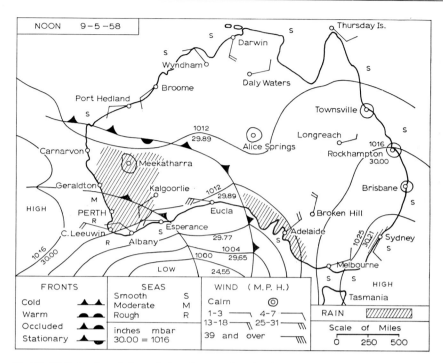

Fig.36. Well developed frontal situation causing the first winter rains to the westerly-exposed shores: the first front reaches the tropic. The centre of the depression is well to the south of the mainland, as usual, and cannot be shown on this map. (West Australian weather map.)

During the winter the closeness of the mid-latitude depressions increases very rapidly to its July maximum, with an average of three cyclonic centres skirting the southwestern coast, and passing over Tasmania during that month (Fig.37). At lower latitudes, the average pressure at the cyclonic centres is about the same on both the western and eastern coasts, namely 1,010 mbar at about 28°S (Fig.38). The pressure decreases rapidly towards the south. As the result of the anticyclonic flow of denser *Tc* air from the continent, the pressure within the cyclones tends to increase slightly as they progress eastwards, so that, e.g., the mean isobar of 1,005 mbar is pushed southeastwards by 7° of latitude, and the 1,000 mbar one by 5°. This effect is felt far to the southeast of the mainland (Fig.38). Apart from the extreme southwest and southeast of the mainland, no cyclonic centre (in the 1952–1963 period) was below 1,000 mbar (KARELSKY, 1965).

On geographical grounds, it was postulated (GENTILLI, 1949) that in winter, frontogenesis would occur weakly off the southwestern coast and more markedly off the southeastern coast. Mean cyclone pressure does not show this, but depression frequency (Fig.37) and minimum pressure (Fig.38, southwestern corner of each 5° slab) both do. By August, the depressions travel further south, and in the whole month only one depression centre is likely to skirt the mainland; the frequency is still slightly greater off the southeastern and southwestern shores (Fig.37) showing that some pericontinental frontogenesis still takes place.

With the end of the winter, depressions travel further south and their centres no longer reach the mainland, although they do pass over Tasmania. The depression centres over the mainland are mostly those of heat-lows and of peri-anticyclonic lows associated with heat-troughs, as is also shown by their general pattern of occurrence (Fig.40).

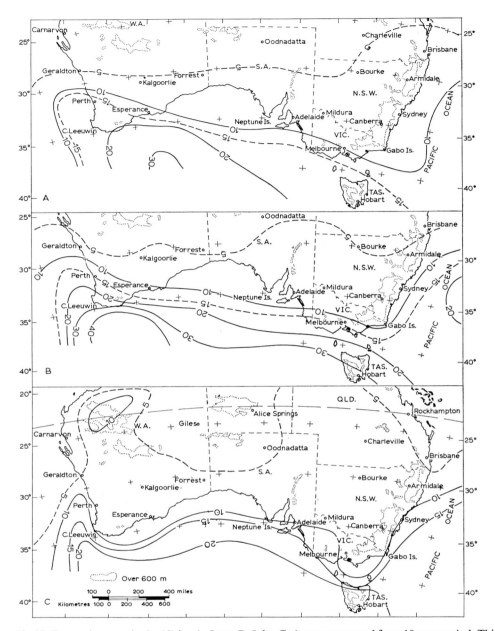

Fig.37. Depression transits ($\times$ 10) in: A. June; B. July; C. August; averaged for a 10-year period. This is the time mid-latitude depressions are closest to the mainland, but even so most of them pass to the south of it; their fronts however penetrate far to the north. Notice the slight repelling effect of the colder southeastern mainland in June and July, resulting in a poleward shift of depression frequencies of about 8° of latitude in June and 9° in July from the first approach to the mainland to the final exit. (Compiled from data in KARELSKY, 1954, 1965.)

A typical transit of a large mid-latitude depression affecting Australia, is shown by Fig. 39. The depression stays well to the south of the mainland throughout the period of 5 days it takes to travel past. On June 15, 1964, it is to the southwest of the mainland, and a strong northwesterly stream associated with an equatorward meander of the jet stream, brings squally gale-force *IsTm* air and heavy showers over the southwestern area. (A local development, not typical, produces a small tornado at Mandurah, Western

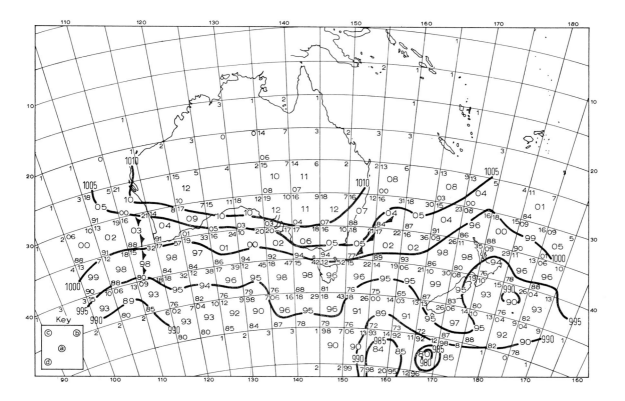

Fig.38. Mean pressure in centres of cyclones in July, 1952–1963. The mean pressure is shown by the large figures at the centre of each slab. In the top right corner is shown the total number of different cyclonic centres crossing the respective 5° slabs in July during the 10 years. In the top left corner is shown the absolute maximum pressure (14 = 1,014 mbar, etc.) and in the bottom left corner the absolute minimum pressure (76 = 976 mbar, etc.). The minor frontogenetic belt to the west and the major frontogenetic belt to the east postulated by the writer in 1949 have been added. *a* = average pressure in mbar; e.g., *a* = 27, i.e., 1027 mbar; *a* = 82, i.e., 982 mbar; *b* = total number of different cyclonic centres in 5° squares in 12 years; *c* = maximum pressure in cyclonic centres; *d* = minimum pressure in cyclonic centres. (After KARELSKY, 1965.)

Australia—50 km south of Perth.) The amount of rain received in 24 h reaches a maximum of 29 mm in the more exposed localities. The maximum temperature of 18.7°C at 12h05 shows that this first front, which is quite typical for this area, is different from the cold fronts of the higher latitudes. The second and third fronts have already been detected and are shown on the weather map.

The jet stream is most strongly developed at the 300-mbar level, ahead of the leading front, but dense cloudiness occurs at the 500-mbar level even beyond the tropic.

The next day these fronts cross the coast, bringing lower temperatures and more showers. By June 17, 1964, the depression has deepened and is off the Great Australian Bight. The northwesterly stream now flows from central Australia to the Eyre Peninsula; the first front has almost disappeared and brings very little rain. Already at Kalgoorlie and on the southwestern coast the jet stream and the surface winds have backed to westerly and the maximum temperature has fallen to 16°C as the third front approaches. The wettest localities in the southwest receive up to 58 mm in the 24 h.

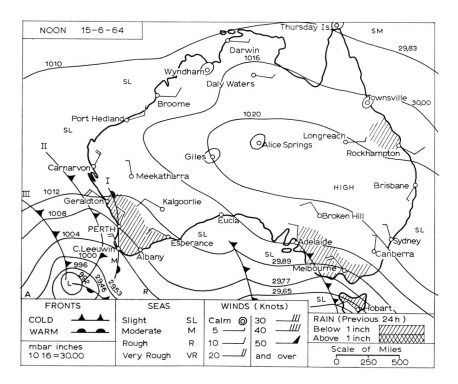

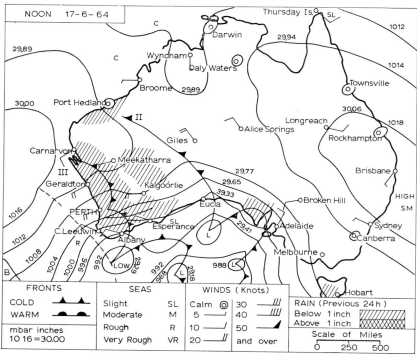

Fig.39. Transit of a complex depression: the centre is close to the mainland at first, but moves progressively further south. On 15.6.64 the first rains are brought by strong northwesterly winds (arrow at Geraldton, W.A.). A tornado occurs at Mandurah, W.A. Other fronts are approaching from the west. Further east, other fronts from the same family have already gone by, bringing coastal rains. On 17.6.64 the first front has disappeared, the second front is causing rain as far north as Port Hedland, the third front, more disturbed, brings heavier rains to a very large area in Western Australia. The de-

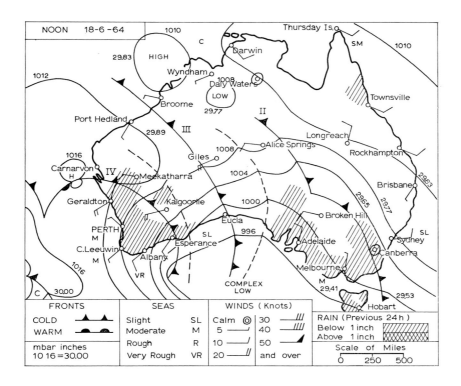

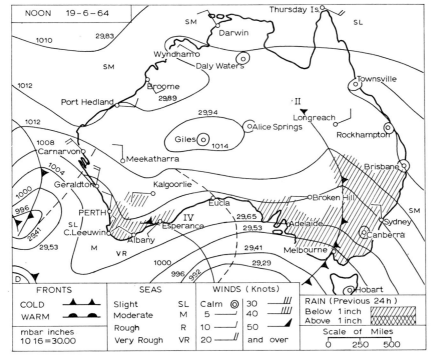

(Fig.39 continued)

pression reveals itself as very complex. By 18.6.64 its rains extend from Perth to Canberra; a fourth front has appeared, and the wind has gradually shifted to southwesterly on the west coast. By 19.6.64 frontal rains are very widespread in eastern Australia, but there still are residual rains over the south coast. To the west, another depression is approaching from the Indian Ocean. (West Australian weather maps.)

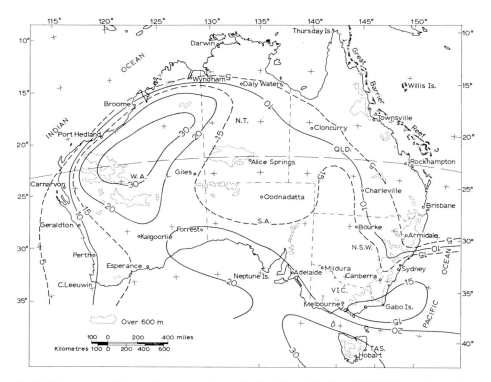

Fig.40. Frequency of depression centres in each 5° slab in October. Mid-latitude depressions travel farther to the southeast now; over the northwest the heat-low reappears with greater frequency. (Compiled from data by KARELSKY, 1954, 1965.)

On June 18, 1964, the wind behind a fourth front has backed to southwesterly, bringing a few showers (max. 26 mm in the 24 h) to the southwestern coast; temperatures remain unchanged. The jet stream now reaches furthest equatorward over central Australia. The second front is extending over eastern Australia, and is being revived by the northeasterly flow of *PsTm* air from the remnant of an anticyclone now over the Tasman–Coral area; some rain is received in southeastern Australia and Tasmania.

The next day (June 19, 1964), a wedge of denser air has moved to central Australia, pushing the second front southeastwards, where it brings widespread rains to that quarter of the mainland. A new and much weaker depression is approaching from the Indian Ocean, and a weaker version of the cycle just described is likely to take place in the following 2–3 days.

The cycle of jet-stream and front transit, from northwesterly to westerly to southwesterly, lasts about 3 days on the western coast, where the system is less disturbed by continental influences; it is usually shortened and less clearly defined towards the eastern coast, partly because of the inflow of *sTc* air, partly because of the revival of some fronts by *PsTm* air from an offshore anticyclone, partly also because of a slight braking effect due to the highlands. QUAYLE (1915) had correctly interpreted the seasonal variations in this pattern, from the observations of cirrus clouds, remarking that the rate of movement of the trough is more rapid in summer than in winter, and that in summer, southwesterly directions are more frequent.

Tornadoes, whirlwinds and waterspouts

These phenomena are more common than is usually thought; the scarcity of reports is partly due to the sparseness of the population over large areas of the interior.

Any statistics of such localized and erratic phenomena would be most misleading, because they would show a close correlation with the distribution of population, so much so that in earlier years, one could have thought of whirlwinds as an urban peculiarity! Australian reports are complicated by the indiscriminate use of the terms "willy willy" and "cockeye bob." BAKER (1945) states that the term "willy willy" is not aboriginal, and comes probably from outside Australia; it may even have been just "willy" at first. It is now used in Western Australian press reports for tropical cyclones, tornadoes and small whirlwinds, but not for coastal whirlwinds originating in winter storms; for these "cockeye bob" or "cockeyed bob" is used (BAKER, 1941).

In the subtropical belt there may arise small but intense *cyclones*, probably as a result of vorticity developed by a sudden acceleration of the subtropical jet stream. Such cyclones are small closed frontal depressions which, because of their dynamic origin, appear along a major trough or front line. The wind is very violent, but the duration of the cyclone is very short. An interesting example, which struck Byron Bay in New South Wales, is described by MOSS (1965). *Tornadoes* are more common, and may approach the frequency they have in the United States (between 100 and 200 per year) but are not as destructive. An excellent review of the information available is given by CLARKE (1962), while WHITTINGHAM (1964) pursues the instrumental and statistical analysis further, and gives maps of the extreme wind gusts that may be expected in 10, 20, 50 and 100 years. Both authors stress the inadequacies of records, and the confused terminology in many reports. The most recent accounts of tornadoes in the extratropical regions come from Mandurah, Western Australia (BROOK, 1965), Adelaide, South Australia (SEAMAN, 1966), Numurkah, Victoria (PHILLIPS, 1965), Smithtown, New South Wales (CARR, 1965). A detailed cinematographic record is available of a fire-induced tornado that occurred in Victoria (KING, 1964): core velocities reached 179 knots vertically, some 30 knots horizontally, and 15–30 revolutions per minute. Nearby parcels of air rose at up to 85 knots without spinning.

An unusual case of twin tornadoes is described by ZILLMAN (1962).

In the early afternoon of Monday, July 9, 1962, Port Macquarie, New South Wales was struck by two small but intense tornadoes which originated as waterspouts at some distance out to sea, then crossed the coast almost simultaneously and about half a mile apart (Fig.41). The more destructive of the two moved almost straight westwards across the town, leaving a trail of devastation about 3 km long and 15–100 m in width. The other, apparently less intense, moved in from the northeast, causing severe damage along a strip which was generally less than 50 m across. The last evidence of its existence as an identifiable tornado was found 1.5 km to the southwest of the spot where it first crossed the coast.

There is strong, if not completely conclusive, evidence to support the belief that although the shorter-lived and less destructive tornado possessed the normal cyclonic sense of rotation, the second one was rotating anticyclonically.

It will be noticed from these accounts that coastal tornadoes tend to occur during or

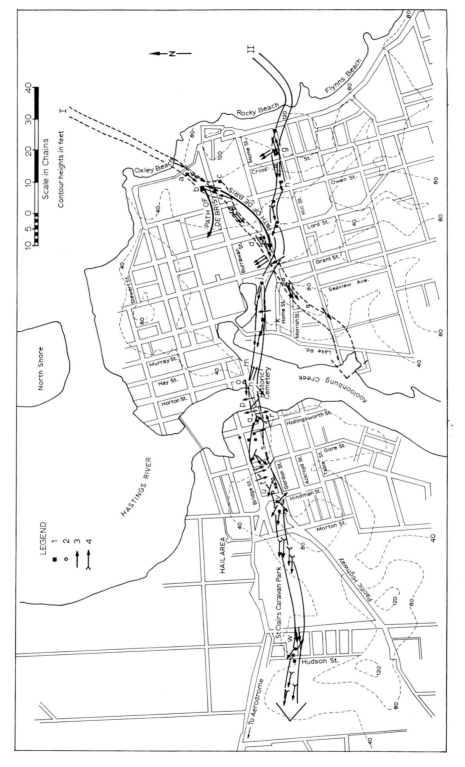

Fig.41. Map of Port Macquarie, N.S.W., showing the paths of the twin tornadoes, damaged areas, and the direction of the wind. *1* = severe structural damage; *2* = minor structural damage; *3* = direction of fall or lean of trees, fences, masts, etc.; *4* = debris strewn in direction of arrow. (After ZILLMAN, 1962.)

after winter, usually along a pronounced front with very humid air on one side. They begin over the coastal waters, and do not proceed more than a few kilometres inland.

The sporadic nature of these phenomena is the main cause of very incomplete records; the main examples however are listed in the early volumes of *Results of Rainfall Observations* and the booklets *50 Years of Weather* published by the Bureau of Meteorology, and once a year in the bulletin *Australia in facts and figures* issued by the Department of the Interior (see References to Chapter 2). Individual case studies appear from time to time in the *Australian Meteorological Magazine*.

Whirlwinds may occur at any time of the year, because of the instability that accompanies some winter frontal surfaces, and much more often, because of convectional instability in summer. There does not seem to be any set geographical pattern for their occurrence. In some cases whirlwinds occur above fierce fires (WHITTINGHAM, 1959).

Some early observers were endowed with great ability to express themselves precisely and some of their descriptions are still very effective. *Wragge's Almanac*, in May, 1903 (1:355) reported the description of a "cyclonette" in Melbourne. "Workmen on the wharves on the Flinders Street extension observed a whirlwind gather up all the coal dust... into one black column" which was estimated to have been 24 m in height and "travelled out to the middle of the river. The river water was rapidly drawn up into the column and the whirlwind became a waterspout..." An iron chimney-stack 24 m high "was seen to rise and remain for a second suspended in the air, then fall with a crash..." Buildings on either side were not damaged.

An eye-witness account of a *waterspout* at Pialba, quoted in 1886, ran as follows: "A heavy black cloud to the northeast... was observed to suddenly drop a huge black tongue, which was met in mid-air by a spiral column of water... The immense black pillar of water thus formed then travelled in a westerly direction, preserving its upward current, for fully a quarter of an hour, during which time the cloud assumed fantastic shapes, and the surface of the sea surrounding the face of the column was studded with a myriad of water jets or splashes..."

The most comprehensive list is that given by RUSSELL (1898), with 38 entries, including the most remarkable sighting of 14 waterspouts within 4 h, off Eden in southern New South Wales, on May 16, 1898.

Mrs. MEREDITH (1861) left this vivid account of *dust devils*, "a phenomenon by no means unusual on the large plains of New South Wales, in dry weather, being a procession across them of tall columns of dust—whirlwinds, in fact, which preserve a nearly uniform diameter throughout their whole length, the upper end seeming to vanish off, or puff away like light smoke, and the lower apparently touching the earth. They move in a perpendicular position, quietly and majestically gliding along one after another, seeming... to be 70–100 ft. (20–30 m) high, and about 20 ft. (6 m) broad. Thus viewed, they do not appear to travel particularly fast, but Meredith tells me he has vainly endeavoured to keep pace with them for a short time, even when mounted on a fleet horse. When they are crossing a brook or river, the lower portion of the dust is lost sight of, and a considerable agitation disturbs the water, but immediately on landing the same appearance is resumed. As some vanish, others imperceptibly arise..."

"I never heard of these gregarious whirlwinds being at all mischievous; they only pick up dust, leaves, little sticks, or other light bodies, which whirl round in them with great velocity..." (STOKES, 1846, had a small boat nearly upset by one of them.)

The "small whirlwinds, or dust whirls as they are commonly called", observed by KIDDLE (1893) in the southern Riverina form quite a representative group. They mostly occurred in the summer and early autumn, travelled at up to 30 km/h, and rotated cyclonically or anticyclonically with almost equal frequency. Their conspicuousness depended on the amount of dust and light debris available. Kiddle observed that those formed in calm weather tended to be more clearly defined and reached greater height. These dust devils ceased in altitude where the force of the wind was greater than their rotational force.

R. R. Farron (personal communication, 1968) observed that at Marble Bar, in tropical Western Australia, dust devils occurred about once a day during hot weather. They were strongly tapered, being several meters wide at the top, and only 1–1.20 m wide just above the ground. They rotated very rapidly, and could easily lift a sheet of corrugated iron. In general, they followed one of the streets; their tendency to proceed straight ahead or follow a bend or a junction depended on their speed.

A general review of *dust storms* in Australia, with special regard to their geographical occurrence, daily period and seasonal frequency, and relation to synoptic situations, was given by LOEWE (1943).

Many dust storms of the interior are manifestations of whirlwinds or tornadoes. We may quote as an example the storm which hit the aerodrome at Alice Springs at 18h45 on November 22, 1964. The township, 10 km away, was not directly hit by the whirlwind, but it was enveloped in swirls of brown dust, while at the aerodrome, at the height of the storm, it was as dark as night. The wind wrecked a light aircraft and badly damaged another, besides damaging several buildings.

Many a dark cloud seen on the distant and parched horizon turns out to be a "Cobar shower" of swirling dust, instead of the eagerly awaited rain!

Land- and sea-breezes

The differential heating or cooling of contiguous land and sea surfaces, causes differences in pressure which, in turn, lead to the sliding of shallow streams of air onshore or off-shore until the balance is restored. At the hottest time of day the air above the land may be 7–8°C hotter than the air above the sea, and the difference in pressure may be considerable.

Any wind that happens to blow at the time will strengthen a breeze which flows in the same direction, and no distinction between wind and breeze would be possible near the surface. Thus a westerly wind in Perth, or an easterly wind at Sydney would strengthen the sea-breeze. (But see discussion of effect of sea-breezes on the cool change, pp.70–72.) A breeze can arise counter to the wind blowing at the time, provided that the breeze be strong enough and the wind not too strong (Fig.42) For instance, at Perth the sea-breeze will arise whenever the air above the land becomes hotter than the air above the coastal waters (Fig.43); it begins as a southwesterly. A strong easterly wind in summer will carry offshore much heat from the land, so that the cool air above the coastal waters is replaced by hot continental air, and the thermal differential fails to occur. On the other hand, a moderate easterly allows the sea-breeze to develop near the surface; the easterly flow will continue immediately above the onshore-flowing breeze, perhaps 550–600 m up.

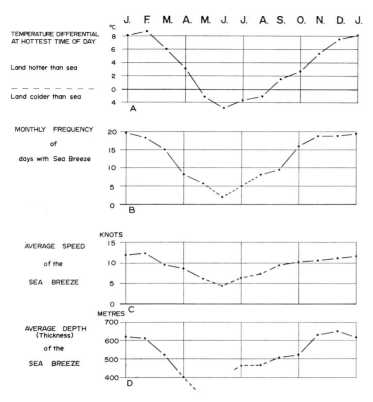

Fig.42. The sea-breeze at Perth. For each month the temperature differential between land and sea at the hottest time of the day is shown, together with the monthly frequency of days with breeze, and the average speed and depth of the breeze flow. Notice the distinct seasonal trend, and the near lack of breeze in the winter. (After HOUNAM, 1945.)

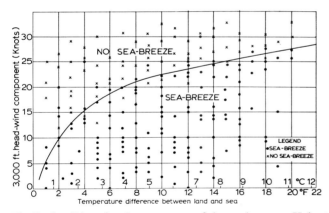

Fig.43. Conditions for the occurrence of the sea-breeze at Hobart, Tasmania, for various values of the head-wind component at 900 m (3,000 ft.) and given differentials of temperature between land and sea-surface. (After WYATT, 1963.)

BARTON (1895) wrote "that occasionally it is possible to fly a kite at Sydney, which rises and is carried away to the southeast, while the sea-breeze below is blowing from the east north-northeast".

In the winter, the general isobaric pattern favours an offshore flow, and land-breezes are consequently strengthened, while sea-breezes fail to occur.

The distance travelled by ordinary breezes is not very great, perhaps 80–90 km for the sea-breeze and 50–60 km for the land-breeze. The average depth of the flow varies considerably, but seldom exceeds 700 m for the sea-breeze, and perhaps 500 m for the land-breeze. There is a definite seasonal variation, with the deepest sea-breezes at Perth occurring in November–December, while the strongest ones come later in January or February (HOUNAM, 1945). From Norseman to Kalgoorlie, the Esperance Doctor blows far more frequently after midsummer than earlier in the season.

The anticyclonic pattern of the southern Australian summer favours the development of thermal contrasts and consequently of sea-breezes in the western half and rather hinders it in the eastern half.

Very similar air masses are often involved in the sea-breeze front as in the cool-change front, and aerological soundings are needed in order to show structural and dynamic differences (BERSON, 1958; CLARKE, 1961). According to CLARKE (1960), sea-breeze fronts "form near the boundary of a strongly heated area, move inland during the afternoon and evening... but are short lived (they do not survive the hours of darkness, and rarely midnight)..." They are similar to inter-anticyclonic fronts, "but tend to be shallower... and do not form in a hyperbolic deformation field, as do the larger scale type".

As a typical eastern coast example, one may quote Newcastle, New South Wales, where the sea-breeze enters the Hunter Valley for some 18 km only; further north or south it is even more limited because of the steep scarp of the highlands. Even the strongest sea-breeze ("black northeaster") does not reach further than 15–30 km inland from Sydney. On the other hand, at Canberra, nearly 100 km inland at an altitude of 550 m, the diurnal pattern of the easterly wind strongly suggests that it is a sea-breeze (CLARKE, 1955). Glider pilots are well aware of its peculiarities and consider it a normal event. CLARKE (1955) established by questionnaire the occurrence of similar breezes at many other points, especially along valleys.

At Hobart, where the sea-breeze occurs on an average of 55 days per year and is usually fairly well marked, so that "at least one major industry has chosen a site such that the sea-breeze provides part of the climate suitable for their operation" (WYATT, 1963), no sufficient correlation was found with other data (head wind at 900 m, air temperature, sea-surface temperature) to enable reliable forecasts to be made. Empirical diagrams

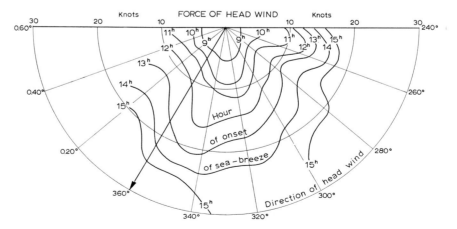

Fig.44. Diagram showing the time of onset of the sea-breeze plotted against the force and direction of the head wind at 900 m at 03h00. (After WYATT, 1963.)

were constructed from the actual records, the first one based on the absence or occurrence of the breeze with a combination of head wind at 900 m and the thermal differential between the maximum air and the sea-surface temperatures (Fig.43), and the
second one giving the time of arrival of the breeze for any combination of force and
direction of the head wind (Fig.44). Good forecasts then became possible (WYATT, 1963).
On the southeastern coast, R. H. Clarke (personal communication, 1965) made a
thorough aerological study of the sea-breeze east of Adelaide (Fig.45). The hindering

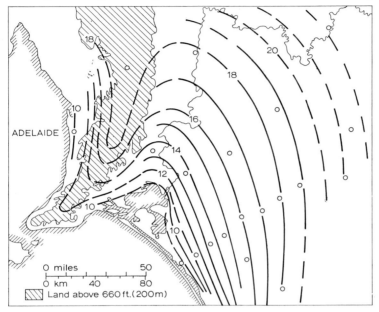

Fig.45. Isochrones of the sea-breeze front in the Lower Murray area, for hourly intervals. Notice the
increasing speed of propagation further inland, and the braking effect of the hills on the shallow wind.
The small circles show the observers' positions. (After R. H. Clarke, personal communication, 1965.)

effect of the highlands and the favourable effect of the very broad and shallow Murray
Valley are obvious. The breeze travelled inland for over 200 km. Cross-sections were
made to a height of 2,700 m (Fig.46); they gave a full picture of the relative air movement
and its characteristics. This investigation provided the information needed for an understanding of the evening wind surges previously reported by REID (1957) at Renmark,
South Australia, 225 km from the coast. These surges were clearly observable 6 times in
November, 7 in December, 13 in January, 14 in February, 8 in March, 5 in April, with
a directional concentration that pointed clearly to the nearest shore.

In the southwest, sea-breezes have long been recognized as a distinctive feature of the
local climate, so much so that they were given local popular names such as (in the likely
chronological order of their adoption) *Fremantle Doctor, Albany Doctor, Geraldton
Doctor, Esperance Doctor,* and more recently even *Eucla Doctor,* in each case from the
name of the coastal locality near which they cross the shore on their welcome way inland (Fig.47).

An extraordinary aspect of these breezes is their deep inland penetration. Cunderdin,
150 km from Fremantle, is reached by the Fremantle Doctor on about 4 days per week,
usually between 20h00 and 21h00. The breeze does not persist, but the temperature is
lowered almost immediately and by up to 12°C. Kalgoorlie is about 350 km from the

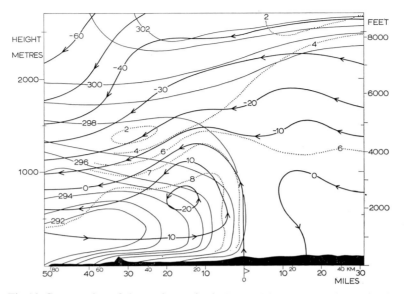

Fig.46. Cross-section of the sea-breeze in the Lower Murray area. The section is taken perpendicularly to the coast, and the streamlines (thick lines) are shown relatively to the advancing sea-breeze front, which at the moment has reached the vertical of *0* at the foot of the diagram. Distances in km (small figures) and miles (large figures) are shown from this point. Isentropes are shown by thin lines and isopleths of the mixing ratio by dotted lines. (After R. H. Clarke, personal communication, 1965.)

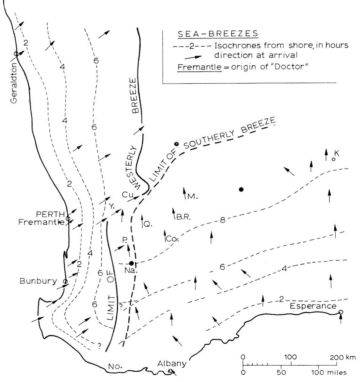

Fig.47. The "Doctor" breezes of the southwest. From the western coastline originate the Geraldton, Fremantle and Bunbury "Doctors", which travel only a little beyond the normal range of a sea-breeze. From the southern coastline originate the Albany and Esperance "Doctors" which appear to travel far beyond the normal range of a sea-breeze. At Pingelly and York either breeze may occur. At the localities shown by black dots no breeze is observed. The initials show the position of Bruce Rock, Corrigin, Cunderdin, Kalgoorlie, Merredin, Narrogin, Nornalup, Pingelly, York.

southern coast; the Esperance Doctor reaches it at about 20h00 after any hot, almost windless day, and produces an immediate drop of temperature. If the breeze began to develop on the shore at about 11h00, it must have travelled at nearly 40 km/h. Quairading, Corrigin, and Bruce Rock are between 320 and 370 km from the southern coast, and are reached by the Albany Doctor between 19h00 and 22h00; this shows an average velocity of 30–40 km/h.

Such determinations of velocity are not wholly representative; these breezes cover such great distances only when conditions are very favourable, because only then do they have the greatest depth and force. When conditions are moderately favourable, the breezes rise later, are more shallow, travel more slowly and do not reach so far inland. The average velocity of the breeze is lower at localities which have it almost every day, because of the many days when it is weak, and does not travel far inland.

A well-developed breeze may continue to blow for 6–7 h near the shore, 3–4 h at 50–100 km inland, and 1–2 h at 200 km inland.

On the western coast, from Shark Bay to Fremantle, the breeze begins near the shore between noon and about 14h00, and moves inland as a southwesterly at about 15 km/h, gradually gaining speed. In the third or fourth hour, it may cover 20–25 km. If it still continues inland, its speed may increase to 30 km/h or more, as mentioned above.

On the southern coast, from Nornalup eastwards, the breeze is even better developed, more regular and more frequent. It comes as a southeasterly or southerly (Fig.47). Its velocity increases as it proceeds, so much so that the radio station at Kalgoorlie can often announce that "the Esperance Doctor passed through Norseman at 6 p.m. and is expected to reach Kalgoorlie at approximately 8 p.m." (A. Carlin, personal communication, 1962), which would give the mean velocity of 80 km/h, far too high for an ordinary sea-breeze.

CLARKE (1955) carried out a thorough study of the Esperance Doctor based on five stations from Esperance to Kalgoorlie for a period of three weeks. Continuous records were obtained, so that the progress of the wind and its effects on temperature and humidity could be followed in detail. From the time–distance charts plotted, it appears that the speed of propagation inland increases slightly but progressively in the 12–14 h taken to cover the distance. From autographic records at Forrest it can be seen that on a good breeze day (21 January, 1955) the Eucla Doctor arrived at about 18h00, and the temperature fell by 2°C almost instantly, while the relative humidity rose from 15 to 30%. By 20h00 the temperature had fallen by 9°C and the humidity had reached 60%.

An interesting conflict develops along the lower western coast, from Fremantle to Nornalup, where the westerly breeze on its way inland soon comes into conflict with the southerly breeze (cf. Fig.47). The westerly breeze does not reach beyond 100–140 km inland, where it finds the cooler air of the southerly breeze, which often arrived there more than 1 h earlier. Thus at Quairading, the southerly breeze mostly arrives between 19h00 and 20h00, perhaps five times per week, after having travelled 330 km from the southern shore, where it began as a southeasterly 6–7 h earlier, while the westerly breeze reaches Cunderdin a little further north and only 150 km from the western shore, between 20h00 and 21h00 and about four times per week.

There is an area of overlap between the westerly and the southerly breezes. York usually gets the Fremantle Doctor (westerly), but occasionally is reached by the Albany Doctor (southerly) instead. Pingelly, further south and a little further inland, gets the Albany

Doctor three to four times per week, and the westerly breeze (unnamed) perhaps once per week. Merredin, much further inland than York, gets the southerly breeze (unnamed) two to three times per week, but very rarely is reached by the Fremantle Doctor instead (unless the southwesterly wind of the cool change associated with an approaching anticyclone is mistaken for a westerly breeze).

In most places, especially those a little further inland, the arrival of the breeze is heralded 10–15 min ahead by loud rustling of trees in the distance, heard as a dull roaring sound or, in dry areas and when the breeze is strong enough, by a swirl of dust. The initial gust is very strong and sudden, then the flow becomes more gentle and steady.

It has been suggested that the exceptional penetration inland of these breezes is due to a broad valley topography, e.g., York receives the southerly Albany Doctor along the Avon Valley, and the southerly Esperance Doctor on its way to Kalgoorlie tends to follow a long line of depressions occupied by salt lakes. It is to some extent true that the cooler air tends to flow along these broad and very shallow valleys because of gravity, but very many localities well away from any depression are reached by the breeze just as effectively. On the other hand, Zanthus, Western Australia, in a shallow depression west of the Nullarbor Plain, often misses the southerly breeze while Cundeelee Mission, on slightly higher ground nearby, benefits almost every day.

Another aspect that may be significant, is the much stronger development of the sea-breeze in the western half of the continent, compared with that in the eastern half. The effect of the eastern highlands is probably sufficient to reduce the thermal differential between land and sea along the eastern shore. In the southwestern quarter on the contrary, there is a rise from sea level to the 300 m or so of the plateau, and then the level rises much more gently to over 400 m. The dry summer air is heated very rapidly and the thermal differential between land and sea soon becomes very great, and more so in the southern belt. Because of the orientation of the land–sea boundary, there arises a *zonal sea-breeze*. There are very few other locations in the world where similar geographical conditions occur, the North African shore being the most conspicuous.

In a general review of the sea-breeze problem, CLARKE (1958) states that "the Kalgoorlie surge appears to have two distinct components: one a degenerate 'seabreeze' of a type probably similar to that found at Renmark, and occurring typically near midnight; the other, of unknown origin, and occurring earlier, say 20h00 (L.M.T.)."

R. H. Clarke (personal communication, 1965) believes that nocturnal cooling is responsible for the air movements which are mistaken for sea-breezes in the interior of the continent; this could explain the apparent and unbelievable speed of travel of the surface stream which reaches, say, Kalgoorlie from the south. This also explains the statements of some informants from the interior who claim that at times, at night, the sea-breeze does travel inland for a great distance, and it is not clear at all where it may be replaced by other air movements, or how it may preserve an apparent continuity over the whole way.

Index to existing literature

Australian tropospheric movements in general: BARTON (1895), BUREAU OF METEOROLOGY (1954, 1956), GENTILLI (1947, 1949, 1955), GIBBS (1946), HUNT (1908), HUNT et al. (1913), KARELSKY (1954, 1956,

1961, 1965), KIDSON (1935), LEEPER (1960), ROYAL AUSTRALIAN AIR FORCE (1942), TAYLOR (1920, 1932, 1940), VAN LOON (1964).

Travelling anticyclones and mid-latitude depressions: BERSON et al. (1957, 1959), CASSIDY (1945a, b), HUNT (1894), KIDSON (1925), LOEWE (1945).

References

BAKER, S. J., 1941. *A Dictionary of Australian Slang.* Angus and Robertson, Sydney, N.S.W., 84 pp.

BAKER, S. J., 1945. *The Australian Language.* Angus and Robertson, Sydney, N.S.W., 244 pp.

BARTON, C. H., 1895. *Outlines of Australian Physiography.* Alston, Maryborough, 180 pp.

BATH, A. T., LLOYD, S. H. and RYAN, P., 1956. Practical techniques used in the Australian region for tropical cyclone forecasting. *Proc. Tropical Cyclone Symp., Brisbane,* pp.121–137.

BERSON, F. A., 1958. Some measurements on undercutting cold air. *Quart. J. Roy. Meteorol. Soc.,* 84: 1–16.

BERSON, F. A., 1961. Circulation and energy balance in a tropical monsoon. *Tellus,* 13: 472–485.

BERSON, F. A. and TROUP, A. J., 1961. On the angular momentum balance in the equatorial trough zone of the Eastern Hemisphere. *Tellus,* 13: 66–78.

BERSON, F. A., REID, D. G. and TROUP, A. J., 1957. The summer cool change of southeastern Australia, I. *C.S.I.R.O., Div. Meteorol. Phys., Tech. Papers,* 8: 48 pp.

BERSON, F. A., REID, D. G. and TROUP, A. J., 1959. The summer cool change of southeastern Australia, II. *C.S.I.R.O., Div. Meteorol. Phys., Tech. Papers,* 9: 69 pp.

BOND, H. G. and RAINBIRD, A. F., 1956. Structure of tropical cyclones, with particular reference to those of 1955–1956 in the Australian region. *Proc. Tropical Cyclone Symp., Brisbane,* pp.159–170.

BRETON, W. H., 1833. *Excursions in New South Wales, Western Australia, and Van Diemen's Land.* Bentley, London, 420 pp.

BROOK, R. R., 1965. Tornado at Mandurah, 15th June 1964. *Australian Meteorol. Mag.,* 50: 26–34.

BRUNT, A. T. and HOGAN, J., 1956. The occurrence of tropical cyclones in the Australian region. *Proc. Tropical Cyclone Symp., Brisbane,* pp.5–17.

BUREAU OF METEOROLOGY, 1956. *Climatic Averages.* Bur. Meteorol., Melbourne, Vic., 107 pp.

CAMPBELL, A. P., 1968. The climatology of the sub-tropical jet stream associated with rainfall over eastern Australia. *Australian Meteorol. Mag.,* 16: 100–113.

CARR, P. E., 1965. Report on tornado investigation—Smithtown (N.S.W.), August 1964. *Australian Meteorol. Mag.,* 48: 23–36.

CASSIDY, M., 1945a. Frontology of subtropical Australia. *Weather Develop. Res. Bull.,* 1: 5–28.

CASSIDY, M., 1945b. Frontal systems peculiar to the Australian continent. *Weather Develop. Res. Bull.,* 2: 5–34.

CLARKE, R. H., 1955. Some observations and comments on the seabreeze. *Australian Meteorol. Mag.,* 11: 47–68.

CLARKE, R. H., 1956. A study of cyclogenesis in relation to the contrours of the 300 mb surface. *Australian Meteorol. Mag.,* 12: 1–21.

CLARKE, R. H., 1958. Midsummer diurnal winds in the southeast of South Australia. *Fire Weather Conference, Melbourne.* Paper 14, 19 pp.

CLARKE, R. H., 1960. The rain-producing potential of cold and sea breeze fronts. In: *Seminar on Rain, Sydney.* Bureau of Meteorology, Melbourne, Paper 11/1: 10 pp.

CLARKE, R. H., 1961. Mesostructure of dry cold fronts over featureless terrain. *J. Meteorol.,* 18: 715–735.

CLARKE, R. H., 1962. Severe local wind storms in Australia. *C.S.I.R.O., Div. Meteorol. Phys., Tech. Papers,* 13, p.56.

COOKE, W. E., 1905. Meteorology of Western Australia. In: *Report of the Eighth International Geographical Congress, held in the U.S.* Edited by the Committee on Printing. Govt. Printing Office, Washington, D.C., pp.386–392.

DEPPERMANN, C. E., 1943. *Upper Air Circulation over the Philippines and Adjacent Regions.* Government Printer, Melbourne, 34 pp.

DE STRZELECKY, P. E., 1845. *Physical Description of New South Wales and Van Diemen's Land.* Longmans, London, 462 pp.

EVESSON, D. T., 1969. Tornado occurrence in New South Wales. *Australian Meteorol. Mag.,* 17: 143–165.

GAFFNEY, D., 1953. The formation of weak troughs off the southwestern coast of Western Australia. *Australian Meteorol. Mag.,* 14: 15–28.

GENTILLI, J., 1947. *Australian Climates and Resources*. Whitcombe and Tombs, Melbourne, Vic., 333 pp.

GENTILLI, J., 1949. Air masses of the Southern Hemisphere. *Weather*, 4: 258–261; 292–297.

GENTILLI, J., 1955. Die Klimate Australiens. *Erde*, 1955 (3–4): 206–238.

GENTILLI, J., 1956. Tropical cyclones as bioclimatic activators. *Western Australian Naturalist*, 5: 82–86; 107–117; 131–138.

GENTILLI, J., 1969. Some regional aspects of southerly buster phenomena. *Weather*, 24: 173–180.

GIBBS, W. J., 1946. Recent developments in weather analysis in Australia. *Australian Geographer*, 5: 47–51.

HARDING, F., 1893. Moving anticyclones in the Southern Hemisphere: a discussion. *Quart. J. Roy. Meteorol. Soc.*, 19: 34–35.

HOUNAM, C. E., 1945. The sea breeze at Perth. *Weather Develop. Res. Bull.*, 3: 20–55.

HUNT, H. A., 1894. An essay on Southerly Bursters. *Trans. Roy. Soc. New South Wales*, 28: 138–185.

HUNT, H. A., 1908. Climate and meteorology of Australia. *Australia, Bur. Meteorol. Bull.*, 1: 27 pp.

HUNT, H. A., TAYLOR, G. and QUAYLE, E. T., 1913. *The climate and weather of Australia*. Bur. Meteorol., Melbourne, 93 pp.

HUTCHINGS, J. W., 1961. Water-vapor transfer over the Australian continent. *J. Meteorol.*, 18: 615–634.

KARELSKY, S., 1954. Surface circulation in the Australian region. *Australia, Bur. Meteorol., Meteorol. Studies*, 3: 45 pp.

KARELSKY, S., 1956. Classification of the surface circulation in the Australian region. *Australia, Bur. Meteorol., Meteorol. Studies*, 8: 36 pp.

KARELSKY, S., 1961. Monthly and seasonal anticyclonicity and cyclonicity in the Australian region— 15 years (1946–1960) averages. *Australia, Bur. Meteorol., Meteorol. Studies*, 13: 11 pp.

KARELSKY, S., 1965. Monthly geographical distribution of central pressures in surface highs and lows in the Australian region, 1952–1963. *Australia, Bur. Meteorol., Meteorol. Summary*, July, 1965, 39 pp.

KIDDLE, H. C., 1893. Small whirlwinds. *J. Roy. Soc. New South Wales*, 27: 91–101.

KIDSON, E., 1925. Some periods in Australian weather. *Australia, Bur. Meteorol., Bull.*, 17: 33 pp.

KIDSON, E., 1935. The analysis of weather charts. *Australian Geographer*, 5: 3–16.

KING, A. R., 1964. Characteristics of a fire-induced tornado. *Australian Meteorol. Mag.*, 44: 1–9.

LEEPER, G. W. (Editor), 1960. *The Australian Environment*. Melbourne Univ. Press, Melbourne, 151 pp.

LOEWE, F., 1943. Duststorms in Australia. *Australia, Bur. Meteorol., Bull.*, 28: 16 pp.

LOEWE, F., 1945. Frontal hours at Melbourne. *Weather Develop. Res. Bull.*, 3: 13–19.

MARES, D. J., 1940. *Know your own Weather*. Angus and Robertson, Sydney, 138 pp.

MARTIN, R. M., 1853. *Australia*. Tallis, London, 576 pp.

McDONALD, W. F., 1938. *Atlas of Climatic Charts of the Oceans*. Government Printing Office, Washington, D.C., 130 pp.

McRAE, J. N., 1956. The formation and development of tropical cyclones during the 1955–56 season in Australia. *Proc. Symp. Tropical Cyclone, Brisbane, 1956*. Bureau of Meteorology, Melbourne, Vic., pp.233–261.

MEREDITH, L. A., 1861. *Notes and Sketches of New South Wales, during a Residence in that Colony from 1839 to 1844*. Murray, London, 2nd ed., pp.86–87.

MORIARTY, W. W., 1955. Large scale effects of heating over Australia, 1. *C.S.I.R.O. Div. Meteorol. Phys., Tech. Paper*, 7, p.31.

MOSS, J. M., 1965. Cyclone in Byron Bay region, New South Wales, 29th December 1964. *Australian Meteorol. Mag.*, 51: 15–28.

NEWMAN, B. W., MARTIN, A. R. and WILKIE, W. R., 1956. Occurrence of tropical depressions and cyclones in the Australian region during the summer of 1955–1956. *Proc. Tropical Cyclone Symp., Brisbane*, pp.25–55.

PHILLIPS, E. F., 1965. The Numurkah tornado of August 1964. *Australian Meteorol. Mag.*, 48: 37–45.

QUAYLE, E. T., 1915. Relation between cirrus directions as observed in Melbourne and the approach of the various storm systems affecting Victoria. *Australia, Bur. Meteorol., Bull.*, 10: 27 pp.

REID, D. G., 1957. Evening wind surges in South Australia. *Australian Meteorol. Mag.*, 16: 23–32.

ROYAL AUSTRALIAN AIR FORCE, 1942. *Weather on the Australia Station*. Air Force Headquarters, Melbourne, I: 641 pp.; II: 470 pp.

RUSSELL, H. C., 1893. Moving anticyclones in the Southern Hemisphere. *Quart. J. Roy. Meteorol. Soc.*, 19: 23–34.

RUSSELL, H. C., 1898. Water spouts on the coast of New South Wales. *J. Roy. Soc. N.S. Wales*, 32: 132–149.

SEAMAN, R. S., 1966. Tornadic squalls in Adelaide on 13th May, 1965. *Australian Meteorol. Mag.*, 14: 24–34.

SPILLANE, K. T. and DIXON, B., 1969. A severe storm radar signature in the Southern Hemisphere. *Australian Meteorol. Mag.*, 17: 134–142.

STOKES, J. L. (Editor), 1846. *Discoveries in Australia*. Boone, London (facsimile edition 1969 by Libraries Board of South Australia, Adelaide).

STURT, C., 1833. *Two Expeditions into the Interior of Southern Australia, with Observations on the Soil, Climate and General Rescources of . . . New South Wales*. Smith Elder, London, 420 pp.

STURT, C., 1849. *Narrative of an Expedition into Central Australia*. Boone, London, I: 416 pp; II: 308 pp.

TAYLOR, G., 1918. *The Australian Environment*. Advisory Council Sci. Ind., Melbourne, Vic., 188 pp.

TAYLOR, G., 1920. *Australian Meteorology*. Oxford Univ. Press, Oxford, 312 pp.

TAYLOR, G., 1932. *Climatology of Australia*. Borntraeger, Berlin, 108 pp.

TAYLOR, G., 1940. *Australia*. Methuen, London, pp.51–74.

TRELOAR, H. M., 1934. Foreshadowing monsoonal rains in Australia. *Australia, Bur. Meteorol., Bull.*, 18: 29 pp.

TROUP, A. J., 1956. An aerological study of the "meridional front" in Western Australia. *Australian Meteorol. Mag.*, September, 1956, 14: 1–22.

TROUP, A. J., 1961. Variations in upper tropospheric flow associated with the onset of the Australian summer monsoon. *Indian J. Meteorol. Geophys.*, 12: 217–230.

TWEEDIE, A. D., 1966. *Water and the World*. Nelson, Melbourne, 317 pp.

VAN LOON, H., 1964. Midseason average zonal winds at sea-level and at 500 mbar south of 25°S, and a brief comparison with the Northern Hemisphere. *J. Appl. Meteorol.*, 3: 554–563.

VEITCH, L. G., 1965. The description of Australian pressure fields by principal components. *Quart. J. Roy. Meteorol. Soc.*, 91: 184–195.

VISHER, S. S. and HODGE, D., 1925. Australian hurricanes and related storms. *Australia, Bur. Meteorol., Bull.*, 16: 54 pp.

WALLACE, A. R., 1893. *Australia and New Zealand*. Stanford, London, 505 pp.

WEINERT, R. A., 1968. Statistics of the subtropical jet stream over the Australian region. *Australian Meteorol. Mag.*, 16: 137–148.

WHIGHT, J. M., 1964. The rainfall patterns associated with cyclone Audrey in Queensland, January, 1964. *Capricornia*, 1: 52–55.

WHITTINGHAM, H. E., 1957. Fire whirlwinds at Imbil. *Australian Meteorol. Mag.*, 25: 59–72.

WHITTINGHAM, H. E., 1958. The Bathurst Bay hurricane and associated storm surge. *Australian Meteorol. Mag.*, 23: 14–36.

WHITTINGHAM, H. E., 1964. Extreme wind gusts in Australia. *Australia, Bur. Meteorol. Bull.*, 46: 133 pp.

WICKHAM, J. C., 1846. Winds and weather on the western and northern coasts of Australia. In: J. L. STOKES, (Editor), *Discoveries in Australia*. Boone, London (facsimile edition 1969 by Libraries Board of South Australia, Adelaide) vol.2, pp.529–543.

WRAGGE, C. L., 1903. A cyclonette in Melbourne. *Wragge's Almanac*, 3 (5), p. 3.

WYATT, R. A., 1963. The sea breeze at Hobart. *Australia, Bur. Meteorol., Work. Papers*, 62–99.

ZILLMAN, J. W., 1962. Report on tornado investigation—Port Macquarie, July 1962. *Australian Meteorol. Mag.*, 39: 28–48.

Chapter 6

The Main Climatological Elements

J. GENTILLI

Pressure

Being a relatively small land-mass surrounded by extensive water-masses, Australia does not exert a drastic influence on the pattern of pressure and winds, but rather is subject to the winds which are part of the general circulation in these latitudes, and which have been discussed in the previous chapter.

It is symptomatic of this predominance of oceanic influences that the monthly mean pressure at Darwin is taken as an indicator of the southern oscillation, the index of which is the normalized mean pressure difference Papeete minus Darwin (TROUP, 1967). The maps of *mean pressure* usually given in textbooks are to a great extent misleading because they show a statistical abstraction which is the computed result of many widely different and continually changing situations. On the average, *annually*, there is a ridge

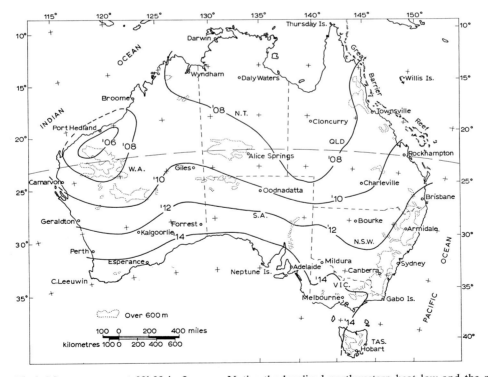

Fig.1. Mean pressure at 09h00 in January. Notice the localized northwestern heat low and the more diffuse northeastern low. ′08 = 1,008 mbar, etc. (Adapted from *Climatological Atlas*.)

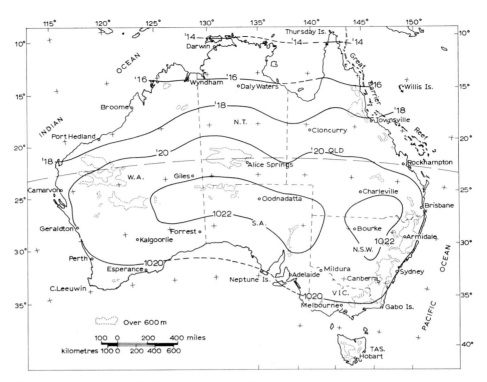

Fig.2. Mean pressure at 09h00 in July: as normally happens, the highest pressures are found in the sub tropical latitudes and more towards the eastern side of the mainland. (Adapted from *Climatologica Atlas*.)

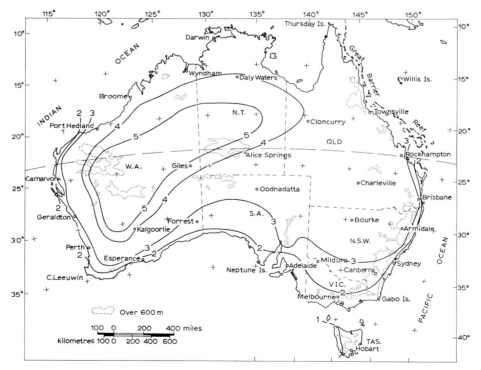

Fig.3. Mean daily pressure drop from 09h00 to 15h00, in mbar, January. The greatest drop is in the hot and dry northwestern quarter, the smallest along the shore, where the sea-breezes are effective.

of high pressure (1,014 mbar) extending west-southwest–east-northeast from Perth to Brisbane. The mean annual pressure decreases in a north-northwestern direction to 1,009 mbar around Darwin and in a south-southeastern direction to 1,012 mbar in the latitude of Hobart.

The mean January pattern (Fig.1). shows a general fall in pressures, with the north-western heat-low down to below 1,006 mbar, and the highest pressure, around 1,014 mbar, being confined to the Great Australian Bight.

The July pattern (Fig.2) shows a rise to 1,020 mbar in the southern half of Australia, from Shark Bay to Rockhampton, and, perhaps, over 1,022 mbar in the interior of southeastern Australia, an increase of some 10 mbar compared with January. In the far north, the mean July pressure is about 1,014 mbar, an increase of 6 mbar over January.

The *mean daily* range of pressure in continental areas is mostly determined by the heating of the ground. In *summer* (Fig.3) the greatest heat intake occurs in the northwestern half, except, of course, along the shore, and a decrease of 4 to over 5 mbar between

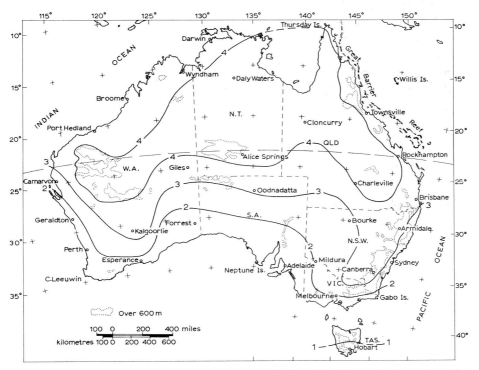

Fig.4. Mean daily pressure drop from 09h00 to 15h00, in mbar, July. The tropical areas have the greatest drop, the coastal belts, especially on the western side, the smallest. The prevalent westerly flow at this time of the year accounts for most of this difference.

09h00 and 15h00 is quite normal. Along the northern coast, the decrease is about 3 mbar; along the southern coast it is down to less than 2 mbar. In *winter* (Fig.4), the pattern is more even and almost symmetrical, but displaced some 500 km to the north.

At low latitudes, the global double diurnal variation becomes apparent. At Innisfail, Queensland, it has the following mean values (LAWSON, 1964):

I (night) minimum 03h30, 1,007.8 mbar

I (morning)   maximum 09h15, 1,010.3 mbar

II (afternoon) minimum 15h30, 1,006.9 mbar

II (evening)   maximum 22h00, 1,009.6 mbar

(mean April, 1963–March, 1964)

The semidiurnal variation is also noticeable on the tropical west coast. MCRAE and MC-GANN (1965) give a barograph trace from Port Hedland which in addition shows a series of 12 waves with a period of 20–25 min and an amplitude of 0.2–0.6 mbar.

Atmospheric pressure thus varies throughout the day, as well as throughout the year, and also over longer periods, but superimposed on these pressure cycles of more or less fixed periodicity, are variable cycles due to the flow of air over the continent and to the general circulation of the atmosphere, and accidental but violent fluctuations due to hurricanes and tornadoes (see Chapter 5).

A study of barograph traces at Perth reveals the superimposition of pressure waves of 5–6 days' periodicity (due to the transit of moving anticyclones, cf. Chapter 5), diurnal and semi-diurnal waves (in varying combinations of amplitude) and short-period waves which only appear very occasionally and resemble those described for Port Hedland above, but may vary in their periodicity. There are also sporadic fluctuations which so far have not been studied. REBER (1967) analyzed 51 months of microbarograph records in Hobart, and found, besides the 24- and 12-h periods, an oscillation with 8-h period which reverses its phase from summer to winter, with "very rapid transitions close to zero during the equinoxes". The tesseral oscillation (6-h period) which moves from east to west with the sun, and has greater amplitude at higher latitude, begins to be noticeable.

Distribution of heat

The higher temperatures

Because of its latitude, Australia may be considered the hottest continental mass, having a mean January (mid-summer) temperature above 28°C over some half of its area. North Africa reaches higher temperatures over a larger area in July (northern midsummer), but the southern part of that continent is much less hot. Higher temperatures are reached at various localities in southwestern Asia, but there is no doubt that, taking the whole continental mass into account, Australia reaches the highest average temperature for its total area. A pioneer study by WALLIS (1914), gave the mean thermal anomaly for each month; from that information a map of the duration of anomalies > 3°C has been prepared (Fig.5), which shows that half of Australia is abnormally hot for over half the year.

POLLAK (1931) correlated the thermal anomalies of Adelaide, Darwin, Perth and Sydney with those of Praha (Prague), Stykkiskolm in Iceland, and Nerchinsk in the U.S.S.R. (Far East); the highest coefficients were negative.

In general there is a tendency for summer temperatures above 40°C to occur most frequently in the northwest, partly because of the transport of heat by the trade winds (Chapter 5, Fig.5 and 23) and partly because of the intense insolation received through the clear, dry air, but such high temperatures may occasionally occur anywhere in Australia, even in Tasmania, depending on the specific pattern of air flow and heat

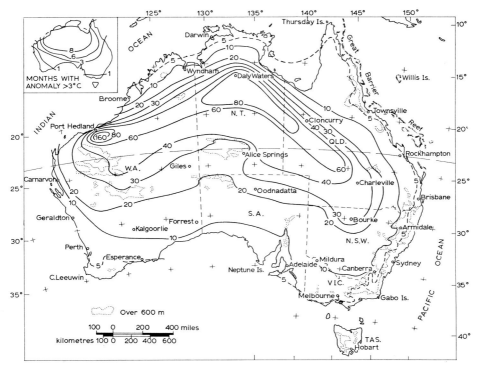

Fig.5. Duration of the longest heat wave, in days. The map shows the maximum number of consecutive days during which the maximum temperature reached 100°F (37.8°C); the record goes to Marble Bar, near Port Hedland, near the centre of the permanent heat low and downslope from the plateau for the trade winds. The inset map shows the mean number of months with an anomaly greater than 3°C; the general patterns shown by the two maps are quite similar. (Main map from Bureau of Meteorology; inset after WALLIS, 1914.)

transport. The source of heat is the cloudless interior (Chapter 4, Fig.1–5); for this reason, the highest temperatures are brought to Perth by north-easterly winds, to Melbourne and Hobart by northerly winds, to Sydney by westerly winds. The Australian *absolute maximum* temperature of 53.1°C was recorded at Cloncurry, Queensland, on January 16, 1889. The *longest hot period* was a series of 160 consecutive days (October 21, 1923–April 7, 1924) during each one of which the daily maximum at Marble Bar, Western Australia reached over 37.8°C (Fig.5). At the same locality, the *mean daily maximum* exceeds 37.8°C from October–March, and 40°C from November–February inclusive (Fig.6). In December, Marble Bar records the *highest mean monthly temperature*, with 33.8°C; in January it is scarcely less hot, with 33.7°C. Its *mean daily maximum for the year*, 35.7°C, is also the highest in Australia.

Nights usually are less hot in dry Marble Bar; it is Wyndham, Western Australia which has the *highest mean daily minimum*, 27.4°C for November, 27.3°C for December, 26.8°C for January (Fig.7) and 24.3°C for the year. As a result of these hot nights, Wyndham has the *highest mean annual* temperature in Australia, 29.1°C. Its hottest month is November, with a mean monthly temperature of 32.2°C.

Darwin, closer to the equator but with cooler nights, has a mean annual temperature of 28°C.

The hottest time of the year is progressively later from north (November) to south (February), corresponding in a general way with the time of most effective insolation.

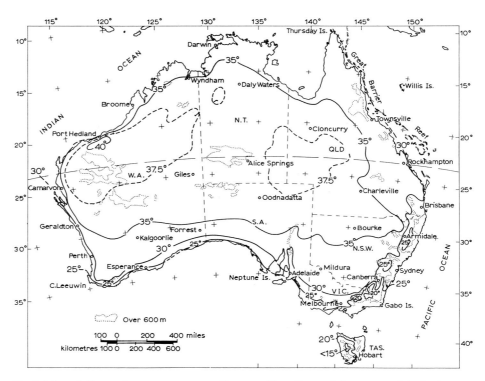

Fig.6. Mean daily maximum temperature, January. The effects of the ocean, of the land mass, of the highlands are all clearly seen, and so is the transport of heat by the southeasterlies. (Simplified from BUREAU OF METEOROLOGY, 1962.)

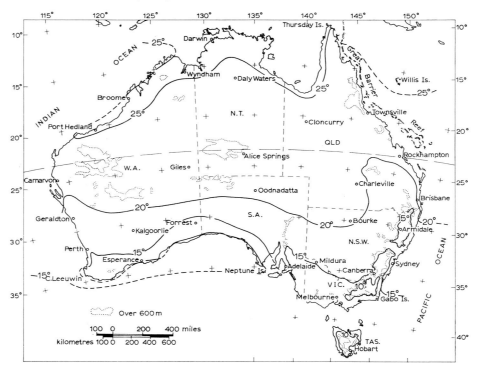

Fig.7. Mean daily minimum temperature, January. Nocturnal radiation lowers inland temperatures considerably, especially in the dry areas. The drainage of cool air downhill smoothes the thermal contrasts in mountain regions. (Simplified from BUREAU OF METEOROLOGY, 1962.)

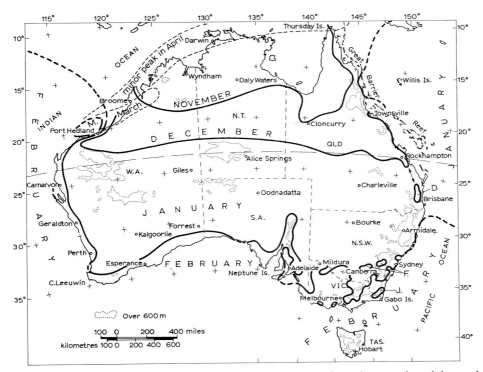

Fig.8. The hottest month of the year, day temperatures. The continental pattern is mainly zonal, with November being the hottest month in the far north, where the sun passes overhead about that time and December just north of the tropic, for the same reason. In the far south the lag is felt and the hottest days are in February. The ocean has a retarding influence, causing a lag of one month along the eastern shore and of over two months along the western shore, with the greatest lag at Port Hedland, probably because of some upwelling of cool water caused by the regular offshore winds. Along the northern shore there is also a secondary maximum in March or April.

There is a lag between the highest day temperatures (Fig.8) and the highest night temperatures (Fig.9), and a tendency for the higher areas to reach their maximum temperatures a little later than the surrounding lowlands. The ocean delays the onset of the highest temperatures in the coastal areas, the more so in the west (over two months at Port Hedland) because of the mild water temperature in the northward drift which prevails at this time of the year.

The daily maxima are of the greatest importance, because in central and northern Australia, most work is done out of doors, where air-conditioning cannot be used. Any high night temperatures, on the other hand, may be avoided at some cost by the use of air-conditioning indoors. Fig.10 and 11 show the number of months, and the actual months, during which the mean daily maximum remains above a given temperature. It will be noticed from Fig.11 that apart from the obvious heat-low, the highest temperatures are reached at the southern margin of the monsoonal belt (see p.82) in the pre-monsoonal months.

On occasional days, under an anticyclonic summer regime, almost any part of Australia may be "subject to hot winds from the interior... resembling the blast of a furnace. The thermometer then rises to 115°F (46°C) and occasionally even higher when extensive bush fires increase the heat... In the desert interior these hot winds, nearer to their source, are still more severe. On one occasion, Captain Sturt hung a thermometer on a

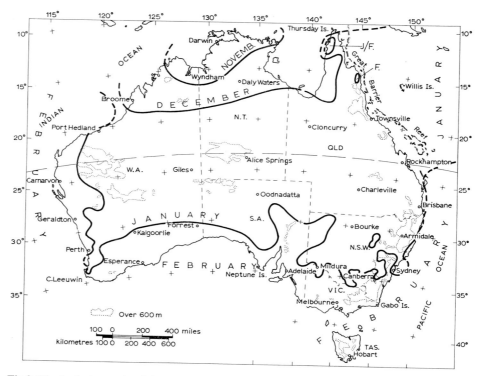

Fig.9. The hottest month, night temperatures. The pattern is similar to that of the day temperatures, but nocturnal radiation causes a lag: for instance, Daly Waters has its hottest days in November but its hottest nights in December. There are some local anomalies, e.g., around the Flinders Range, and on Cape York Peninsula.

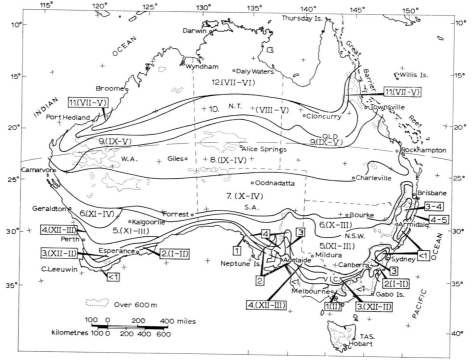

Fig.10. Months with mean daily maximum greater than 80°F (26.6°C). This is mainly a zonal pattern, modified by the ocean along the shore, and by altitude.

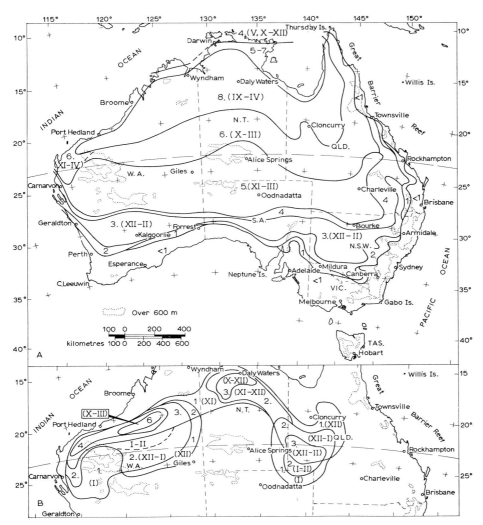

Fig.11. Months with mean daily maximum: A. above 90°F (32.2°C); B. above 100°F (37.8°C). Heat transport by wind begins to be noticeable in the upper map and is very obvious in the lower one. The cooling effect of the trade winds on the northeast coast is also very marked.

tree shaded both from the sun and wind. It was graduated to 127°F (52.8°C), yet the mercury rose till it burst the tube!" (WALLACE, 1893).

The lower temperatures

The middle-to-low latitude and the very modest altitude of Australia (Mount Kosciusko, the highest peak, reaches 2,234 m and the mean altitude of the continent is well below 300 m), are not conducive to extreme heat loss and consequently to very low temperatures. Southeastern Australia is the coolest part throughout the winter, because of its combination of greater land-mass, slightly higher latitude, and greater altitude. For analogous reasons, a much smaller cool area occurs in the southwest (Fig.12 and 13). July is the coldest month, on the average, throughout Australia, but occasionally, in some years, June is colder than July.

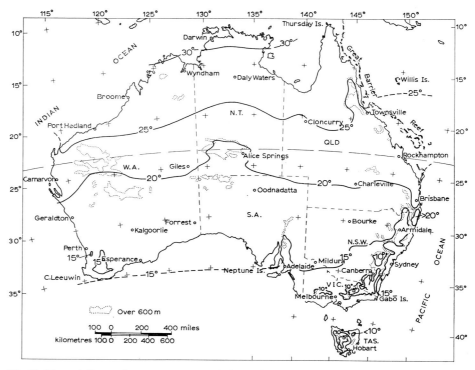

Fig.12. Mean daily maximum temperature, July. Altitude is practically the only factor which alters the zonality of this pattern. (From Bureau of Meteorology, 1962.)

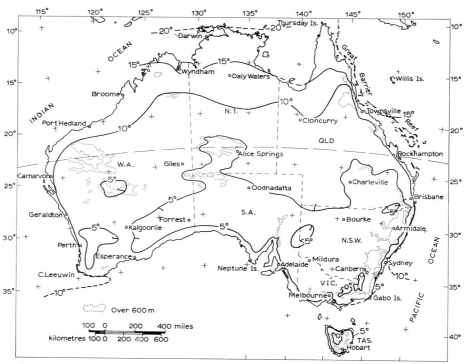

Fig.13. Mean daily minimum temperature, July. Cold-air drainage during the night spreads the effect of the highlands to much wider areas, but at the same time subdues their pattern considerably. (From Bureau of Meteorology, 1962.)

The *lowest temperature* so far recorded is −22°C, at 1,830 m above sea-level, just below Charlotte Pass in the Australian Alps, on July 14, 1945, and August 22, 1947.

On Hotham Heights, Victoria (1,860 m), the mean annual temperature is 4.7°C. In July, the coldest month, the mean daily maximum is 0.2°C, the mean daily minimum, −4.2°C. On the average, the mean daily minimum is below freezing from May–September, inclusive, the daily mean from June–mid-September. The month with the highest mean daily maximum, 15.8°C, is January. February has the highest mean daily minimum, 6.5°C, and the highest daily mean, 11.1°C.

At Lakeside Inn (formerly Kosciusko Hotel), New South Wales, (1,529 m) the mean annual temperature is 6.3°C. In July, the coldest month, the mean daily maximum is 3.8°C, the mean daily minimum −3.9°C. In the hottest month, February, the mean daily maximum is 18.8°C and the mean daily minimum, 6.1°C. The mean daily minimum is below freezing from May to September, inclusive.

Miena, at 1013 m on the Tasmanian plateau, has a mean annual temperature of 6.1°C. In July, it has a mean daily maximum of 4.6°C, and a mean daily minimum of −1.7°C. In February the mean daily maximum is 16.5°C, the mean daily minimum 5.8°C, the daily mean 11.2°C.

If a frost is defined as a temperature below 0°C, it may be stated that most of eastern Australia, south of 25°S, including the lowlands, is subject to occasional frosts during the winter (Fig.14), even though the actual frequency of frost may be very low. Mac-Kinlay the explorer noted "with surprise the hard frosts that were encountered in the

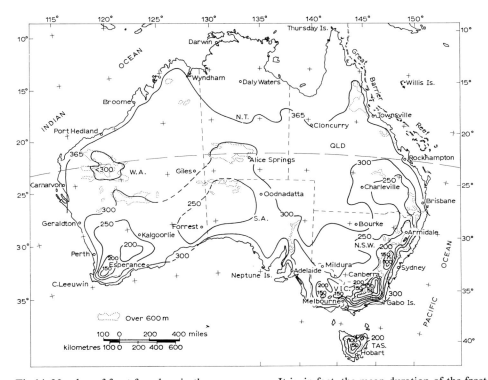

Fig.14. Number of frost-free days in the average year. It is, in fact, the mean duration of the frost-free period; many of the days excluded from the count may have had only a very occasional frost. (From FOLEY, 1945.)

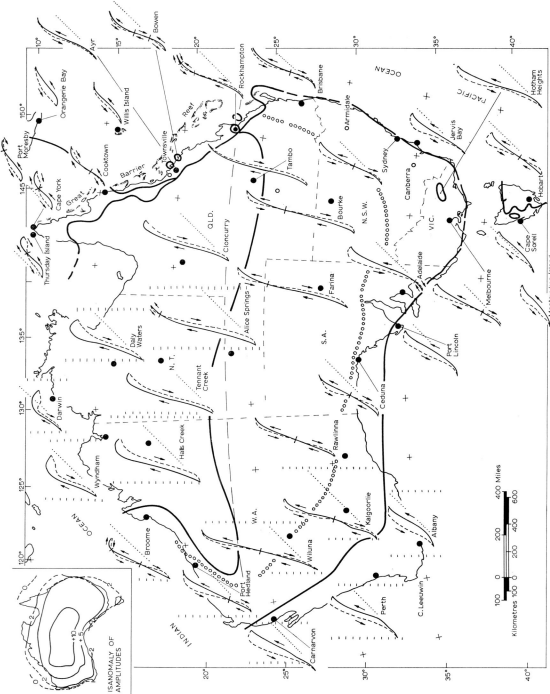

Fig.15. Thermal regime, as shown by the monthly cycle of mean temperatures. The continuous lines in the small graphs show the ascending thermal phase, from July onwards; the dashed lines show the descending phase after January, irrespective of whether the hottest time falls earlier or later. In the tropical continental type, the ascending phase is always warmer than the descending one, i.e. spring is always warmer than autumn, if such terms have any meaning in those latitudes. In the oceanic type the reverse happens, and autumn is always warmer than spring; the prevailing westerly drift carries this type further inland in Western Australia. In the subtropical continental type, shown between the two thick lines on the map, the pattern is mixed; the beaded line shows where the thermal year is equally divided into four quarters, with October and April equally warm. The slope of the graphs shows the thermal amplitude; the gap between ascending and descending phases shows the seasonal lag, which is much more pronounced on the western side, e.g. Carnarvon. The inset map of isanomalies of thermal amplitudes is due to Loewe[?]

130

valley of the Upper Burdekin, about latitude 20° and at a height of certainly not more than 1,000 ft. (300 m) above sea-level" (BARTON, 1895). Spring and autumn frosts may only occur at greater altitudes, usually above 300 m from Ballarat, Victoria, to Stanthorpe, Queensland. Land above 600 m may have an occasional summer frost. Spring frosts may cause serious damage to susceptible crops, especially when accentuated by concave topography, e.g., in the Kooweerup market gardening area near Melbourne (LEEPER, 1960). In subtropical areas, even light frost may have a bad effect on some pasture plants (COLEMAN, 1964).

Oceanic influences and thermal amplitudes

The effect of the ocean is most conspicuously seen in the bending of the mean maximal isotherms towards the north in the summer (Fig.6) and of the mean minimal isotherms towards the south in the winter (Fig.13); this is more marked on the western coast, where, in the winter, the pattern of the ocean currents is reversed. A similar effect is seen on the map of frost-free period (Fig.14) all along the coastline, both on the mainland and in Tasmania.

TABLE I

CONTINENTALITY AND LAG OF THERMAL PHASE
(After JOHANSSON, 1931)

Latitude (S)	Continental	Dev. from normal		Lag	Asymm.
	K	v	h	d	m
15°	38	39	17	−11	28
20°	41	11	13	1	12
25°	41	2	6	2	4
30°	30	0	3	2	2
35°	16	−6	7	6	0

Another result of the ocean's influence is the smallness of the *autumnal thermal gradient*, i.e., of the mean drop in temperature for the period March–June (GENTILLI, 1960). It is largely a function of latitude and continentality, being lowest at Cape York (0.4°C/month), rising to a maximum of 4.7°C at Nullagine, Western Australia, to decline to 1.2°C at Cape Leeuwin, Western Australia, and 1.1°C at Maatsuyker Island, Tasmania. These very moderate values show that Australia is almost free from excessive thermal contrasts during the cycle of its seasons. The thermal gradient is shown by the steepness of the graph lines in Fig.15, continuous lines for the vernal gradient and dashed lines for the autumnal one.

The seasonal phase and amplitude of Australian temperatures were subjected to harmonic analysis by PRESCOTT (1942), with interesting results.

The *annual thermal amplitude* (Fig.15) depends mostly on latitude and continentality.

It varies from less than 5°C in the far north to just over 17°C in the interior, small values indeed. The reversal of the ocean drift off the western shore has a slight moderating effect on coastal temperatures, and the annual amplitude is not as great there as might be expected, e.g., it is practically the same, 10.5°C, at Perth as at Sydney (Fig.15). In the south, the amplitude is considerably reduced by oceanic influences and all along the southern coast it does not exceed 11°C.

The localities with the greatest annual thermal amplitude are Wiluna, Western Australia (17.9°C), Thargomindah, Queensland (17.9°C), and White Cliffs, New South Wales (17.8°C). The smallest annual thermal amplitudes recorded on the mainland coast are those of Cape York, Queensland (2.9°C), Mapoon, Queensland (4.5°C), and Darwin, Northern Territory (4.7°C), all at low latitudes. At latitudes higher than 13°S, even on the coast, the annual amplitude on the mainland exceeds 5°C. Willis Island, off the Queensland coast at 16°18′S, has an annual amplitude of 4.4°C. Maatsuyker Island, at 43°40′S and the southernmost meteorological station in Australia, exposed to continuous westerly winds between two oceans, has the extremely small annual amplitude of 4.8°C. Cape Sorell, on the Tasmanian mainland but also fully exposed to the ocean westerlies, has a mean annual amplitude of 6°C (Fig.15). The isanomaly of annual amplitudes is also shown in Fig.15 (inset).

JOHANSSON (1942) analyzed Australian temperatures. *Phase lag* is obtained by subtracting spring temperature from autumn temperature and dividing by 2: values range from 15 at Geraldton on the west coast, through moderate negative values in the interior, to 6 at Sydney on the east coast. From north to south one meets —7 at Darwin, —4 at Alice Springs, 10 at Cape Borda on Kangaroo Island. *Asymmetry* (skewness) is measured by taking the average of spring and autumn temperatures: it ranges from 27 at Darwin through 10 at Alice Springs to —6 at Cape Borda, and from —2 at Geraldton to 5 at Sydney. *Continentality* k is given by:

$$k = (1.6 \, A : \sin \text{lat}) - 14$$

where A is the annual amplitude. Values range from 54 at Mundiwindi, in tropical Western Australia, and 54 at Alice Springs, to less than 10 along the southeastern fringe, 8 at Hobart, and 2 at Wilson's Promontory. Values calculated by JOHANSSON (1931) for each parallel within Australia are shown in Table I.

The writer recently computed continentality values using the same formula and more recent data. The results in general agree with Johansson's, and show the importance of continental influences in Australian thermal climates. It is also clear that the latitude factor should be modified by a suitable coefficient; the values found for characteristic islands from north to south are: Thursday 10.8; Willis 11.4; Rottnest 9.9; Breaksea 3.1; Kangaroo 2.0; Maatsuyker —2.9.

The *thermal regime* shows some characteristic variations (Fig.15). North of the tropic, the vernal semester is constantly warmer than the autumnal semester, except on the eastern and western coasts. In all the coastal belt south of the tropic, it is the autumnal semester that is warmer. In the extratropical interior the regime is mixed, with the autumnal temperatures dipping very early in the north, and very late in the south. Broome, Western Australia, Bowen, Queensland, and Rockhampton, Queensland, have transition regimes (Fig.15). The mid-latitude oceanic regime, extends inland into Western Australia but is strictly littoral in eastern Australia. Northeastern Queensland already shows a subequatorial regime.

The *daily thermal amplitude* is affected by topography as much as by continentality and atmospheric moisture. The greatest daily amplitudes are found on concave landforms, with semi-arid climate, 200 or more km away from the sea on the lee side of the highlands: Tambo, Queensland, 17.0°C; Yeulba, Queensland, 16.9°C; Kynuna, Queensland, 16.8°C; Canowindra, New South Wales, 16.7°C; Nullagine, Western Australia, 16.4°C; Winning Pool, Western Australia, 16.2°C. At Alice Springs the mean daily amplitude is 15.3°C. On the Nullarbor Plain the smooth, dry, bare limestone surface readily loses at night, the heat received during the day; at Cook, South Australia, the mean daily amplitude is 15.7°C.

On convex surfaces the mean daily amplitude is reduced: on Hotham Heights, Victoria (1,860 m), it is only 6.8°C against 13.7°C at Omeo, Victoria, not very far away but in a valley at 643 m. Even when the difference in altitude is almost irrelevant, topographical forms affect the mean daily amplitude: at Guildford, Western Australia, on river flats, the amplitude is 12.8°C, while at Kalamunda, a little farther inland but on a strongly convex spur of the plateau, it is 10.6°C.

All along the coast, the daily amplitude is small: Sydney, on the coast, 8.2°C; Melbourne, on enclosed Port Phillip, 9.8°C; Perth, Western Australia, some 10 km inland, 10.4°C, but Fremantle, on the coast, 7.9°C, and Rottnest Island, several kilometers offshore, 6.3°C.

The mean daily amplitude is reduced by the advection of maritime air, e.g., in the belt of westerly winds (Cape Sorell on the western coast of Tasmania 5.5°C, St. Helen's on the eastern coast 10.8°C) and in the belt of trade winds (Cooktown, Queensland, on the eastern shore of Cape York Peninsula, 6.9°C, and Mapoon, Queensland, at a lower latitude but on the western shore of the same peninsula, 9.0°C).

Advection of dry continental air increases the daily amplitude, e.g., average for the year at Derby, Western Australia, 11.6°C, and Broome, Western Australia, 10.6°C. Where the seasons bring an alternation of maritime and continental air, the daily amplitude reflects this very clearly even over whole monthly periods. Thus at Perth, Western Australia, it is 8.2°C in July (maritime winter) and 12.0° in February (anticyclonic continental summer). At Darwin, Northern Territory, it is 10.4°C in July and August (continental trade wind) and 7.0° in January (maritime monsoon). At Sydney, where maritime and continental air may alternate at any time, the month with the greatest mean daily amplitude is September (9.4°C) and that with the smallest, February (7.3°C). Detailed tables of maxima and minima for 20 localities, in tenths of a degree Fahrenheit with frequencies in tenths of a day throughout the year, have been prepared by ASHTON (1964).

The water balance

The rain-bearing factors

The tropospheric movements discussed in Chapter 5 bring rain to different parts of Australia at different times during the year (Fig.17); thus, monsoonal rains in the north come mostly during the summer (November–March), and frontal rains in the southwest mostly during the winter (May–September). Rain from tropical cyclones comes in the

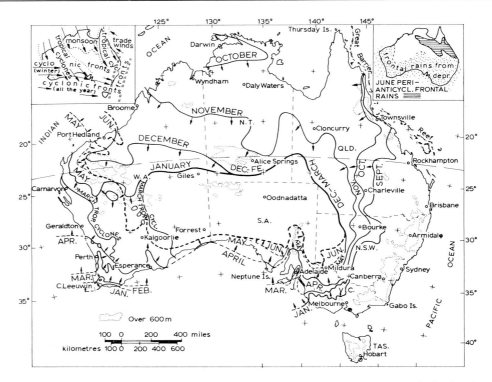

Fig.16. The advance of summer and winter rains; only the months which bring rain farther than the respectively previous months are shown. The thick black line shows the farthest advance of monsoonal rains, the thick dashed line that of mid-latitude frontal rains. The inset map at the left shows the main rain-bearing factors with their generalized location, that at the right shows where the peri-anticyclonic frontal rains are a noticeable feature of the climate.

summer or very early autumn (December–March), is not very frequent, but when it comes it may reach almost any part of the mainland (Chapter 5, Fig.28 and 30). The trade winds bring orographic rain to a very small part of the northeastern coast (Fig.16, inset).

FITZPATRICK (1964) analyzed Australian rainfall by applying harmonic analysis to monthly means from 192 stations. Amplitudes, phase angles, dates of the maximum points, and absolute and percentage variance values for 6 harmonics were obtained. The first harmonic (single cycle within the year) accounts for most of the seasonal cycle; it is due to the latitudinal position of the anticyclones.

"This zone represents a major separation or 'climatic divide' between the summer-active rainfall mechanisms of northern Australia and those which are predominantly influential during the winter season in southern Australia." Tropical cyclones and thunderstorms account for most of the residual variance. A preliminary survey of the frequency of selected monthly totals in stations along a meridional transect done by the writer shows bi- and tri-modal distributions for the summer months, more pronounced in the northern and central stations. They are probably due to multiple incursions of *Tm* air brought about by tropical cyclones or monsoonal advances.

Much research on the actual processes of rain formation has been done by the Division of Radiophysics of the Commonwealth Scientific and Industrial Research Organisation. Frontal rains mostly fall from clouds which have become super-cooled from $-12°$

to −35°C as a result of uplifting. Some of the super-cooled droplets freeze on suitable nuclei, thus beginning the actual rain-forming process.

Condensation of vapour around large salt nuclei can produce droplets large enough to facilitate coalescence and the subsequent fall of rain at above-freezing temperature (warm rain). Much showery rain in coastal districts is due to this cause. At lower levels, the salt content of the air may be very high; research by the Royal Society of Western Australia (ANONYMOUS, 1929) showed that, with strong onshore winds, rain-water contains 10–90 parts of Cl per million (in weight).

YAMAGUCHI (1963) gives tables of the occurrence of cloud types, frequency of cloud-top temperature (warmer than 0°C, between 0° and −10°C, colder than −10°C) and of associated rain, for 14 stations well distributed throughout the Australian mainland. This information is also correlated with the presence of maritime, continental or transitional air. Pioneering research on the artificial stimulation of rain has been carried out in Australia during the last 20 years, but there is no conclusive evidence so far that rainfall averages have been affected by the experiments, of which a good general account is given by SMITH (1966).

The rains of northern Australia come nearly entirely from monsoonal situations (see p.82) but their immediate cause is the instability of *Eq* and *Tm* air spreading over continental locations, and the consequent thunderstorms. While over the whole season the area watered by these rains is enormous (over 4,000,000 km²), a day-by-day analysis shows them to be mostly highly localized and patchy, as may be expected from their convectional nature (Chapter 5, Fig.18). There are, however, instances of important southward invasions of *Eq* air with very widespread and plentiful rains, caused by streamline convergence, advectional cooling, and at the monsoonal front, frontal uplift. In February 1955, an extraordinary transgression of *Eq* air covered all of Western Australia, reaching as far as latitude 32°S at the surface and at least 35°S in altitude, bringing very high humidity and heavy rains everywhere, causing a multitude of local floods; it is also likely that the same transgression affected eastern Australia (see Chapter 3).

Fig.16 shows the monthly positions of the 25 mm isohyet during the phase of spreading rains, i.e., from spring to mid-summer in the north and from autumn to mid-winter in the south; it will be noticed that in the east the rains remain practically uniform throughout the year, and in the west the most extreme alternations of rains and drought occur in the northern and southern portions, respectively, at opposite times of the year.

In October, the first monsoonal rains appear near Darwin, but are adequate only over less than 90,000 km²; at the same time an advance of the inter-anticyclonic rains occurs along a meridional belt passing through Charleville and Townsville, adding some 120,000 km² to the 950,000 or so watered in September. During November, monsoonal situations normally extend east-southeastwards at a rate of about 30 km per day and near the tropic in central Queensland, part of the 3,000,000 km² of country which they water overlaps with some of the 1,500,000 km² watered by the inter-anticyclonic front and upper trough (Fig.19). In December, monsoonal rains extend over 3,750,000 km² and usually reach beyond the tropic everywhere except in Western Australia where, however, some very heavy but occasional falls may then be brought by tropical cyclones.

During January, the frequency of monsoonal situations reaches its peak, watering over 4,000,000 km² (Fig.16), and during February there is little change but for a slight retreat in the west and advance in the east. By March, the rains tend to decrease a little, and

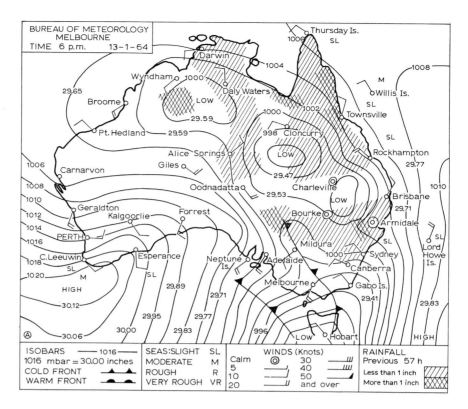

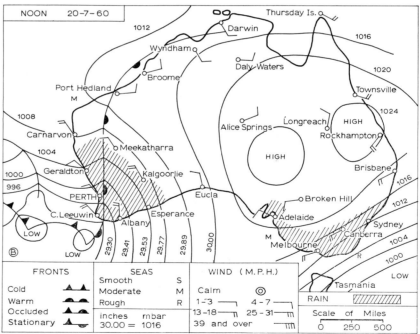

Fig.17. Days with good rains: A. from widespread monsoonal lows which link up with an active inter-anticyclonic front and a mid-latitude depression to the south; B. from the first mid-latitude front of a complex depression, with strong northwesterly flow of *ITm* air over the southwestern quarter, while the last front of a preceding depression still brings rain to the southeastern coast, but with a southwesterly flow; C. from an inter-anticyclonic front over eastern Australia, obviously with a development in altitude; D. from a peri-anticyclonic front over the tropic. (Age and West Australian weather maps.)

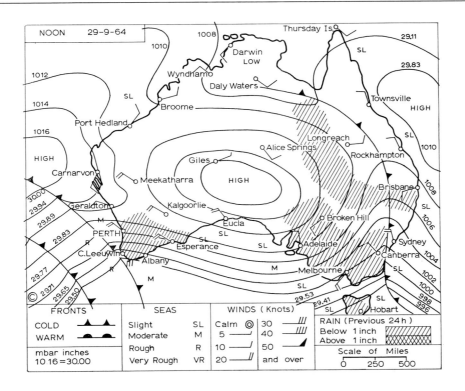

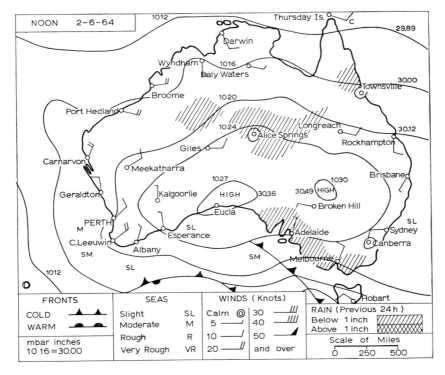

Fig.17C, D (Legend see p.136)

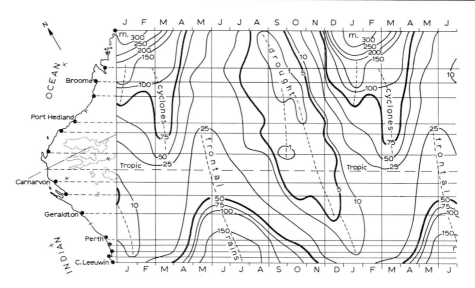

Fig.18. Monthly rainfall on the west coast (in mm) from Kunmunya (Port George IV) in the north, to Cape Leeuwin in the south. The contributions of the monsoon *(m.)*, the tropical cyclones and the frontal rains are clearly distinguishable, except for the overlap of monsoonal and tropical cyclonic rains. The extreme drought of the spring months is a salient feature.

more noticeably so in the west, where, on the other hand, tropical cyclones become less infrequent, and more than compensate for the decreased monsoonal activity. Over the western part of the mainland, rain from tropical cyclones extends over an average of nearly 2,000,000 km² and is enough to alter the normal seasonal trends (Fig.18 and 28). Tropical cyclones bring at least as much rainfall to the eastern part of the mainland (Fig.19), but this fact is hidden by the rains brought by monsoonal situations and above all by the inter-anticyclonic front (Fig.19). From an examination of the pattern of tropical cyclonic rainfall (Fig.27 and 28), it seems that the heaviest rains occur while the cyclone is still north of the tropic, when it is very slow (4–5 km/h) and its enormous load of moisture is discharged over a relatively small area. A peculiar pattern of drought prevails immediately to the south, irrespective of topography, between the tropic and latitudes 26°–27°S. Further south, the cyclone draws in local air, and a frontal surface develops, which changes the structure of the cyclone and causes widespread and usually heavy rains, the amounts being the more remarkable, because the whole cyclone then travels at 100–150 km/h. It may also be a general rule that, while in the tropical tract, the rain occurs more to the east of the cyclone's path; in the extratropical tract the (then frontal) rain occurs more to the west, i.e., on the side where the cooler air is being drawn into the system. Fig. 27 and 28 give an idea of the large areas affected by cyclonic and cyclofrontal rains.

The trade winds (Chapter 5, Fig.18 and 21) bring some orographic rain to the coast between Rockhampton and Cooktown, but very little of this rain penetrates inland, and only along the 350 km of steeper land between Port Douglas and Cardwell is it really important. The area affected may be some 30,000 km² (Fig.19).

Over northeastern Australia, during the early winter, *Tm* air from the Coral Sea is lifted by the cool *Tc* or *sTc* air issuing from any large anticyclone, and the very extensive, if fairly stable, peri-anticyclonic front that it forms, brings light rains to a very large area

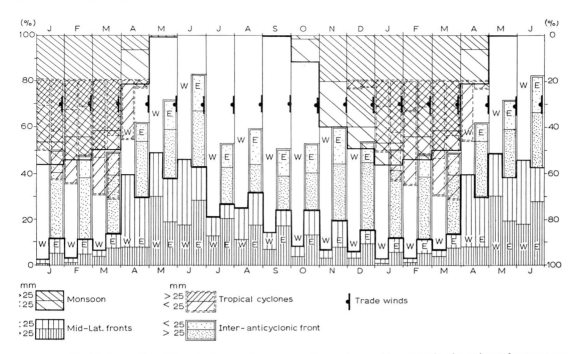

Fig.19. Area affected by rain from various causes. For each monthly rectangle, the estimated percentage of the area receiving rain from each factor is shown by different shading (close for over 25 mm per month, open for less); a distinction is also made between the western and eastern parts of the mainland. Tasmania is counted with the eastern part. The most sudden break is the one taking place after June, with the sudden contraction of the mid-latitude frontal rains and the spreading of the tropical drought to the eastern half.

(Fig.16 and 17). With a strong meridional circulation, the likelihood of rain is increased because of the progressive cooling of the southward-spreading *Tm* stream and the strong convergence in the northward spreading *sPm* stream, while a strong zonal circulation reduces or even prevents the development of the peri-anticyclonic front.

In the summer coastal front, which usually arises in inter-anticyclonic situations along the western shore (see p.60), the air masses are stationary or very nearly so, relatively to each other, and bring no rain, although some cloud lines may become apparent.

Inter-anticyclonic fronts (see Chapter 5, Fig.6) are rain bearing along the eastern shore (Fig.19) because they lead to the formation of an upper trough or cold pool. The amount of rain can be much greater when the inter-anticyclonic situations produce a particularly effective upper trough by reviving a pre-existing front (Fig.16 C). Variations in the temperature of the sea-surface affect such fronts, by increasing or decreasing the thermal contrast of the overlying air; such variations usually come about through changes in the trajectories of oceanic currents and drifts, rather than in the actual temperature of large bodies of water. The approach of the warm (southward) current to the eastern shore in winter causes an intensification of fronts and an increased rainfall; the intromission of cooler water, either as a northward drift or as an upwelling due to offshore winds, causes coastal fogs and stratus but reduces the intensity of the rain.

Along inter-anticyclonic fronts, there may also arise small secondary waves which become small but fully developed depressions; they rapidly bring in air from higher latitudes (GIBBS, 1946) and may cause most intense, if not very widespread, falls of rain. This cause of rain was clearly stated by J. C. Foley (personal communication, 1948) who said that "widespread rains may as a rule be expected with trough formations in the 500-mbar contours extending northward (including northwest or even west-northwest) over the continent in such a manner that a well marked air stream from some direction north of west persists for at least a short period ahead of the trough. Divergence in the stream is an important factor in its relation to convergence in the lower strata of the atmosphere, but this may not always be apparent in the contour pattern at the 500-mbar level... A cut-off low aloft may not be accompanied by any marked feature on the surface isobaric chart as it traverses eastern Australia, but a cyclonic development on the east coast, particularly the coast of New South Wales, may be expected as the upper low approaches this area... From the point of view of climatology... it will be of interest to note that extensive rains in the inland can usually be traced to deep streams of moist air flowing into the interior from tropical regions to the northeast, north or northwest." Subsequent work on the jet stream has confirmed all this, albeit with a changed emphasis on the cause–effect relationship. This is the dynamic development studied by CAMPBELL (1968) in connection with significant rainfall episodes in the Brisbane and Newcastle areas (see Chapter 5).

Obviously jet stream maxima, upper air divergence and induced destabilization below could not produce rain without adequate supply of moisture and an attainable dew point, hence the frequency of rainfall from these causes only well after the autumn equinox, and the general alignment of the affected belt within reach of the maritime air.

The effectiveness of the inter-anticyclonic front and its occasional cyclonic vortices is shown as far south as Hobart. TROUP and CLARKE (1965) show that about as much rain reaches Hobart with easterly winds as with westerly winds, in each season (slightly less in spring, slightly more in summer). The rain from the west is brought by distant mid-latitude depressions (see below), the rain from the east by cyclonic vortices formed off the east coast. The falls of rain caused by eastern depressions are often much heavier (over 25 mm in 24 h) but also less frequent.

The rains from the inter-anticyclonic fronts show very little variation throughout the year, covering a maximum area of nearly 1,500,000 km² in January and February (when they overlap with the monsoonal and tropical-cyclonic rains in the north) and a minimum

TABLE II

MEAN ANNUAL RAINFALL ON WESTERN COASTS (mm)

Latitude	Australia	North America	South America	Southern Africa	North Africa and Europe
20°	300	875	0	25	100
22°	300	750	0	25	50
24°	225	100	0	25	25
26°	250	100	25	25	75
28°	400	100	50	50	175
30°	575	200	100	100	200
32°	700	250	275	175	375
34°	1,000	375	500	750	375
36°	625	500	950	–	750
38°	650	1,000	1,875	–	500

------- separates the rainfall of the freely exposed Western Australian coast from that of the partly sheltered South Australian coast.

of some 950,000 km² in September, which is the driest month for the mainland as a whole (Fig.19).

Frontal rains from mid-latitude depressions (see p.94) only skirt the southern shore during the summer (December–February) and come mostly from a southwesterly direction. In March, they advance northwards by some 100–150 km, covering about 800,000 km² (Fig.16). They are increased even by a modest orographic uplift, as can be seen above the Darling Scarp, Western Australia (Chapter 4, Fig.9) and the Mount Lofty Range, South Australia. The frequency and effectiveness of the fronts increases rapidly in April, with a northward advance of over 200 km to cover a total of some 2,500,000 km²; the prevailing direction is still from west-southwest. Both frequency and effectiveness increase overwhelmingly in May, when the advance along the western coast passes 1,200 km, a progress of 40 km a day, at a rate which can only be explained by a change in the behaviour of the jet stream, with a sudden increase in the equatorward swing and a tightening of its meanders, and a strong west-northwesterly flow. The area watered by these frontal rains in May reaches some 3,250,000 km².

In June, the position of these frontal rains is consolidated along the same line, except for small but characteristic changes near Port Hedland, Western Australia, around the Flinders Range, South Australia and north of Mildura, Victoria (Fig.16), which can only be explained with a further swing and tightening of the jet stream's meanders, so that the initial rain-bearing flow is from the northwest. This pattern may also occur in July (Fig.17).

A peculiarity of the winters frontal rains is their extremely rapid northward spread and increase in quantity throughout May (Fig.18). Their southward retreat and simultaneous decrease during August, September and October, are slower and at an almost constant rate throughout this period. The area watered on the western side of the mainland (Fig.19) is 1,250,000 km² in June, 800,000 in July, 700,000 in August and 550,000 in September; on the eastern side, including Tasmania, over 2,200,000 km² in June, 1,000,000 in July (this great reduction is due to the predominance of anticyclonic situations because of the braking effect of the highlands), 1,250,000 in August and 950,000 in September. During October the summer pattern sets in, the fronts become infrequent and travel further to the south, and the rains are much lighter and erratic; the area watered is reduced to some 400,000 km² in the west and still remains about 950,000 km² in the east, but a very large portion of it gets less than 25 mm during this month (Fig.19). From November–March, these rains are significant only over the southeastern mainland and Tasmania.

The annual march of the rainfall has been described by HUNT (1914) and may be shown by a series of monthly rainfall maps, e.g., as is done in the *Atlas of Australian Resources*, or by a movable sliding or concentric chart. HUNT (1914) visualized the advance and retreat of the rains as an apparent "oscillatory movement... about a centre in the vicinity of Forbes, in New South Wales ... the amplitude of oscillation exactly equals 72° of arc or one-fifth of a circle." He designed an effective "rain clock". TAYLOR (1920) designed a sliding chart, i.e., he visualized the seasonal swing as a latitudinal displacement and not as an oscillatory alternation.

Our Fig.16, 20 and 21 as well as FITZPATRICK's (1964) isochronal map of the first harmonic's maximum shown in 2-weekly intervals vindicate HUNT's (1914) concept and his placement of the nodal point in the vicinity of Forbes.

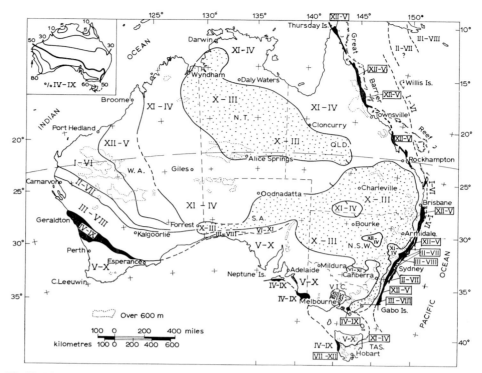

Fig.20. The wettest semester and (inset) the percentage of the yearly rains falling in the April–September semester. Notice the October–March ridge (stippled), the gradual concentric spreading of later rains towards the southwest (with the joining of the opposed systems from the north and from the southwest), and the razor-sharp alignment on the east coast. The inset map has been slightly modified from PRESCOTT (1931), because later statistics do not show the high percentages of winter concentration originally shown; the spot with the highest proportion of April–September rains (the Abrolhos Islands) is shown by the arrow.

The graphic representation of the various aspects of this complex cycle requires a time-chart along a meridional coastal transect, preferably on the western side in order to show the seasonal variations with the least interference by topography and continentality (Fig.18), a map of the advance or retreat of the rains in order to show the range of spatial transgression or regression as the case may be (Fig.16), and a continental time-chart in order to introduce the areal dimensions which are essential in the study of a continental land-mass (Fig.19).

The seasonal incidence of the various rain-bringing factors causes a variation in the seasonal incidence of rain. The "time ridge" runs approximately from Wyndham, Western Australia, to Sydney, with October–March as the wettest semester (Fig.20). North of the tropic, the October–March period may be taken as that of the most undisturbed monsoonal influence. In November, December and January there is a general tendency for the monsoonal fringe to spread southwestwards (see also Fig.16) and for tropical cyclones to increase in frequency and affect the western area (see p.88). This tends to make November wetter than October, December wetter than November, and January wetter than December, as one progresses further west, as far south as the tropic. On the other hand, beginning from the extreme southwest, frontal rains begin to advance (Fig. 16) and reach their farthest extension in May–June, enough so as to cause June to exclude December from the wettest semester already at Port Hedland, Western Australia

(20°23′S), July to exclude January on the coast near the tropic, August to exclude February from Carnarvon to Kalgoorlie, Western Australia. The "all frontal rain" semesters then appear, with the early winter semester (April–September) from Geraldton to Esperance, Western Australia, and the late winter semester (May–October) further south; this is the classic winter rain area south of latitude 30°S.

The eastward cline from the "time ridge" is easily explained in the continental area, by the average extension of monsoonal rains well into April from Darwin to Rockhampton, Queensland, April being wet enough to cause the exclusion of October. What happens along the coastal fringe, however, is probably linked with oceanic water circulation: over a few miles there is a rapid change from December–May, which is the normal wettest semester on the littoral, to January–June (Willis Island and Sandy Cape, Queensland), to February–July (Sydney, Jervis Bay, New South Wales) and even, in the extreme fringe (Orangerie Bay, New Guinea at 10°22′S; Port Stephens, New South Wales, at 23°42′S; Gabo Island, Victoria, at 37°34′S) to March–August. The eastern highlands exclude any decisive effect by the westerly fronts (notice the "subtropical corridor" of October–March maximum all along the eastern slopes from Canberra, Australian Capital Territory to Sale, Victoria).

That part of the yearly rainfall which comes between April–September (Fig.20, inset) is of decisive importance for the winter–spring crops, and has some significance in pedogenesis (PRESCOTT, 1931); it depends on the mid-latitude frontal rains, and reaches maxima of 92% at the Abrolhos Islands (years 1891–1895), 86% at Greenough and 85% at Geraldton and Dongara, Western Australia (years 1911–1940).

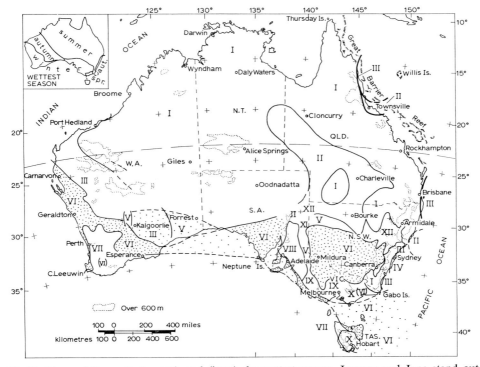

Fig.21. Maps of the wettest month and (inset) the wettest season. January and June stand out, but around the respective areas lies a complicated pattern of transitions and overlaps. February and March (tropical cyclones) and July (mid-latitude frontal rains) dominate smaller areas. The most complex pattern is shown around Adelaide. The inset map is from ANDREWS (1932).

The most common wettest months are January in the monsoonal area and the interior, and June in the frontal-rain belt (Fig.21). May is the wettest month on the frontal-rain fringe, and March in parts of both tropical cyclone belts (western and eastern, the latter only fragmentary). The most complex node occurs in South Australia and western Victoria, where the cold air pool of late June and July reduces the rainfall of those months and indirectly makes August or September the wettest month.

The overall pattern of daily recurrence of rain at Melbourne was analyzed by WATTERSON and LEGG (1967), who found that "previous history only influences a given day via the immediately preceding day". Longer dry and wet spells tend to follow a "generalized negative binomial distribution", but the months leading into and out of winter showed some anomalies.

The amount and variability of the rain

The inadequacy of its rainfall is Australia's major problem: this inadequacy is obvious in the geographical distribution (Fig.22) which leaves 37% of the land with less than 250 mm of rainfall per year, 57% with less than 375 mm, and 68% with less than 500 mm (cumulative percentages). Things are worse if single areas, e.g., states, are considered: on the average, 83% of South Australia, 58% of Western Australia, 25% of the Northern Territory, 20% of New South Wales and 13% of Queensland receive less than 250 mm per year. Only 0.5% of South Australia, 5.5% of Western Australia, 16% of New South Wales, 17% of the Northern Territory, 23% of Queensland and 27% of Victoria receive more than 750 mm/year. Only in Tasmania is rain plentiful, with more than half the island receiving over 1,000 mm/year (BUREAU OF METEOROLOGY, 1962).

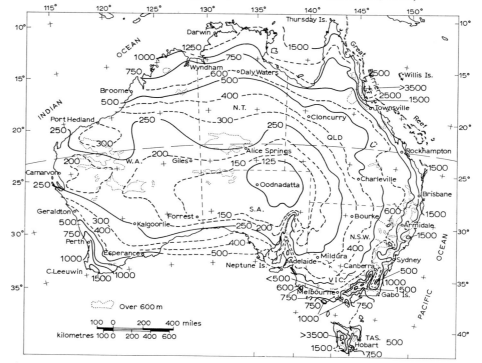

Fig.22. Mean yearly rainfall based on the 1911–1940 period. (From BUREAU OF METEOROLOGY, 1962.)

The driest area is located in South Australia, around Lake Eyre, 2,000 km from the western, 1,400 from the eastern and 1,000 from the northern shores (Fig.22). It is very low, the shores of the lake actually being below sea-level. It is at the latitude where the anticyclonic situations are most frequent and best developed. In addition, its position in respect to the extensive western plateau, which is about 400 m high, causes a slight subsidence to occur in the westerly streams (winter frontal and summer monsoonal, respectively), so that very little precipitation can take place, on the average less than 125 mm/year.

The lowest, long-term mean yearly total, 104.9 mm, is found at Troudaninna, with a partly interrupted record extending over 42 years (1893–1936). Only one calendar year exceeded 230 mm (the year 1894, with 281 mm). On several consecutive years the rainfall was well below average, e.g., for the calendar periods 1895–1903 (yearly mean 74 mm) and 1818–1929 (yearly mean 67 mm). Over the five years from December 1924–November 1929, the yearly mean total was only 43 mm.

Another station, Mulka, has received less than 75 mm per calendar year on 17 years out of 39; only on two years (1920 and 1955) did it exceed 250 mm. The mean yearly total for the four years between October 1926–September 1930, was 32 mm.

Kanowana recorded 11, 57, 22 and 24 mm, respectively in each of the years from 1896 to 1899, inclusive, the mean being 28.5 (Bureau of Meteorology, 1962).

Another aspect of normal rainfall deficiency which reaches an extreme degree of incidence in Australia is the seasonal drought. In summer, the southwestern corner and in

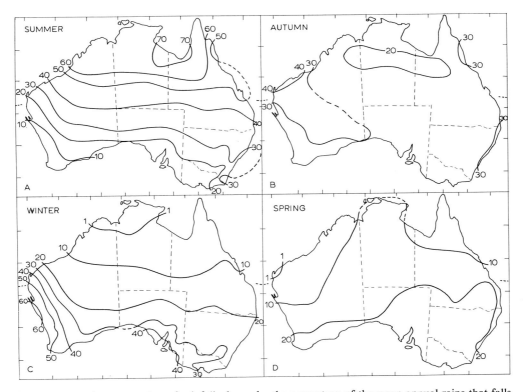

Fig.23. Seasonal concentration of rainfall, shown by the percentage of the mean annual rains that falls within each season. Notice the monsoonal pattern of the summer, the effect of tropical cyclones in the autumn, the distinctly westerly alignment in the winter, the southwesterly bias of the spring. (From Andrews, 1932.)

winter the northern third of the mainland, receive less than 10% of their respective yearly rainfall (Fig.19 and 23). Most of the northern area is also badly deficient during the spring (ANDREWS, 1933). When combined with high temperatures, seasonal droughts have more severe effects on all organisms; the relevant concept of effective rainfall is discussed in the next section (p.172).

Non-seasonal droughts are very critical in Australia because so much of the country is near the lower threshold of usable rainfall. The drier limit of wheat farming extends over a length of nearly 4,000 km, and the slightest deficiency in rainfall in its vicinity has immediate economic repercussions. In the drier pastoral districts the average climatic conditions must be classified as marginally arid, and a drought reduces pasture growth and precipitates a crisis in what is an already precarious equilibrium.

Many droughts have occurred in Australia's brief history, e.g., in 1864–1866, 1888, 1880–1886, 1888, 1895–1903, 1908, 1914, 1918–1920, 1929, 1939–1945. In a year of widespread but not catastrophic drought, about half the mainland may receive less than 250 mm of rain, and in a catastrophic drought year the proportion might reach about two-thirds (WINTERBOTTOM, 1945), and the rainfall may be less than half the normal amount over large areas. There are droughts which are catastrophic because of their duration as well as their intensity, e.g., the drought that struck the northwestern division of Western Australia in the periods 1922–1927 and 1934–1941 because of the failure of cyclonic rains to reach that area for several years, or the inadequacy of monsoonal rains over parts of northern Australia during two or three consecutive summers,

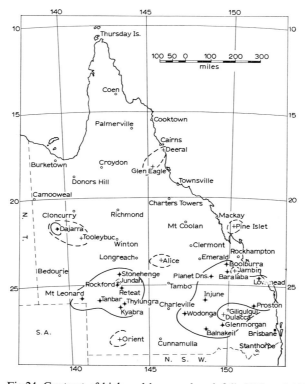

Fig.24. Contrast of high and low yearly rainfall, 1950 and 1951. The stations shown by a diamond and situated in the areas circumscribed by the continuous line, had their record wettest year in 1950 and their record driest year in 1951. The stations shown with a cross and circumscribed by a dashed line had their driest year in 1951. (From Australian News and Information Bureau.)

e.g., in 1911–1916, 1925–1929, 1937–1938, 1951–1952. A thorough review of droughts recorded in Australia is given by FOLEY (1957). That droughts are quite unpredictable is shown by Fig.24; over some 250,000 km² of Queensland territory, two consecutive years were the wettest (1950) and driest (1951) ever recorded.

GIBBS and MAHER (1967) made a statistical comparison of mean and median values, and evaluated the rainfall of every year from 1885 to 1965 in terms of deciles from the median. The three lowest deciles indicate various degrees of relative drought intensity.

As extreme examples of scarcity of rain over a complete calendar year, one could quote from ASHTON and MAHER (1951), the 0.5 mm received at the 510-mile station on the Rabbit Proof Fence, Western Australia (No. 510 R.P.F.), 1.3 mm at the De Grey station, Western Australia (when no tropical cyclones crossed the area) and in the Eyre dry bowl, 9.9 mm (spread over 5 days) at Mungeranie, South Australia (BUREAU OF METEOROLOGY, 1962).

The rainiest part of Australia is the steepest section of the trade wind coast of Queensland, between Port Douglas and Cardwell (Fig.22 and 25). The wettest locality is Tully (Tully Mill), with a mean yearly total of 4,445 mm, a maximum of 7,897 mm in 1950 and a minimum of 2,667 mm in 1943. In about one year out of three, Tully receives more than 5,000 mm of rain.

Further north, at Harvey Creek, the mean annual rainfall is 4,242 mm; the heaviest yearly total on record, 6,472 mm (in 1921) and the lightest, 2,044 (in 1903); a yearly rainfall of 5,000 mm is exceeded once in seven or eight years. Better known localities in this rainy coastal area are Port Douglas (16°30'S, 2,024 mm), Cairns (16°55'S, 2,205 mm), Innisfail (originally called Geraldton, 17°32'S, 3,912 mm), and Cardwell (18°15'S, 1,956 mm).

That these heavy falls are due to the effect of topography there is no doubt: on Willis Island, situated over 200 km offshore at 16°18'S 149°59'E, the mean yearly rainfall is only 780 mm, and yet Willis Island is exposed to the same trade winds, tropical cyclones and occasional monsoonal thunderstorms as the coastline opposite; but the island, a

TABLE III

VARIATIONS IN RAINFALL QUANTITY AND REGIME[1]

Station	Jan.	Feb.	Mar.	Apr.	May	June	July	Aug.	Sept.	Oct.	Nov.	Dec.
Northeast Qld.												
Willis I	89	106	**107**	88	76	65	56	33	32	**29**	38	59
Innisfail	490	602	**687**	468	323	187	119	106	**81**	94	134	243
Mareeba	**223**	220	179	58	17	19	9	6	**4**	14	33	110
Mt. Garnet	160	**165**	125	34	23	19	12	6	**3**	15	49	112
Eyre Lowland												
Oodnadatta	**17**	15	8	*5*	7	**13**	**4**	6	*5*	*14*	8	11
Tasmania												
Queenstown	171	*126*	162	*223*	221	227	245	**257**	255	232	212	179
Clarendon	*30*	32	*30*	*40*	33	*40*	39	*36*	39	**50**	*38*	47
Maatsuyker I	83	*68*	88	104	106	112	**114**	111	101	102	93	90

[1] Bold face numbers = wettest and driest months.
Numbers in italics = secondary wetter or drier months.

coral reef, is very flat and low above the ocean. It should be noted, however, that the effect of topographic obstacles on tropical air streams is to cause an immediate release of precipitation on the lower slopes, not on the upper slopes as happens in cooler climates, so that one may speak of "orographic depletion" of the rainfall.

The rainfall gradient towards the interior (Fig.22) is extremely steep, because of the abrupt effect of the mountain barrier on the tropical air streams: over the 36 km from Cairns to Mareeba, the yearly rainfall decreases from 2,205 mm to 891 mm, an average of 36 mm/km; over the 110 km from Tully to Mount Garnet (17°52′S 145°07′E) it decreases from 4,445 mm to 772 mm, an average of nearly 34 mm/km. The regime also changes completely (Table III and Fig.25).

Constant exposure to the westerly fronts gives a high rainfall to the western coast and slopes of Tasmania; at Lake Margaret the mean yearly total is 3,685 mm, and the highest yearly total on record is 4,504 mm in 1948. The steepest rainfall gradient is southeastwards, where Clarendon, about 120 km away in the sheltered Clyde Valley, gets only 455 mm per year; the gradient is nearly 27 mm/km.

Besides the quantity of the rain, its duration and frequency are most important. The finer measure of this aspect of the rainfall is the number of days in which it rains ("wet days", "rainy days"). In Australia these are the days with a precipitation >1 point = 0.01 inches = 0.24 mm. It is doubtful whether such a small amount of rain falling in widely spaced drops on a warm metal gauge-funnel, would run into the gauge at all, so

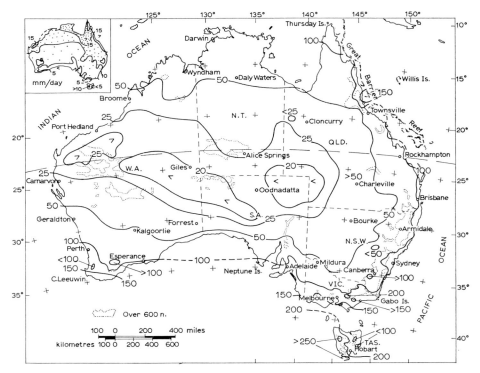

Fig.26. Mean annual number of days with rain and (inset) mean rainfall per day with rain. The subtropical belt, with the greatest frequency of anticyclones, has the lowest number of days with rain. Monsoonal rains are heavy but usually do not occur on many days. Mid-latitude frontal rains on the other hand persist for much longer. Notice that the effect of topography on the number of wet days is much more subdued than it is on the actual amount of rain (cf. Fig.5.22). (Map of number of days from BUREAU OF METEOROLOGY, 1962.)

it must be assumed that some hot "wet" days go unnoticed, since the reading is usually done once in 24 h. Secondly, no accurate comparison is possible with frequencies of "wet" days otherwise defined, e.g., ≥0.1 mm, because with the Australian definition quite a number of days with very light drizzle, frequent in Tasmania, will not be counted. Fig.26 shows the annual number of days with rain; only the areas exposed to mid-latitude fronts have a high frequency of wet days, 150 or more per year, reaching a maximum of just over 250 on the slopes above Zeehan and on Maatsuyker Island. The inset map shows the mean rainfall per wet day; the tropical cyclonic rains bring the heaviest falls, with an average of more than 15 mm/day, while the mid-latitude fronts bring persistent but light rains of less than 5 mm/day, unless topographic uplift causes an intensification of the fronts, as happens in western Tasmania.

A much larger map of the mean precipitation per wet day is given by JENNINGS (1967), who found the highest values at Reid River near Townsville (25.1 mm per wet day) and Innisfail (23.7 mm per wet day), and the lowest values on the mainland at Koppio at the southern tip of Eyre Peninsula (2.4 mm) and in Tasmania at Oatlands (3.1 mm). The highest intensity in Tasmania is found at Gormanston, on the western slopes: 12.4 mm per wet day. As was pointed out by PRESCOTT (1931a), a rainfall intensity below 5 mm is likely to lead to the accumulation of salt in the soil, and the isohyet of 5 mm is thus specially significant.

The actual incidence of the rainfall as distinct from the average can be most uneven.

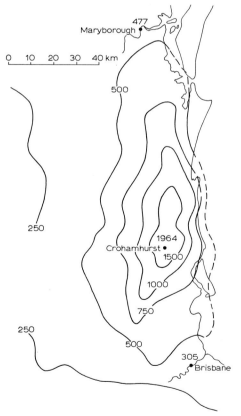

Fig.27. The Crohamhurst cyclone and the rainfall of the period 1–4 February 1893, heaviest recorded in Australia. (Adapted from BRUNT, 1958a.) These are the four wettest days: rain actually fell from January 30 to February 5.

Rain from tropical cyclones may be extremely heavy, e.g., on the northeastern coast on February 3, 1893, a pressure trough, associated with a tropical cyclone, brought the Australian record fall of 907 mm to Crohamhurst, Queensland, and 739 mm to Mooloolah and 638 to Landsborough nearby. The rain brought by this disturbance amounted to 1,715 mm at Mooloolah and 1,690 at Crohamhurst (Fig.27). That month, Crohamhurst received a total of 2,733 mm. This rainfall remained a classic in Australian records and was the subject of a detailed aerological and hydrological analysis (BRUNT, 1958). A similar event happened in 1958, when on February 18, a tropical depression brought 878 mm to Finch Hatton, 760 to Mount Charlton, 707 to Calen and 671 to Mount Jukes, Queensland (BRUNT, 1958). Fully developed cyclones can bring falls which are just as overwhelming: on April 1, 1911, a cyclone brought 800 mm to Port Douglas, Queensland, and 617 to Kuranda, and next day another 732 mm to Kuranda and 778 to Yarrabah. It is likely that in some instances some of the rain failed to be recorded because the gauge overflowed between inspections.

On the northwestern coast there have been similar downpours: the cyclone that brought the western record fall of 747 mm to Whim Creek on April 3, 1898, had brought 355 mm to Pilbara the previous day. A more northerly cyclone brought 356 mm to Roebuck Plains on January 5, 1917, and next day another 568 mm to Roebuck Plains and 356 to Broome, drenching Derby with 418 mm on the third day. Even when the amounts are less spectacular, the rainfall can be truly remarkable (Fig.28).

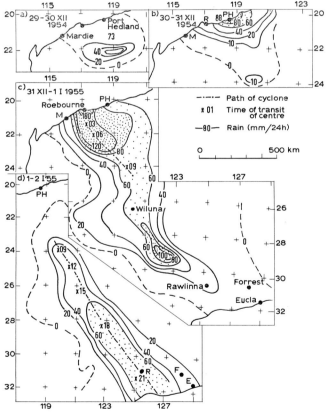

Fig.28. The Western Australian cyclone of New Year 1955, shown in its daily rainfall and track. Notice from *(a)*, *(b)* and *(c)* that good rainfall preceded the cyclone's centre (position every 3 hours shown by small crosses along the track, with the hour shown). (From data by W. A. Divisional Weather Bureau.)

On the northern coast, the moisture load of cyclones is lighter, and no daily totals above 365 mm have been recorded so far.

The contribution of water to dry areas is a most important effect of tropical cyclones. In 1950, parts of western Queensland which are usually arid, were flooded. In 1964, on part of Cape York Peninsula, where the rainfall is usually just enough to make the climate subhumid, rain from cyclone Audrey alone was some 20% of the yearly mean. South of the Gulf of Carpentaria, the proportion rose to 40%. Further south, the usually arid Cunnamulla and Thargomindah districts received in 24 h little less than their yearly mean rainfall (Fig.29). At Eulo, where the mean yearly rainfall is 230 mm, this cyclone brought 234 mm, of which 180 fell within 12 h (WHIGHT, 1964; BRUNT, 1966).

In dry regions the cumulative effect of tropical cyclones is also particularly noticeable. Over a large area, which corresponds to the greatest frequency of western cyclonic tracks, the rainfall increases from February to March, to decrease again slightly in April (Fig.30). Since the summer rainfall of many localities in the western wheatbelt is only 25–50 mm, and the rainfall for February or March perhaps 10–15 mm, it takes only one tropical cyclone bringing 100 mm in 48 h to raise the mean rainfall for the month by 2 mm if the record goes back 50 years, by 4 mm if the record goes back 25 years, etc. Normally May or June is the wettest month in the Eastern Goldfields, Western Australia, and yet because of the torrential downpours brought by a few tropical cyclones the wettest month there is now March (GENTILLI, 1955).

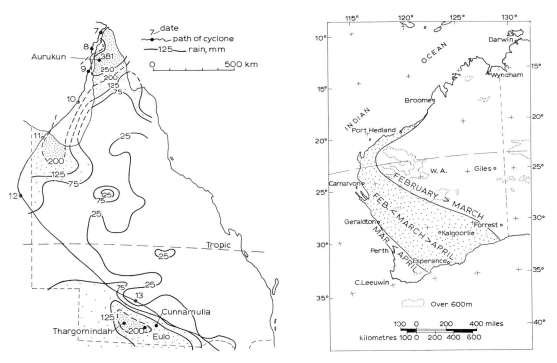

Fig.29. Cyclone Audrey of 1964: position on each successive day shown by the black dot and the date along the track, total rain over the whole period shown by the isohyets with the respective amount. (Simplified from WHIGHT, 1964.)

Fig.30. Effect of tropical cyclones on the mean rainfall: in the belt beaten by cyclones (shown stippled) the month of March gets more rain than February on the summer-rain side and more than April on the winter-rain side, thus upsetting the normal zonal pattern.

A few of the heaviest falls recorded are due to monsoonal developments; on February 23, 1955, an enormous transgression of *Eq* air extended over Queensland and New South Wales (BOND and WIESNER, 1955); during its rapid progress a cyclonic centre developed over the dry country near Bourke, New South Wales. In the 24 h ended at 09h00 on February 23, more than 100 mm had fallen over 75,000 km², with some parts of the area receiving 150 mm or more, and a few places up to 250 mm. The rains continued during the day, causing floods by the western rivers (Macquarie, Castlereagh, Namoi, Gwydir, MacIntyre), as well as the worst floods on record by the eastward-flowing Hunter River. This event was a unique monsoonal outburst (see p.85) rather than a cyclonic manifestation.

Some very heavy winter downpours have also been recorded on the coast of New South Wales: on June 24, 1950, Dorrigo received 635 mm of rain, and Upper Orara, 503. It is very likely that warm water offshore contributed most effectively to the load and instability of the air stream.

The southernmost excessive rains were all recorded on or near the eastern coast: those of February 27, 1919, reached 472 mm at Candelo, and 454 at Bega, New South Wales, and those of April 5, 1929, brought 33υ mm to Mathinna, 282 to Cullenswood, and 281 to Riana, on the eastern coast of Tasmania.

Frontal rains from the west, although more persistent and widespread, bring lighter daily falls, even the heaviest ones being below 250 mm in the 24 h and even where orographic conditions are most favourable. For example, the greatest daily total at Mount Buffalo, Victoria (1,271 m) fell on June 6, 1917, and was only 217 mm.

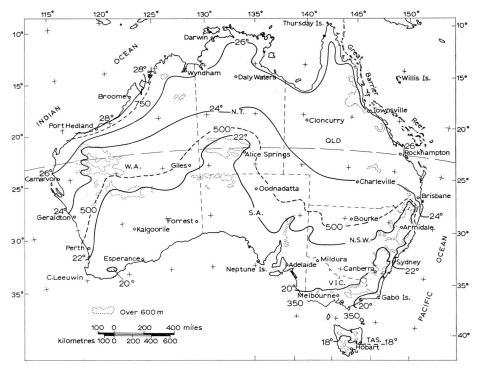

Fig.31. Extreme dew point persisting at least 24 hours (continuous lines) and maximum rainfall possible (dashed lines) over relatively large areas, from the maximum thunderstorm development. (From WALPOLE, 1958.)

Needless to say, nearly all the eastern coast is subject to frequent floods: in New South Wales alone, in the 1946–1955 period, 298 floods were reported, varying in frequency between 13–75 in a year. In 1949–1950, serious floods affected New South Wales and Queensland; in 1952–1955, those two states plus Victoria; and in 1956, these three states and Tasmania.

The knowledge of the maximum amount of rain that may fall within any given period, is of the greatest importance for the prevention of flood damage to many structures, such as bridges, dams, canals, culverts and roads, and stormwater channels. WALPOLE (1958) lists the outstanding falls of rain in Australia, among which are some of short duration, such as 121 mm in 15 min at Fremantle, Western Australia, on June 20, 1895; 229 mm in 2 h at Florence, Queensland (situated at the margin of the arid region, 20°43′S and 140°30′E); 457 mm in 6 h at Maroochydore, Queensland. Several of the falls of longer duration (24 h or more) have been quoted above. The highly localized downpours are usually due to thunderstorms or to small cyclones or whirlwinds that arise peripherally to or within larger systems (see pp.103–105). The amount of precipitable water may be calculated from the dewpoint for any given area. Fig.31, simplified from WALPOLE (1958), shows the highest dewpoint temperatures recorded during 24 h, and the estimated maximum rainfall possible in that time over 1,295 km² (500 sq. miles).

JENNINGS (1967) calculated the maximum probable rainfall in 24 hours within a period of one year, using the constants given by the STORMWATER STANDARDS COMMITTEE (1958). On his map the localized distribution and the variable amount of the east-coast rainfall stand out quite clearly. Apart from the east coast, the mainland probable 24-hour maximum in one year ranges from over 115 mm near Darwin to under 25 mm on the Nullarbor Plain. The maximum probable rainfall at given points, with special reference to Melbourne, was examined by KARELSKY (1960), who gives valuable references.

The knowledge of the possible deficiencies of the rain is also very important, and has been the object of much research. The first step was to express the variability or the reliability of the rainfall in simple statistical terms. TAYLOR (1920) chose the mean deviation from the mean, and expressed it as the percentage of the same mean:

$$100 \frac{\Sigma d}{N} : X$$

in a greatly generalized map; the same expression was used in a more detailed map by GENTILLI (1947) (Fig.32A). WADHAM and WOOD (1939) preferred the standard deviation $\sigma = \sqrt{\Sigma d^2/N}$ which is capable of better mathematical treatment than the mean deviation, but it is doubtful whether the map of the coefficient of variation:

$$100 \sqrt{\frac{\Sigma d^2}{N}} : X$$

thus obtained (Fig.32B) constitutes a sufficient improvement on a continental scale to warrant the additional labour. ANDREWS (1932) having preferred the mode (most frequent value) to the mean as an average, decided to express reliability as the ratio between the frequency of modal values ($\pm 10\%$) and the total number of years on record, i.e., 100 Mo/N. This gives a map in which the highest values correspond to the most reliable rainfall, while in the other maps the highest values correspond to the most variable, i.e., least reliable rainfall. To make the maps in some way comparable, the

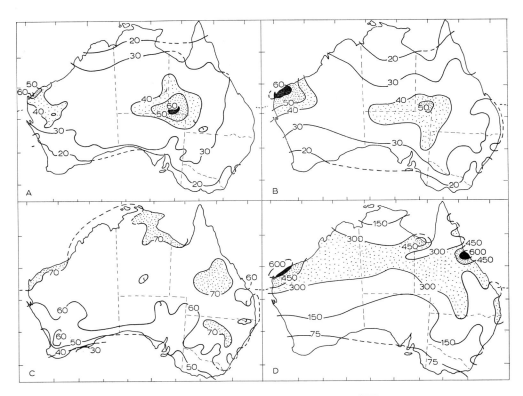

Fig.32. Maps of the rainfall variability, $100 \dfrac{\Sigma d}{N} : X$ (A), variation, $100 \sqrt{\dfrac{\Sigma d}{N}}^{2} : X$ (B), modal unreliability (C) and Maurer variability (D). In A. is shown the mean deviation from the arithmetic mean or common average as a percentage of the same average. In B. is shown the standard deviation from the same average, also as a percentage; notice the different effect on monsoonal rains. In C. the degree of unreliability is shown by the percentage of years on record in which the modal (most frequent) amount of rain fails to occur. In D. special coefficients are used to equalize light and heavy rainfalls so that variability downwards receives the same weight as variability upwards. (A. from GENTILLI, 1947; B. from BARKLEY, 1931; C. modified from ANDREWS, 1933; D. from LOEWE, 1948.)

complementary values have been taken instead, i.e., $100 (1 - Mo/N)$, in order to obtain the map in Fig.32C. It is quite obvious that there are considerable discrepancies in the assessment of the reliability of the rain of tropical Australia. According to Andrews, it shows very little variability. The reason is to be sought in the nature of the mode and in the way it is determined; if a class-interval of 25 mm is used to determine the mode for a station with a mean yearly rainfall of 250 mm, a class interval of 100 mm must be used to determine the mode for another station with a mean yearly rainfall of 1,000 mm. If the smaller class-interval is used, say 25 mm, then the higher rainfall will appear as much more variable than it really is.

The Bureau of Meteorology experimented also with formulas based on the median as obtained from cumulative frequency distributions:

$$100 \ \frac{\text{semi-interquartile range}}{\text{median}} \qquad \text{and also:} \qquad 100 \ \frac{95\% \text{ range}}{4 \times \text{median}}$$

The maps thus obtained (unpublished) differ only slightly from the simple variability map of Fig.32A, except for the fact that the southern stations (those with a greater frequency of mid-latitude frontal rains) show a subdued variability compared with the monsoonal stations, when the semi-interquartile range is used.

As mentioned above, GIBBS and MAHER (1967) have now evaluated the annual rainfall throughout Australia for each year from 1885 to 1965 in terms of deciles. They have also prepared maps showing the annual rainfalls never reached and never exceeded in 90% of the years, and the median annual rainfall. It should be stressed, however, that a greater frequency of lower deciles simply shows a greater degree of skewness; given enough time (unless there is a definite trend in climatic change) all deciles have an equal frequency if the distribution is normal.

LOEWE (1948), used the Maurer index of average variability:

$$S = \frac{100}{n} \, n \, \sum_{1} (s - \bar{s})^2$$

where n = number of values of s; and s is obtained from $r = a \, (b^s - 1)$, where the constants, a and b are determined empirically and, for rainfall in inches, are $a = 6.476$ and $b = 1.18$. The advantage of this index is that it eliminates the exaggeration of high-rainfall variations caused by the class-interval of the mode, while at the same time ensuring that no "damping" takes place where heavy falls are more frequent than light falls, as happens on the northeastern coast. A good example is obtained from the following comparison of yearly rainfalls, in mm:

	1932	1933	1934	1935	1936	1937	1938
Onslow, W. A.	109	434	716	58	66	279	76
Townsville, Qld.	813	1,245	991	279	1,600	610	1,041

In the case of Onslow, there is a normally low rainfall with sporadic wet years; at Townsville, on the contrary, there is a normally rainy climate with sporadic dry years. The mean deviation and the standard deviation stress the variability of the climate of Onslow but fail to stress that of Townsville, while Maurer's formula (Fig.32D) stresses both equally well.

Over smaller areas and for shorter periods, the coefficient of variation is to be preferred because of its complete objectivity. In Fig.33 is shown the variation of the monthly rainfall for each month from March–June in the Western Australian wheat-belt. The decrease in variation from March (early autumn) to June (early winter) and, within each month, from north–south, is clearly shown.

For the assessment of the probability of receiving a given rainfall at a given station, use is frequently made of cumulative frequency curves. The rainfall of the station is plotted on a graph (Fig.34) from which it is then easy to read either the amount of rain which is likely to be received a given proportion of the time (e.g., in 75% of the years) or the probability of receiving a given amount of rain (e.g., 1,000 mm or more).

MAHER (1966), while accepting annual rainfall at most Australian stations as normally distributed, pointed out that many monthly rainfall distributions are heavily skewed because of the combination of many low or zero totals with sporadic heavy rains. The confidence limits of any averages are therefore much farther apart for stations with such irregular rainfalls. Since the formula for the calculation of the confidence limits includes the size of the sample, it is clear that even the "normal" period of 30 years leads to an

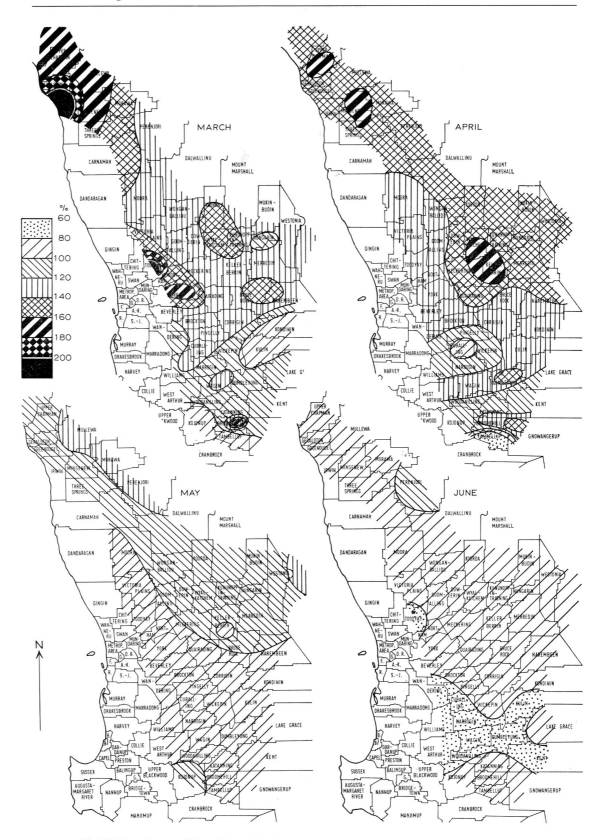

Fig.33. Variation coefficient (100 σ/X) of the monthly rainfall in the Western Australian wheat belt.

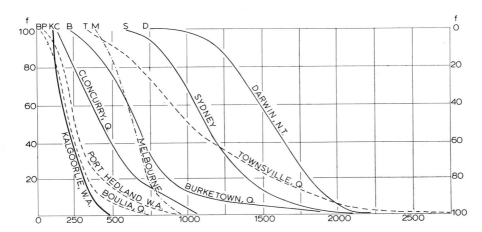

Fig.34. Cumulative frequencies of the rainfall at various localities. Frequency in percentage form shown upwards to the left, and downwards to the right; rainfall in mm shown at the bottom. (From J. C. FOLEY, personal communication, 1948.)

exaggerated estimate of variability in terms of confidence limits. Records extending over the longest possible period must be used.

Particularly good or bad rainfall years in any part of Australia are usually due to the effective or ineffective establishment of the respective rain-bearing factor: in Fig.35 it is possible to single out in Fig.35A the year 1950 for its good monsoonal and inter-anticyclonic rains and weak mid-latitude frontal rains; in Fig.35B the year 1948 for its good tropical cyclonic rains in the west, poor monsoonal rains in the north and normal inter-anticyclonic rains in the southeast; in Fig.35C the year 1958 for its very good mid-latitude frontal rains and good inter-anticyclonic rains in the east (extending much further north than in 1948); in Fig.35D a generally "patchy" year, with more or less normal amounts of rain almost everywhere. Since, as is well known, a relatively low or falling pressure is associated with most rain-bearing situations, it follows that a year of high and widespread cyclonicity (e.g., 1947, see Fig.36) is a year of good general rains. Many attempts have been made at correlating the rainbearing factors with other manifestations, in order to arrive at long-term forecasting or at least foreshadowing. Thus QUAYLE (1918) following the method tried by Todd in Adelaide, in 1896, correlated the May–October rainfall of northern Victoria with the number of "monsoonal troughs, dips and cyclones" experienced in the preceding November–April period; the correlation seemed good at first, but subsequently failed.

The relationship between rainfall and some other climatological element (mostly preceding Darwin pressure, sometimes also preceding temperature, or preceding pressure at Darwin and various Pacific or South American stations) was studied by various authors between 1928–1953 (TRELOAR and GRANT, 1953) but every promising avenue turned out to be closed, e.g., while Quayle in 1928 found that the correlation between preceding Darwin pressure and following northern Victorian rains was —0.81, Treloar found that for the period 1929–1948, the value had fallen to an insignificant —0.22. This research, however, confirmed that some negative relationship exists, and that it extends across northern Victoria and the Riverina, New South Wales; it also proved that district rainfall is more significant than single-station rainfall, and that a good statistical indicator is May–June Darwin pressure for July–August, Victoria–Riverina

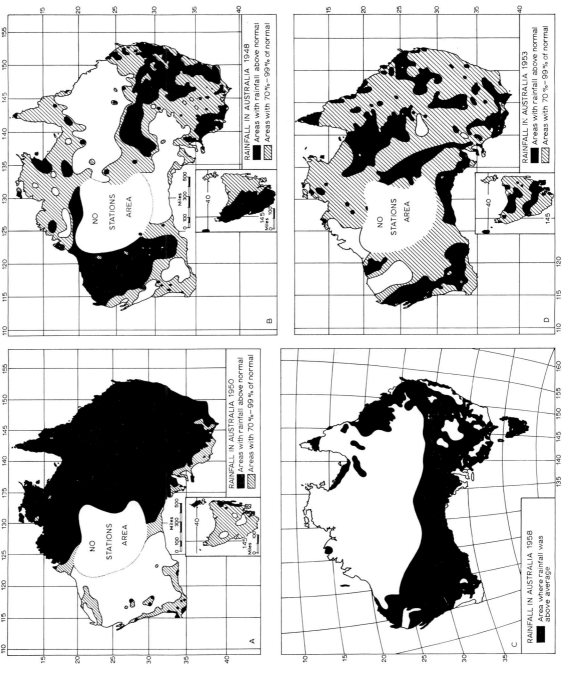

Fig.35. Examples of years with good rains from: A. monsoon and inter-anticyclonic front; B. western tropical cyclones; C. mid-latitude fronts. D. The pattern of rainfall in a year with normal amounts almost everywhere. (From Australian News and Information Bureau.)

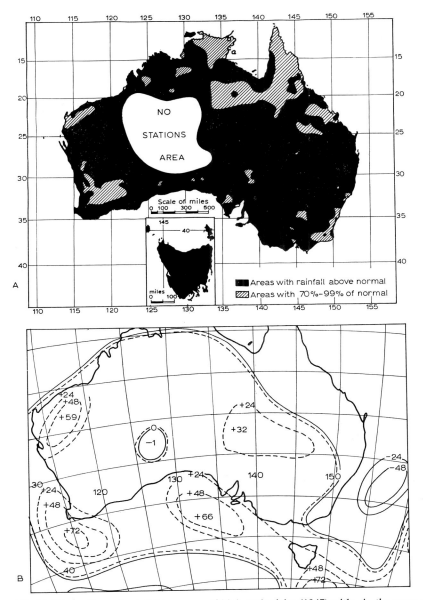

Fig.36. The pattern of rainfall in a year of high cyclonicity (1947) with: A. the areas with rainfall above normal shown black; and B. the deviation in the frequency of cyclonic centres from the normal for each 5° slab. In the case of frontal rains, the rain-bearing fronts extend well to the north of the respective cyclonic centres. (A. from News and Information Bureau; B. from KARELSKY, 1954.)

rainfall, although there is little forecasting value in this because of the closeness in time. The addition of data from other stations brought no improvement.

Evaporation, humidity and the water balance

Although it had been felt for many years that a knowledge of evaporation in Australia was necessary in order to estimate the water balance in general, and the water stress on plants and animals in particular, little progress was possible until suitable measurements could be obtained. Evaporating tanks were installed at Melbourne (circular, 1860), Adelaide (square) and Dubbo, New South Wales (shape unknown) in 1870; Perth, Western Australia, installed a 20.3-cm evaporating dish in 1876; circular evaporating tanks were installed in New South Wales at Sydney (1880), Walgett and Young (1886), Cataract River (1888). In 1891, the South Australian state government had a circular tank installed at Alice Springs. In 1892, a square slate tank was installed at Coolgardie, Western Australia, and in 1899 a similar one in Perth, where observations in the dish were continued side by side for some time. The first installations this century were a circular tank at Douglas, New South Wales, in 1901, and 20.3-cm dishes at Carnarvon, Cue, Laverton, Wiluna in the dry regions of Western Australia. HOUNAM (1961), in reviewing the history of evaporation measurements in Australia, points out that data from evaporating containers of different type cannot be readily or accurately compared. HUNT (1914) gave a tentative map of evaporation in Australia, but the values for tropical Australia were later on found to be grossly underestimated.

While better measurements of evaporation were being sought, indirect methods of estimating the hydric stress had to be found, by combining some of the climatological elements for which data were more readily available.

A pioneer of Australian thought in this field was WILLS (1887), who showed on a map of Australia those areas with over 20 inches (about 500 mm) in the summer semester and those with over 10 inches (about 250 mm) in the winter semester, pointing out that much more summer rain is needed in order to produce the equivalent amount of plant growth to that fostered by much less winter rain.

ANDREWS and MAZE (1933) studied the application of De Martonne's index of aridity to Australia, and realized that annual values were not wholly significant. They computed monthly values, on the basis of $r/(t + 10)$ where r is the monthly rainfall and t the mean monthly temperature, while DE MARTONNE (1926) had suggested $12r/(t + 10)$, as was still preferred by LAUER (1952). Since Andrews and Maze took a limiting value of 1 and De Martonne and Lauer a limiting value of 20 for the arid zone, a comparison may be useful:

ANDREWS and MAZE (1933)	DE MARTONNE (1926); LAUER (1952):
$r/(t + 10) = 1$ (original)	$r/(t + 10) = 1.6$ (equivalent)
$20r/(t + 10) = 20$ (equivalent)	$12r/(t + 10) = 20$ (original)

Andrews and Maze pointed out that "the arid border is a zone which fluctuates from month to month. In bad seasons, it may extend into whole divisions of any of the states, and signalize crop failure. In good seasons the wheatlands flourish, and far into the interior grass and water are abundant...From careful examination of the statistics supplemented by our knowledge of the behaviour of some crops under certain climatic conditions, a tentative selection of the index $R/(T + 10) = 1$ was made as the arid boundary" (ANDREWS and MAZE, 1933).

Having chosen the monthly aridity index 1 as a significant indicator of arid conditions, Andrews and Maze drew a map to show the number of arid months throughout the average year (Fig.37). "This brings us to the consideration of the *duration* of the arid period and its position in the year, a point which we regard as of primary importance." The relevant features of the map may be summarized as follows:

(*a*) There is an area in the central southern portions of the continent which has practical-

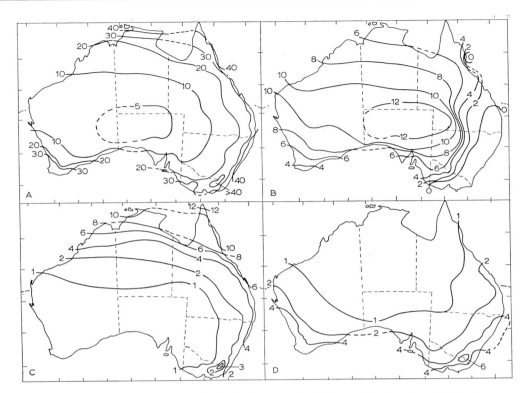

Fig.37. The index of aridity $r/(t + 10)$: A. for the year; B. in its monthly duration; C. for January; D. for July. (After ANDREWS and MAZE, 1933.)

ly continuous arid conditions. This extends from southwestern Queensland to the Western Australian boundary.

(*b*) The steepest gradient (from 12–2 arid months) occurs in southwestern Queensland.

(*c*) The main belt with no arid month, some 350 km wide, extends from just south of Bundaberg, Queensland, to west of Portland, Victoria; there is also a very small coastal area in northeastern Queensland, east of the Atherton Tableland. Most of northern Australia has five or more arid months.

Andrews and Maze did not attempt to define other climatic regions besides the Australian desert. Their criterion for the definition of monthly aridity was taken up by MILES (1947) who classified the hydric state of each month in Queensland as:

agriculturally suitable if $r/(t + 10) > 1$ in $> 70\%$ of the years;

pastorally suitable if $r/(t + 10) > 1$ in > 51–70% of the years.

The suitability of the climate is thus assessed on the basis of probability of recurrence of $r/(t + 10) > 1$ conditions; it presupposes that such degree of wetness is equally significant for any kind of crop or livestock. The adoption of a probability basis in conjunction with a climatic index was a definite contribution.

The next map of Australian climates based upon the length of period with given values of de Martonne's index of aridity was not published until much later, as part of a world map by Troll and Paffen (TROLL, 1964).

The 09h00 relative humidity (Fig.38) reflects the characteristics of the prevailing airmasses much more closely than temperature. It ranges in January from 60% in the

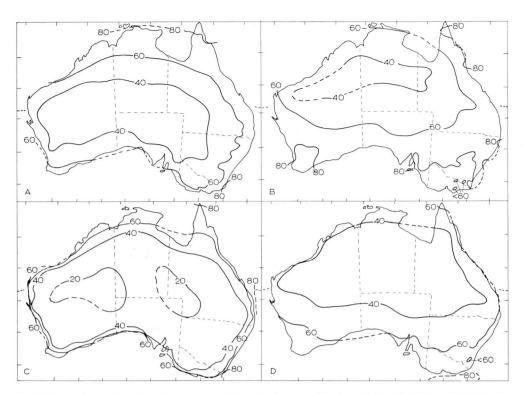

Fig.38. The relative humidity, in percentage form, in January (A, C) and July (B, D) at 09h00 (A, B) and 15h00 (C, D) respectively. The daily rhythm is very conspicuous, and the dryness of the interior in summer very obvious. (Simplified from *Climatological Atlas*.)

southwest to 80% in the north (monsoon time) and in July from 60% in the north to 80% in the southwest (mid-latitude depression time). The eastern coast remains more humid, around or just below 80% throughout the year. Diurnal heating acts rapidly, and in January by 15h00 the humidity in the interior may fall to less than 20%. In the coastal areas the decrease in humidity is moderated by the moisture brought in by the sea-breeze. In July the diurnal heating is less pronounced, but a fall in humidity of some 20%

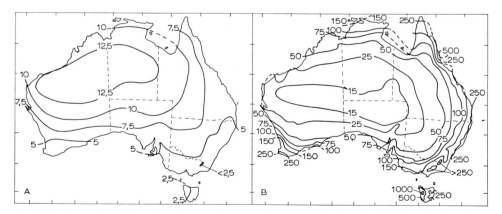

Fig.39. A. Saturation deficit for the year in mm of mercury. B. Meyer's *R/S.D.* ratio for the year. (After PRESCOTT, 1931.)

is quite general (Fig.39). It was obviously desirable to devise an index of hydric effectiveness based on humidity rather than on temperature.

The Waite index is related to TRANSEAU's (1905) R/E ratio (originally P/E) and to MEYER's (1926) $R./S.D.$ ratio (originally N-S) which, for Australia, gives the values shown in Fig.39; it is based on the principle that "the vapour pressure deficit of the air (saturation deficit) is the major factor influencing evaporation. It is a function of the temperature and relative humidity of the air and can be calculated for stations where these data are available. From Dalton's law of evaporation, it is known that the intensity of evaporation in a controlled atmosphere is almost proportional to the saturation deficit of the air" (DAVIDSON, 1934b).

It was PATTON (1930) who showed the approximate general relationship $E = 0.8\ S.D.$, where E = yearly total evaporation in inches of water, and $S.D.$ = mean annual saturation deficit in mm of mercury. The relationship between the yearly (or monthly) total evaporated height of water E (or e) and the mean annual (or monthly) saturation deficit $S.D.$ (or $s.d.$), expressed by the height of a column of mercury measured in the same linear units, as given by various sources is shown in Table IV.

TABLE IV

RELATIONSHIP BETWEEN THE YEARLY (OR MONTHLY) TOTAL EVAPORATED HEIGHT OF
WATER E (OR e) AND THE MEAN ANNUAL (OR MONTHLY) SATURATION DEFICIT $S.D.$
(OR $s.d.$)

	Year $E/S.D.$	Month $e/s.d.$
PATTON (1930) (homogeneous units)	203	16.9
PRESCOTT (1931a)	230	19.2
PRESCOTT (1931b)	263	21.9
PRESCOTT (1949)	258	21.5
PRESCOTT et al. (1952)	252	21.0
Bur. of Meteorol. evaporation maps (1954)	254	21.2

The purpose for which the Waite Institute climatic index was gradually evolved was "the assessment of the efficiency of rainfall in any month for agriculture, biological activity or soil leaching by taking into account also for that month, the evaporating power of the air as measured either by the evaporation from a free water surface, or by the atmospheric deficiency in humidity." (PRESCOTT and THOMAS, 1949.)

The need for a monthly hydric index became apparent when DAVIDSON (1933), while investigating the distribution of the lucerne flea, *Smynthurus viridis*, in South Australia, found that "under South Australian conditions the ratio of monthly rainfall to evaporation is a most useful index to the conditions at the soil surface during each month. Temperature does not indicate the amount of moisture in the air, and with *S. viridis*, when temperature is favourable, saturation deficit may be the factor limiting activity and population increase... When the numerical value of the ratio of mean monthly rainfall to evaporation falls below one (November–March), the environment may become too dry for the survival of the active stage of the insect..." In its initial form, therefore, the Waite index was $r/e > 1$, where r is the monthly rainfall and e the monthly evaporation from the Australian tank.[1]

Davidson produced a map showing "where the mean rainfall exceeds evaporation in any particular month, and, therefore, the areas and periods in which *S. viridis* can exist in the active stage. The basic information used in preparing this map was the mean monthly rainfall at 220 stations, and the evaporation (calculated) at 21 stations.

"General hatching of the oversummering eggs of *S. viridis* may be expected at the end of the dry season, in the month in which the ratio falls below a value of approximately 1.0... The permanent establishment of the species in an area will depend upon (*a*) the number of successive months in which the R/E exceeds 1; and (*b*) the duration and severity of the dry period as affecting the viability of the eggs ..." (DAVIDSON, 1933). Extending the investigation to the remainder of Australia, DAVIDSON (1934a) prepared similar maps for the other states, and also (DAVIDSON, 1934b) for the Commonwealth. The *duration* of given conditions of moisture balance was given more emphasis.

"The moisture conditions on the surface soil... are also important in relation to the germination of seeds and the growth of seedlings. In general, however, plant growth depends upon the moisture in the deeper layers of the soil. Where temperature is favourable, vigorous growth of certain plants, for instance, grasses, will occur in an area where the surface conditions are dry, so long as moisture is available in the lower depths of the soil. This is a feature of the grassland areas of Queensland and northern Australia. The rate at which the soil moisture is used up, will depend upon the transpiration activity of the plants and the intensity of evaporation in relation to the rainfall. Therefore, the duration of the periods having particular values for R/E is important when considering the ratio as an index of aridity" (DAVIDSON, 1934b).

The map given by DAVIDSON (1934b) shows an enormous area where no month has $r/e > 1$, surrounded by an irregular "humid crescent" which, between Townsville and Fraser Island, is only 120–130 km wide, broadening out progressively to reach over 900 km in the southeast and 600 in the north. There are important differences between the broader humid areas: in the southeast (Newcastle, Bega) there are eight humid months, from February to September; in the southwest (Cape Leeuwin, Albany) seven, from April to October; in the north, four, from December to March.

Later on DAVIDSON (1935) considered the monthly value $r/e = 0.5$ "as the lower limit at which adequate moisture will be available for plant growth. Fluctuations about this mean will result in lower values occurring from time to time in certain areas, thereby

---

[1] The Australian evaporation tank is made of metal (galvanized iron or copper), cylindrical, 3 ft. (914 mm) in diameter and 3 ft. deep. It is fitted concentrically in another cylindrical metal tank 4 ft. (122 cm) in diameter and 2 ft. 10 inches (864 mm) deep. The outer tank is sunk into the ground with its rim just above the surface. Both tanks contain water: the outer one is kept full in order to attract birds and small mammals that might be tempted to drink from the measuring tank—this is why it is called the *water jacket* or *guard ring*. Further protection may be given by a nylon mesh over the inner tank (wire netting is still widely used but it may affect the rate of evaporation). The level of water in the inner tank is kept at least 1 inch (25 mm) below the rim, to prevent loss because of wind or heavy rain splashing, but it must not be allowed to fall below 4 inches (102 mm) from the top, otherwise evaporation will be hindered. The measuring is done by a hook gauge in a still well fitted to the inside of the tank (HOUNAM, 1961). The overall amounts evaporated do not differ greatly from those of the U.S. Weather Bureau, Class A pan, but Australian values are greater in winter (because of heat transfer from the soil) and U.S. values in summer (because of heat transfer from the air). The Class A pan, as recommended by the W.M.O., has been used since 1967.

producing temporary drought conditions; the intensity of desiccation will depend upon the value of the ratio and its duration... The period when moisture is effective may be considered as the growing period.

"Australia may be mapped into moisture zones having different degrees of favourableness for vegetation, according to the number of months in which moisture is effective for plant growth... For instance, in the area bordering on the arid central portion... moisture is effective for one month only. The number of effective months increases progressively towards the coast. Over a large belt along the eastern coast and in the highlands of Victoria and Tasmania, the... ratio exceeds 0.5 in every month of the year."

DAVIDSON (1935) also stresses that the average rainfall must be considered with some caution because of the great variability of the rainfall from year to year.

DAVIDSON (1936) then pointed out that "owing to the mild climate and marked seasonal rainfall in Australia, moisture is to be considered as the major influence affecting the distribution and seasonal activity of insects... In dry regions, moisture restricts or limits the permanent establishment of insects and in those areas having a definite dry season, their activities may be restricted to the favourable months of the year (wet season); the insects survive the dry months by aestivation in particular stages of their development." By means of the r/e ratio (r = monthly rainfall; e = monthly tank evaporation), Davidson defined the intensity of monthly wetness or dryness as shown in Table V (his original terms in brackets).

TABLE V

THE INTENSITY OF MONTHLY WETNESS OR DRYNESS

Values of r/e	Degree of dryness or wetness
<0.25	arid
0.25–0.5	semi-arid
0.5–2.0	subhumid ("semi-humid")
2.0–4.0	humid
>4.0	perhumid ("wet")

The map given with the same paper shows the duration, in months, of the period(s) with $r/e = 0.5$ or more. Adopting a duration concept, Davidson calls "arid zone" the area where 1–6 months have a r/e ratio of 0.5 or less, and "humid zone" the remainder. Because of the halving of the r/e ratio, from 1 to 0.5, the "humid zone" shown on the 1936-map is far wider than that on the 1934a-map. The eastern Queensland humid corridor widens from 120–130 to nearly 400 km. The broadest "humid" stretch, in the southeast of the continent, increases from 900–1,450 km. Newcastle and Bega have no dry season, having all 12 months with $r/e>0.5$, while they have eight months with $r/e>1$. At Cape Leeuwin up to seven months with $r/e>1$ correspond 8 with $r/e>0.5$. However, at Albany, Western Australia, the sudden onset and rapid falling-off of the frontal rains, causes the same seven months to have ratios over 0.5 and over 1, the transition from dry to wet and vice-versa taking place within a few weeks. At Darwin, Northern Territory, the lowering of the threshold r/e ratio from 1 to 0.5 allows the inclusion of both November and April within the humid season, which thus extends over six months. In a substantial area of southeastern Queensland, over 250,000 km², there are two distinct "humid" seasons, a major one in summer, a minor one in winter, mostly confined to June. When the period with $r/e>0.5$ lasts one month or less, it is reasonable to count it as "humid" because there will be moisture stored in the soil, thus making the two "humid" seasons continuous.

PRESCOTT (1936) still preferred the more sensitive $r/s.d.$ index as a test of aridity, taking the lowest monthly $r/s.d.$ ratio in the yearly series of twelve, as an indicator of the most arid conditions in Australia. On the other hand, TRUMBLE (1937, 1939) found that a monthly ratio $r/e = 0.3$ is the minimum for agricultural crops in southern Australia, and called it the "influential rainfall". The period during which the monthly values of the ratio exceed this value (period of influential rainfall) corresponds almost exactly with the length of the agricultural season. TRUMBLE (1939) used a simple and most effective graphic device, by plotting on the same graph the monthly rainfall and one-third of the monthly evaporation in order to assess the length of the rainfall season. A similar device was later used by Gaussen and by Walter.

With better evaporation measurements available for longer periods, the conversion of $r/s.d.$ to r/e or vice versa became acceptable.

PRESCOTT (1938a) stated that near Adelaide he found that soil moisture remained above the wilting point of plants during those months in which $r/s.d.>5$ (or alternatively, $r/e>0.24$). On the other hand, "full moisture conditions for optimum biological activity" prevailed during those months with $r/s.d.>13.2$ or alternatively, $r/e>0.62$.

A ratio $r/s.d. = 35$ was taken (PRESCOTT, 1938b) as the limit for the maintenance of soil saturation or an actively growing dense plant cover. In the same paper are given iso-hygromenal maps for $r/s.d. = 5$ (or $r/e = 0.24$) and $r/s.d. = 35$ (or $r/e = 1.68$), as well as single monthly maps with the $r/s.d.$ ratio in multiples of 5. "This series of maps reveals that the greater part of tropical Australia is subject to seasonal drought for eight months of the year, and that very wet conditions...prevail only for three months of the year along the northern coastline, and in small areas along the eastern coast for up to eight months as at Innisfail in northern Queensland."

The Waite index was soon used by LAWRENCE (1941) in her study of tropical Australia and later on by FARMER et al. (1947) for an analysis of pasture conditions in south-western Queensland; the latter authors felt that values of r/e between 0.2–0.3 provided a more realistic threshold than the rigid 0.25 previously used.

MOLNAR (1948) in a comparative study of East African and tropical Australian hydric climates also used $r/s.d.>5$ (PRESCOTT, 1938a, b) as the limiting value for each month in order to determine the *mean length of the growing season* in consecutive months (in eastern Queensland the June rains are often sufficient to make the month wet, but this wet period remains isolated from the sequence of summer wet months). In addition, MOLNAR (1948) determined the *reliable growing season* in terms of the *consecutive* months which have a ratio $r/s.d.>5$ twice out of every three years (66.6 probability, which in view of the more stringent *s.d.* criterion, makes it comparable with the $>70\%$ probability proposed by MILES (1947) for agricultural suitability determined by the $r(t + 10)$ index).

The Waite index was to be further improved by better estimates of evaporation and by the introduction of an exponential coefficient.

PRESCOTT (1940) verified that the slight time-lag "between the combined forces controlling evaporation and those of temperature and humidity on which the values for satur-

170

ation deficit are based", could be reduced by introducing solar radiation as a correcting factor to evaporation. In 1942, he published a study of the phase and amplitude of the march of temperature in Australia, with a map showing the lag of temperature behind solar radiation. This in turn makes it possible to correct the relationship between temperature and evaporation by allowing for the lag in temperature (PRESCOTT, 1943a). TRUMBLE (1945) confirmed a ratio $r/e = 0.3$ as the most significant for the maintenance of topsoil moisture in months with total tank evaporation ranging from 75–150 mm, such as April–May and September–October in the southern agricultural areas. With lower evaporation rates, such as prevail in the cooler and wetter months of June–August, a ratio closer to 0.5 would be required to maintain an adequate amount of soil moisture needed to ensure continuity. With the higher evaporation of summer, or for the assessment of influential rainfall for xerophilous plants, ratios ranging from 0.25 down to 0.1 would be suitable because of drought adaptation. Observations in the field indicated a value more nearly approaching 0.1 for conditions northwest of Port Augusta, South Australia, where summer evaporation ranges from 200 to 350 mm per month.

Studies of evaporation from the soil show that the top 10 cm "tends to lose a fifth to one-half the moisture which evaporates from a water surface as exposed in the standard Australian evaporimeter, over the time between wetting by rain and drying... The ratio is highest at a low rate of loss, as in the winter months, and is lowest when the rate of evaporation is high, as in summer. For the critical months at the commencement and termination of the effective rainfall season, the evaporation from an exposed soil surface is about one-third that from the evaporimeter" (TRUMBLE, 1948).

In its present form the Waite index is:

$$r/(s.d.)^m$$

where r = monthly rainfall (i.e., water received) or evaporability (i.e., water needed), and $(s.d.)$ = mean monthly saturation deficit at 09h00, expressed by the height of a column of mercury, or:

$$r/e^m$$

where e = total monthly evaporation from a free water surface (usually the Australian pan evaporimeter), and m = "a constant varying from 0.67 to 0.80 with a probable mean of 0.73" (PRESCOTT, 1949); but more simply 0.75 "in order to avoid an impression of too great precision" (PRESCOTT et al., 1952).

PRESCOTT (1949) could also give a general climatic equation for yearly normals:

$$\log R = 0.7 \log (S.D.) + \log K$$

Or:

$$R/(S.D.)^{0.7} = K$$

where R = the height of the rainfall; $S.D.$ = the height of a column of mercury expressed in the same linear units, corresponding to the vapour pressure of the saturation deficit; K is the annual climatic index-value.

Critical annual values of K corresponding to zonal soil boundaries are:

Desert/grey-and-brown soils	23
Grey-and-brown soils/red-brown or black earths	45
Red-brown or black earths/podzols	83

Table VI gives the monthly values of the index found significant for $m = 0.75$ (PRESCOTT and THOMAS, 1949; PRESCOTT, 1956); corresponding values of the non-exponential

TABLE VI

MONTHLY INDEX VALUES FOUND SIGNIFICANT FOR m = 0.75

	$r/e^{0.75}$	r/e	$r/(s.d.)^{0.75}$	$r/(s.d.)$
Break of growing season	0.4	0.25	4.0	5.7
Bare avoidance of wilting point	0.5	0.33	5.0	7.5
Vegetation of low transpiration, start of drainage through bare soil	0.8	0.56	8.0	12.2
Vegetation of average transpiration, start of soil leaching	1.2	0.92	12.0	19.7
Vegetation of high transpiration, soil saturation on rain forest	1.6		16.0	
Rice fields	2.0	1.68	20.0	35.0

earlier indices are also given (PRESCOTT, 1941; PRESCOTT et al., 1952; WHITE, 1955). It might be mentioned that PRESCOTT (1951) gave slightly different values for some of these examples; he also used the simplified form $r/e^{0.7}$. The two formulas may be interchanged by use of the relationship $e = 21.2$ *s.d.* (for a month of 30 days) but one may still prefer the saturation deficit version where evaporimeters are not stardardized; the actual depth of water evaporated depends on the size of the evaporimeter and the height of the free rim (PRESCOTT, 1938; HOUNAM, 1956, 1961). The smaller the evaporimeter, the greater the depth of water lost—hence, the abnormally high evaporation shown by the portable dish evaporimeter installed for several years at Laverton, Western Australia.

PRESCOTT et al.(1952) showed the rainfall efficiency for every month, in terms of $r/(s.d.)^{0.75}$, so that the monthly changes can be easily followed throughout the year.

HOUNAM (1955) computed rainfall–evapotranspiration daily decline curves for several stations (converting the monthly $r/e^{0.7} = 0.54$ to daily values) and assumed that the end of the growing season took place after soil moisture had remained at about zero for at least two days. He also drew comparable time curves of the mean monthly rainfall and the mean monthly effective rainfall $r/e^{0.7} = 0.54$ and from their graphic intersection determined the date of the end of season. Comparison of the daily decline graphs with an observer's field reports showed them to be quite satisfactory. WHITE (1955) assessed the *initial effective rainfall* in drier New South Wales and found a similar pattern of variation as was mentioned by TRUMBLE (1945), with the drier areas needing less than $r/e^{0.7} = 0.54$.

VOLLPRECHT and WALKER (1957) tested various statistical ways in which the beginning ("break") and the end of the growing season could be determined with the Waite index from rainfall and evaporation data. The effective rainfall of each month is computed from $r/e^{0.7} = 0.54$, and is then plotted on a monthly time graph on which the actual rainfall is also entered. The intersections of the two lines show the beginning and the end of the growing season. Tests on a daily basis and by the least-square method gave very similar results.

The Waite Institute index was never worked into a system of climatic classification; the reason may be implicit in TRUMBLE's (1945) statement that "the main pitfalls to be avoided in using agro-climatological measures lie in (*a*) the exclusive employment of means, yearly values, abstract expressions generally, and (*b*) the treatment of climatic data without experience and understanding of the biological phenomena which climatic data have been utilized in part to interpret."

There has not been much further work on the Waite index. CROWE (1957) compared evaporation estimates obtained with the methods of Prescott, Penman and Halstead respectively. He also concluded that a consistently satisfactory conversion of saturation deficit values into evaporation values is not possible. He also pointed out that with Prescott's formula $1.2\,e^{0.75}$ for evapotranspiration from vegetation of average water needs, evapotranspiration will exceed evaporation from a tank whenever the latter is less than about 50 mm for the month. After discussing Penman's and Halstead's methods, Crowe suggested a formula for the computation of evapotranspiration from the maximum temperature of the hottest month:

$$e_{tr} = 7.3\,(T_{max} - 6°C)$$

subject to slight corrections for precipitation and length of day in each month.

FITZPATRICK (1963) preferred to combine thermal and hygrometric data in the estimate of evaporation, as follows (in our notation):

$$e = a + b\,(v_{sT} - v)$$

where e is the estimated evaporation from the standard tank; a and b are constants, $a = 1$ and $b = 10$ for monthly data if e is in inches; v_{sT} is the saturation vapour pressure at a synthetic temperature obtained from $0.9\,T_{max}\left(1 + \log\dfrac{h}{12}\right)$ in which T_{max} is the mean maximum temperature (°F) and h the mean length of the day in hours; v is the saturation vapour pressure at the average of the 09h00 and 15h00 dewpoints.

A statistical test of correlation of 754 actual station–month tank evaporation measurements from 13 Australian stations (including tropical ones) and Port Moresby, with evaporation estimates according to the THORNTHWAITE (1948) and the Halstead formulas as well as mean and maximum temperatures, saturation deficit, and $v_{sT} - v$ difference, showed that the best estimates and no serious error would result from use of the $(v_{sT} - v)$ difference. Slightly poorer results would come from using the saturation deficit or the Halstead formula, especially in low-latitude stations. The use of vapour pressure is preferable to that of saturation deficit derived from relative humidity (as in some of Prescott's formulas) because vapour pressure is less easily affected by short-term factors. STERN and FITZPATRICK (1965) also compared daily and monthly evaporation records, estimates and parameters for the wet and dry parts of the year, confirming the value of Fitzpatrick's method.

Following upon the installation of many new evaporating tanks at selected stations (HOUNAM, 1961), the BUREAU OF METEOROLOGY (1963) published new detailed evaporation maps based primarily on actual Australian tank data, and secondarily on data based on humidity and cloudiness. The Australian tank is sunk into the ground, and evaporation from it shows some lag compared with the United States tank which is shallower and exposed clear of the ground. Fig.40 and 41 show the great difference between January and July in the amounts evaporated, Australian evaporation rates in January being among the highest in the world.

NIMMO (1964) made a comparison of evaporation rates from sunken circular pans of various size and a U.S.W.B. class A pan, all installed at Kirkleagh, Queensland, for a period of 6 years. He compared these rates with those from tanks of the same size in-

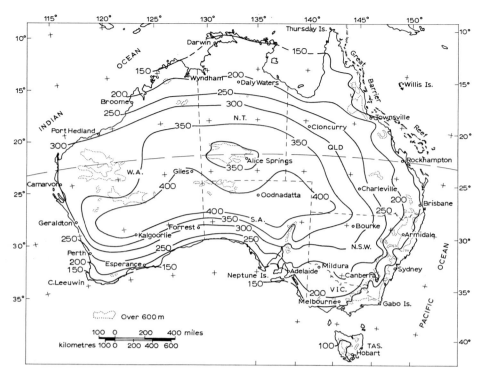

Fig.40. Mean total evaporation in January, from tank measurements and estimates. (From BUREAU OF METEOROLOGY, 1963.)

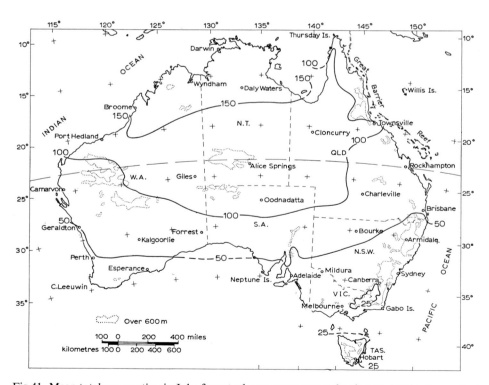

Fig.41. Mean total evaporation in July, from tank measurements and estimates. Notice the much simpler pattern and the much lower values compared with January. (From BUREAU OF METEOROLOGY, 1963.)

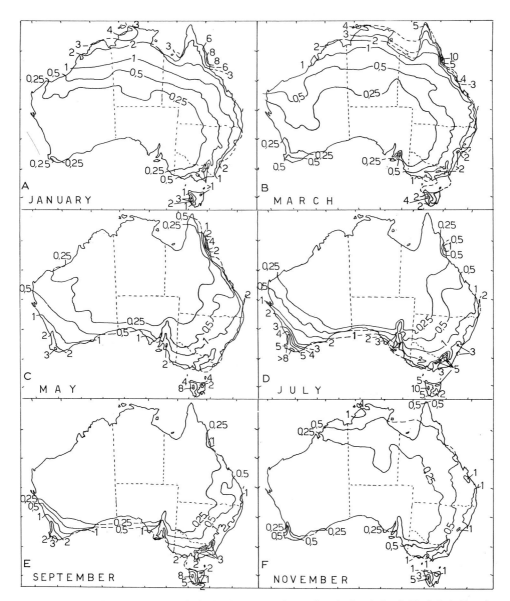

Fig.42. Bi-monthly hydric values of $r/e^{0.7}$. The values chosen for the isopleths (0.25, 0.5, 1, 2 etc.) correspond approximately to the hydric conditions needed to start plant growth, to maintain vigorous growth, to begin run-off, etc. Other values may easily be interpolated. The individual maps should be compared with ordinary rainfall maps.

stalled at Griffith, New South Wales, and with data from American observations, and obtained conversion and regression equations. The overall scatter of individual data remains rather large (HOUNAM, 1964; NIMMO, 1964).

With such information available, it was possible to compute a new set of $r/e^{0.7}$ data, from which the bi-monthly maps of Fig.42 and the frequency maps of Fig.43 have been prepared. These maps in turn are the basis for a new evaluation of climatic differences within Australia which will be discussed in the following section. The bi-monthly maps show the water balance in its seasonal cycle, from the large surplus in the northern areas in summer to the less widespread winter surplus in the southern areas. The frequency

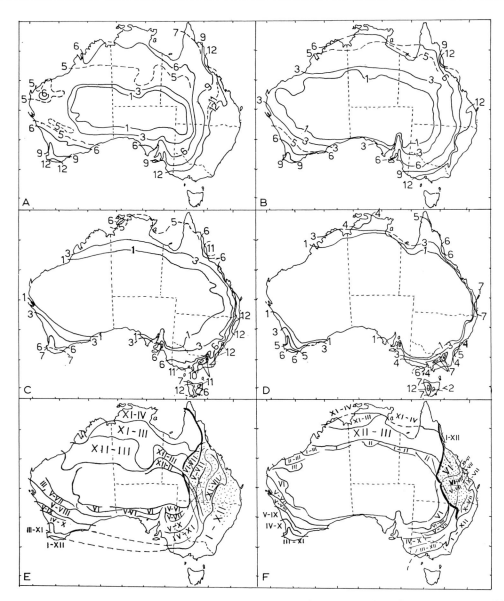

Fig.43. Length of the effective hydric season, as determined from the number of months with: A. $r/e^{0.7}$ > 0.25; B. $r/e^{0.7}$ > 0.5; C. $r/e^{0.7}$ > 1; D. $r/e^{0.7}$ > 2; E. and F. show the actual months during which the respective hydric conditions prevail; E. months with $r/e^{0.7}$ > 0.25; F. with $r/e^{0.7}$ > 0.5). Notice the stippled area in Queensland where summer rains and winter rains overlap, usually leaving a dry period in between.

maps have been prepared for ratios of 0.25, 0.5, 1 and 2, all close to the various critical thresholds mentioned by Prescott, and should provide a good indication of the length of season for various types of land use.

Climatic regions

A rather bewildering array of pragmatic (objective) systems of climatic classification is at the disposal of the public. The system which has been most widely used is the one

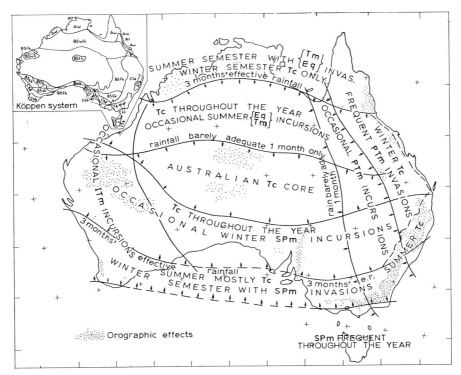

Fig.44. Regions of air-mass frequency, as shown by the more frequent invasions or the occasional incursions by each type of air mass; the more characteristic seasons are also noted, with some indication of the adequacy of the rainfall.

devised and repeatedly modified by Köppen (Fig.44 inset). It was first specifically applied to Australia by ANDREWS and MAZE (1933) and later, with slight modifications, by GENTILLI (1948) and DICK (1964). While it gives significant regional boundaries in the tropical zone, it is quite inadequate for the delimitation of subtropical climatic regions, if these are at all to correspond to vegetational or edaphic regions. Already in 1947, the AUSTRALIAN INSTITUTE OF AGRICULTURAL SCIENCE stated that "maps of this type are of very limited value to agricultural scientists, largely because they are too broad and fail to take adequate account of the very different requirements of different crops."

Forestry workers in Queensland and New South Wales made extensive use of a simple but finely subdivided method of climatic classification proposed by SWAIN (1928) and very easily applied to local ecological conditions and for the evaluation of world homoclimes. It is mainly based on the character of the mean annual rainfall (light, medium, heavy), the rainfall of the driest month, and the mean temperature of the coldest month.

The Waite index, in its latest form $r/e^{0.7}$ or $r/e^{0.75}$, provides a completely flexibile measure of the hydric efficiency of any climate; it has proved far more accurate than THORNTHWAITE's (1948) method, which gave unrealistic results (LEEPER, 1950; GENTILLI, 1953). However, as was pointed out in the preceding section (p.172), it was never worked into a system of classification, although TRUMBLE (1939) hinted at the possibility, and the classification stage was very nearly reached in the latest publications (PRESCOTT et al., 1952).

In recent years, there has been a growing demand for a genetic approach to the classifi-

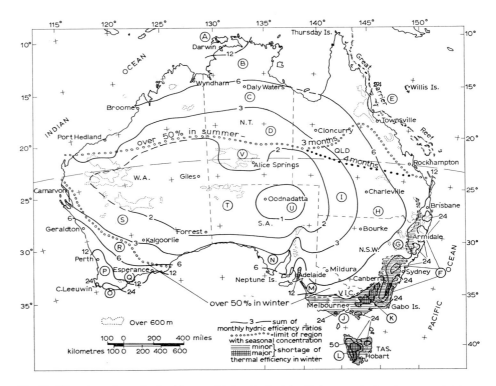

Fig.45. Climatic regions obtained by combining the hydric efficiency ratio for the year, with its seasonal concentration and any shortage of thermal efficiency. The boundaries of the 22 regions shown are only tentative, but they agree very closely with the belts of natural vegetation.

cation of climate. The period 1930–1940 witnessed the spreading of the belief that an air-mass classification would provide the answer, but during World War II the concept of air-mass lost its value (FLOHN, 1957). There still is a demand for an air-mass classification of Australian climates, especially among teaching geographers, and (while recalling the section "Air masses of Australia" on p.54 for the characteristics of the air masses themselves) Fig.44 is suggested as a tentative subdivision of Australia into air-mass frequency regions. It must be understood that such clear-cut boundaries are merely convenient abstractions. It is significant, however, that the lines showing the limits of three months with effective rainfall $r/e^{0.7} > 1$ and of one month $r/e^{0.7} > 0.25$, if the effect of orography is ignored (which one cannot, in fact, afford to do), coincide with these air-mass-frequency boundaries.

A reference to Fig.16 (inset left), will show that the *Tm-Eq* invasions are due to the monsoon, the *PTm* invasions to anticyclonic interplay, the *sPm* invasions to mid-latitude fronts, which also bring occasional *ITm* streams to the western shore. Occasional *ITm* and *PTm* incursions are due to tropical cyclones.

The writer prefers a genetic classification of climates based more definitely on the atmospheric circulation, with its zonal, seasonal, and zonal–seasonal changes. It is understood that, wind being moving air, the concept of airmass is retained, but more emphasis is given to its movement than to its physical characteristics. Within the airstream, there arise gradual physical changes as it proceeds along this or that kind of surface. In Australia, a very interesting problem of classification is offered by the climates of the eastern coast: *PTm* air comes in from the ocean, as part of anticyclonic circulation or trade wind, and it was dry upper air, then it descended and became moist travelling over the ocean—but how far inland does it have to travel before it loses its newly acquired maritime characteristics? Even if one prefers to call this an anticyclonic or a trade wind climate, the problem remains the same, except that it seems more logical then to proceed to an assessment of the hydric effectiveness of the air, which can now be done with the Waite index, with the aid of the evaporation maps now available. The weakness of the airmass system is, that while near the surface, the air has certain characteristics, it may be quite different higher up (and most of the rain is developed "higher up") and will be different a little further on, so that some kind of hydric measurement or evaluation must be done anyway, and the direction of movement assumes a fresh importance if one is to know the precedents and the immediate future of that air; this in turn brings one back to the concept of wind and it might be just as easy then to place the wind first in the list of criteria.

Briefly then, the system of classification advocated here takes the zonal belts with regular winds throughout the year, or with regular convergence or divergence, and distinguishes these as *core* climates. It then recognizes the *alternate* climates, where two winds or stream patterns alternate during the year. Within each zone thus classified, the system evaluates the hydric and the thermal degrees of efficiency by using the $r/e^{0.7}$ formula and the number of day-degrees (or month-degrees), respectively.

A first outline of this classification was given by GENTILLI (1955) and the main criteria followed are still valid, although perhaps the nomenclature may vary slightly. On a world scale, the closest classification to this one is that given by KUPFER (1954), based on Flohn's work of 1950–1952, discussed by FLOHN (1957); however, Kupfer did not make use of any climatic index.

PRESCOTT and THOMAS (1949) state that "most of the work on these (Waite climatic) indices has been based on experience in South Australia, where there is a relatively sharp break in the season, and where a value of 4 for the new index $(r/(s.d.)^{0.75})$ is followed up in the months immediately following by much higher values. This applies with greater emphasis in the southwest of Western Australia, as well as in the Kimberley region of that state and in the Northern Territory. In much of Queensland, however, a monthly value of this index of 4 for the break of season, may not be followed up immediately by substantially higher values, so that it does not follow that the growing period of five months will be adequate for agriculture, ...although it may be on the average for pastoral pursuits. It is for this reason that account has been taken also of the period during which the index has a value of 8 or over." In this case it is clear that a hydric balance just adequate for the break of season is not adequate for its safe continuation. On the other hand, BREMER (1965) pointed out that in northern Australia the concentration of the rainfall was generally greater than in Africa or South America, so that a classification based on the number of humid months (in this case, Lauer's classification based on De Martonne's monthly index $12r/(t + 10) > 20$) recognizes no humid month and, therefore, no possibility of plant growth in some areas with scant rainfall, where in fact there *is* plant growth. It seems clear from Bremer's analysis, that months which are not rainy enough to be deemed "humid" by Lauer's method, may still have enough moisture for well adapted xerophytes.

The two Queensland centres of Emerald and Charters Towers provide an example of the problem discussed by Prescott and Thomas. Taking $r/e^{0.7}>0.5$ as roughly equivalent to $r/(s.d.)^{0.75}>4$, we find that Emerald has five months with $r/e^{0.7}>0.5$, from November to March; then the winter rains bring conditions just as humid to June and most of July. However, the April–May and August–October droughts make the short wet season in winter of very little use. No month, not even the wettest summer one, has $r/e^{0.7}>1$. At Charters Towers, on the other hand, only the four months, December–March, have $r/e^{0.7}>0.5$, but January and February have $r/e^{0.7}>1$, conditions which permit more vigorous growth.

The natural vegetation at Emerald is a layered scrub of brigalow (*Acacia harpophylla*) while at Charters Towers grows a mixed tropical woodland with various species of *Eucalyptus* and grass. In overall volume of plant-matter the two localities do not differ very much. The summation of their monthly $r/e^{0.7}$ ratios gives almost the same value, 6.27 for Emerald and 6.22 for Charters Towers, but it is undoubted that far more adaptability is needed by plants in the Emerald climate.

Coming to the problem raised by Bremer, and following the same method, we find that $\Sigma^{12} r/e^{0.7}$ gives 15.94 for Darwin (marginal tropical layered forest), 8.94 for Katherine (mixed tropical woodland) and 5.74 for Daly Waters (mixture of sclerophyll low-tree savanna and low arid woodland). The index reduces the plentiful rainfall of Darwin to a rather moderate value, but gives it a hydric ratio similar to those of the *Eucalyptus* forests of the southwest (Mundaring, Western Australia, 20.04; Perth, 17.72; Dinninup, Western Australia, 13.85) and the southeast (Grafton, New South Wales, 14.53; Lithgow, New South Wales, 15.55; Alexandra, Victoria, 15.71). This is the only index which gives truly comparable results over such a span of latitude and with such different rainfall regimes. Use of the De Martonne–Lauer index is inadequate under uniform rainfall regimes:

	Darwin, N.T.	Perth, W.A.	Lithgow, N.S.W.	Alexandra, Vic.
$\Sigma^{12} r/e^{0.7}$	15.94	17.72	15.55	15.71
Lauer No.	6.7	6.7	12.0	10–11.0

It is, therefore, suggested that $\Sigma^{12} r/e^{0.7}$ be taken as the hydric index, supplemented where necessary by other criteria.

It seems desirable to give a hydric grade to each month as well, in order to arrive at a finer distinction between months than is afforded by the mere "humid" or "arid" alternative.

The following limits are suggested:

Month's hydric grade	$r/e^{0.7}$ *ratio*
Perarid	<0.12
Arid	0.12–0.25
Semiarid	0.25–0.50
Subhumid	0.50–1.25
Humid	1.25–2.50
Perhumid	>2.50

It is also suggested that values above 4 be neglected in the assessment of the hydric bioclimate, because the excess water is likely to run off too rapidly to be of any benefit.

As to thermal conditions, and following TRUMBLE (1939), we suggest the following evaluation:

Type of plant growth	Mean monthly temperature
None; full vernalization	$<7°C$
Minimal; some vernalization	$7-10°C$
Moderate; minimal vernalization	$10-15°C$
Active; no vernalization	$>15°C$

Thermal efficiency is more than adequate over most of the mainland, with the exception of the highlands. In Tasmania, during the winter, thermal efficiency is low and, on the higher mountains, it is non-existent. It is, therefore, necessary to introduce a correcting factor to the abnormally high hydric summation values that are obtained for such areas; it could be done simply by not counting the hydric values of months with mean temperature below freezing, and counting only part of the values of months with mean temperature between 0°–6°, or, more simply, taking the period with month-degrees above 6° and counting only the hydric efficiency for that period.

Pending confirmation by observations and experiments, we recommend that each month's *hydric grade* obtained from $r/e^{0.7}$ be reduced by 10% of every 1°C by which the mean monthly temperature falls below 13°C. It is thus assumed that any hydric condition is of no advantage to plants when the mean monthly temperature falls to 3°C or less. The use of the *coefficient of thermal adversity* would then be as follows:

Mean monthly temperature	Hydric biopotential
13°C or more	$r/e^{0.7}$
12°C	$0.9 (r/e^{0.7})$
11°C	$0.8 (r/e^{0.7})$
...	...
4°C	$0.1 (r/e^{0.7})$
3°C or less	0

It must be stressed that the *hydric potential* takes full account of monthly ratios above 4, and is not adjusted for temperature, while the *hydric biopotential* is treated as suggested above. We call the yearly summation of the 12 monthly adjusted values the *phytohydroxeric index*:

Annual phytohydroxeric index $\Sigma^{12}r/e^{0.7}$	Zonal plant formation
>24	rain forest
12–24	forest
6–12	woodland, tree savanna
3–6	grassland, dry woodland
1–3	dry grassland, arid scrub
0–1	sandy or stony desert

Seasonal index values, and the number of dry months of various grades of aridity, are highly significant and cannot be disregarded.

Fig.45 shows the general outline of the climatic regions thus obtained. Needless to say, such regions are suitable for studies of the plant environment, but are only very broadly indicative of the environment of homeotherm animal species.

Minor climatic elements

It is not possible to do justice to the many minor climatic elements here. Whenever possible, their mean or normal monthly values have been included in the climatological tables (pp.269–381); there one may find, for instance, data on wind and cloudiness. A few elements better known from descriptive records (apart from frequency data) are briefly discussed below.

Dew

Dew is more common than most people realize, but is seldom measured. Throughout the southern areas, on many winter nights, the minimum temperature falls below the dew-point temperature, but this is not invariably so, and fails to happen on several successive nights during an anticyclonic spell. On the other hand, local conditions can easily affect this phenomenon; a rapid check of data shows that dewpoint temperatures are abnormally high in localities situated near watercourses, as for instance Guildford, Collie and Bridgetown in Western Australia. Soon after the end of winter the western part of the mainland becomes much drier, and nightly minima remain well above the dewpoint. In tropical Australia conditions may be affected by the enormous disparity in dewpoint temperatures between the air masses that may alternate within a short time; here again, a sheltered site close to some water (as for instance at Wittenoom, Western Australia, 22°12′S, 118°10′E) brings the dewpoint above the minimum on very many nights. Some Duvdevani dew gauges were installed in 1956, but the results of the observations were not published because of the localized nature of the phenomenon, and because observations at fixed times did not allow the recording of variations due to the time of sunrise. In general, dew is more frequent near the ground than at 1 m or more above it. At Perth the total amount of dew recorded was highest at 06h00 in most months, but nights with dew are more frequent at 03h00 or even at midnight. Little dew evaporates from the gauges between 06h00 and 07h00 in the winter, but by November about half the amount is lost from the gauge within that hour, and in December most of it. At Perth, during the year 1957, the dew gauge above the ground recorded a total of 19 mm of dew, and the gauge at 1 m recorded 15 mm; in the same year, the standard rain gauge had recorded only 0.75 mm of dew, almost a negligible fraction (R. Vollprecht, personal communication, 1965).

Fog

Fog is neither common nor widespread. On a winter morning, the writer observed fog from the aircraft at the mouth of every valley on the western slopes of New South Wales, while the intervening slopes and spurs were perfectly clear. Because of the prevailing location of settlements, an unduly high number of observers would have recorded fog on that particular morning. LOEWE (1944) points out that in coastal locations, winter fogs are common inside inlets and embayments, while summer fogs, far less common, occur only at more exposed sites such as capes and promontories, which on the other hand have very few fogs during the winter. As with dew, the presence of concave topography and sheltered water surfaces provides the most favourable conditions for the development of fog. All the frequency statistics available for coastal stations have been collated by LOEWE (1944), and detailed data for the very infrequent fogs observed at eight Western Australian aerodromes have now been assembled by MAINE (1968). There is a very interesting trend in the proportion of reduced visibility (below 3 miles = 4,827 m) due to fog and to other causes, among which dust or smoke *haze* is most frequent, or heavy precipitation in tropical cyclonic situations. Thus at Broome in February all reduced visibility was due to heavy precipitation, in April there was no reduction from any cause, but in June 95% of the reduced visibility occasions were due to fog. At Port Hedland heavy rain or dust haze may reduce visibility in January, when no fog has ever been recorded. In May and June fog is the only cause of reduced visibility, in July it is the cause of nearly 90% of the occurrences. At Carnarvon the rare fogs restrict visibility more in August and September, heavy precipitation in February and March. At Geraldton fogs account for some 80% of the September and 90% of the October restrictions but only for less than 10% of those occurring in June or July. At Perth, dust or smoke haze may be the main factor in January and February, heavy rain in May–August, fog in September–December. In all cases the actual frequency of fogs is quite low, mostly below 2% for half-hourly observations, and usually in the very early morning only, before or immediately after sunrise.

Snow

Snow is unknown in northern Australia, not having been recorded north of Charters Towers, Queensland (20°02′S, altitude 311 m). It is rare, at sea level, throughout southern Australia. It was a memorable event when, on 28th June 1836, it snowed in Sydney for about half an hour; snow lasted on the ground, in places, for up to one hour. At Melbourne there were noticeable snowfalls in 1849 and 1899; recently, in 1951 (twice) and 1969. Snow was seen a few times on Mount Lofty, South Australia. It falls more often (every 2–3 years) on the Stirling Ranges, Western Australia. On the Australian Alps it falls every year above 1,000–1,500 m, very irregularly, at times with falls of more than 1 m in one day. There now is a small but thriving tourist industry based on skiing, ski-lifts and snowmobiles, especially around Mount Kosciusko.

Snowfalls are always associated with a strong southerly flow which alone can cool the ground and the lower troposphere sufficiently to prevent a premature melting of the crystals during their descent. Such flow is more likely to occur when a deep mid-latitude depression is immediately followed by a very large anticyclone. Some falls of snow, although of brief duration, may extend to large areas in southeastern Australia, south of Toowoomba on the highlands and south of the Lachlan on the plain. On the higher tablelands snow comes in sudden and tremendous storms, because of the relatively large moisture content of the air exposed to sudden cooling. On the night of 21 December 1851 the rev. W. B. Clarke, a well-known amateur climatologist, slept under the summit of Kosciusko on 12 m of hard, dry, crystallized snow; he remarked that "it seems strange that perpetual snow... should be met with so near the Equator at an elevation of only 1,800 m." Cattle, sheep and human beings trapped in the snow have died of exposure. In July 1834 a snowstorm lasted 3 weeks; snow accumulated to a depth of 1.5 to 5 m on the mountains, and herds of cattle were buried. On 14 January 1964, midsummer, a very deep trough formed by the former tropical cyclone Audrey (see pp. 92,155) and a very large depression south of Tasmania brought strong southwesterlies and enough snow to trap 740 cattle for 17 days only 65 km from Canberra.

Thunderstorms and hail

High moisture loads and frequent atmospheric instability produce conditions which often lead to heavy falls of hail, especially when a moist air stream invades the interior of the mainland. Hailstones the size of an egg are reported at least once a year from somewhere, with damage to iron roofing and cars, occasionally with deaths of birds. Damage to crops can be serious and widespread. While most hailstorms are associated with diurnal atmospheric instability, there have been several instances of nocturnal hailstorms due to frontal uplift of unstable air.

Information on thunderstorms was uneven and inadequate in earlier years. After 1940, thanks to closer cooperation between the Bureau of Meteorology and the Electricity Supply Association (which maintains its own network of observers), better information has become available, and maps of the annual incidence of thunderstorms have been published. The annual average ranges from less than 5 in eastern Tasmania, around Tarcoola in South Australia and Wilcannia in New South Wales, to over 60 around Normanton, Queensland, and the western coast of the Kimberleys, Western Australia, with a maximum of over 80 at Darwin. The east coast has 20–40 thunderstorms per year, against 10–20 along the west coast, with sharp increases along the respective scarps. Thunderstorms are most frequent in summer, except in the extreme southwest, where 6.8 out of the annual 13 come between May and August inclusive. The greatest incidence is in November in the southeast, in December in the east, in January in the north, in February in the northeast. The association with hail is not very close: in Brisbane the record year for both phenomena was 1922, when there were 64 days with thunderstorms but only 8 with hail.

Records of hail vary considerably. The most comprehensive listing was the Western Australian one for the years up to 1926 (BUREAU OF METEOROLOGY, 1929). It shows maps of the most extensive hailstorms: those of 27–31 December 1912 affected 60,000 km²; the one of 30 November 1913 radiated over 44,000 km², devastating over 1,000 km²; those of 8–10 December 1925 affected over 42,000 km². Sydney had its worst hailstorm on 1 January 1947, with hailstones up to 1,800 g in weight, but its most expensive thunderstorm with hail came in winter (6 July, 1931) and caused over $ 400,000 worth of damage.

References

ANDREWS, J., 1932. Rainfall reliability in Australia. *Proc. Linnean Soc. N. S. Wales*, 57: 95–100.

ANDREWS, J., 1933. Seasonal incidence and concentration of rainfall in Australia. *Proc. Linnean Soc. N. S. Wales*, 58: 121–124.

ANDREWS, J. and MAZE, W. H., 1933. Some climatological aspects of aridity in their application to Australia. *Proc. Linnean Soc. N. S. Wales*, 58: 105–120.

ANDREWARTHA, H. G., 1944. The distribution of plagues of *Austroicetes cruciata* SAUSS (Acrididae) in Australia in relation to climate, vegetation and soil. *Trans. Roy. Soc. S. Australia*, 68: 315–326.

ASHTON, H. T., 1964. Meteorological data for air conditioning in Australia. *Australia, Bur. Meteorol., Bull.*, 47: 15 pp.

ASHTON, H. T. and MAHER, J. V., 1951. *Australian Forecasting and Climate*. Published by the authors, Melbourne, Vic., 72 pp.

AUSTRALIAN INSTITUTE OF AGRICULTURAL SCIENCE, 1947. Climatic mapping. *J. Australian Inst. Agr. Sci.*, 13: 138–140.

BARTON, C. H., 1895. *Outlines of Australian Physiography*. Alston, Maryborough, Queensland, 180 pp.

BIGG, E. K. and MILES, G. T., 1964. The results of large scale measurements of natural ice nuclei. *J. Atmospheric Sci.*, 21: 396–403.

BOND, H. G., 1954. Northern Territory drought, 1951–1952. *Australian Meteorol. Mag.*, 5: 5–19.

BOND, H. G., 1960. The drought of 1951–1952 in Northern Australia. In: S. BASU et al. (Editors), *Symposium on Monsoons of the World*. Meteorological Office, New Delhi, pp.215–222.

BOND, H. G. and WIESNER, C. J., 1955. The floods of February, 1955 in New South Wales. *Australian Meteorol. Mag.*, 10: 1–33.

BREMER, H., 1965. Klima und Vegetation im Australischen Nordterritorium. *Petermanns Geograph. Mitt.*, 109: 183–193.

BRUNT, A. T., 1958. The Crohamhurst storm of 1893 and notes on the Mackay storm of February, 1958. In: *Conference on Estimation of Extreme Precipitation*. Bur. Meteorol., Melbourne, Vic., pp. 190–228.

BRUNT, A. T., 1966. Rainfall associated with tropical cyclones in the northeast Australian region. *Australian Meteorol. Mag.*, 14: 85–109.

BUREAU OF METEOROLOGY, 1929. *Results of rainfall observations made in Western Australia*. Government Printer, Melbourne, Vic., pp.108–113.

BUREAU OF METEOROLOGY, 1940. *Climatological Atlas of Australia*. Bur. Meteorol., Melbourne, Vic., 76 pp.

BUREAU OF METEOROLOGY, 1945. *Maps on Average Monthly and Annual Rainfall*. Bur. Meteorol., Melbourne, Vic., 13 pp.

BUREAU OF METEOROLOGY, 1956. *Climatic Averages*. Bur. Meteorol., Melbourne, Vic., 13 pp.

BUREAU OF METEOROLOGY, 1958. *Conference on Estimation of Extreme Precipitation*. Bur. Meteorol., Melbourne, Vic., 271 pp.

BUREAU OF METEOROLOGY, 1960. *Seminar on Rain*. Bur. Meteorol., Melbourne, Vic., 1378 pp.

BUREAU OF METEOROLOGY, 1962. Climate and meteorology of Australia. *Official Year Book of the Commonwealth of Australia*. Bur. Meteorol., Canberra, A.C.T., pp.29–60.

BUREAU OF METEOROLOGY, 1963. *Average Evaporation in Inches*. Bur. Meteorol., Melbourne, Vic., 13 pp.

COLEMAN, R. G., 1964. Frost and low night temperatures as limitations to pasture development in subtropical eastern Australia. *C.S.I.R.O., Div. Tropical Pastures, Tech. Papers*, 3: 8 pp.

CROWE, P. R., 1957. Some further thoughts on evapotranspiration: a new estimate. *Geograph. Studies*, 4: 56–75.

DAVIDSON, B. R., 1969. *Australia Wet or Dry?* Melbourne University Press, Melbourne, Vic., pp.6–36.

DAVIDSON, J., 1933. The distribution of *Sminthurus viridis* L. *(Collembola)* in South Australia, based on rainfall, evaporation and temperature. *Australian J. Exptl. Biol. Med. Sci.*, 11: 59–66.

DAVIDSON, J., 1934a. The "Lucerne Flea" *Smynthurus viridis* L. *(Collembola)* in Australia. *Commonwealth of Australia, C.S.I.R., Bull.*, 79: 66 pp.

DAVIDSON, J., 1934b. The monthly precipitation–evaporation ratio in Australia, as determined by saturation deficit. *Trans. Roy. Soc. S. Australia*, 58: 33–36.

DAVIDSON, J., 1934c. Climate in relation to insect ecology in Australia, 1. Mean monthly precipitation and atmospheric saturation deficit in Australia. *Trans. Roy. Soc. S. Australia*, 58: 197–210.

DAVIDSON, J., 1935. Climate in relation to insect ecology in Australia, 2. Mean monthly temperature and precipitation–evaporation ratio. *Trans. Roy. Soc. S. Australia*, 59: 107–214.

DAVIDSON, J., 1936. Climate in relation to insect ecology in Australia, 3. Bioclimatic zones in Australia. *Trans. Roy. Soc. S. Australia.*, 60: 88–92.

DE MARTONNE, E., 1926. Aréisme et indice d'aridité. *Compt. Rend.*, 182: 1395–1398.

DICK, R. S., 1964. Frequency patterns of arid, semi-arid and humid climates in Queensland—a climatic-year analysis based on Köppen's classification. *Capricornia*, 1: 21–30.

DROESSLER, E. G., 1964. A note on ice nucleus storms. *J. Atmospheric Sci.*, 21: 701–702.

DURY, G. H., 1964. Some results of a magnitude-frequency analysis of precipitation. *Australian Geograph. Studies*, 2: 21–34.

FARMER, J. N., EVERIST, S. L. and MOULE, G. R., 1947. Studies in the environment of Queensland, 1. The climatology of semi-arid pastoral areas. *Div. Animal Ind. Bull.*, 1: 39 pp.

FITZPATRICK, E. A., 1963. Estimates of pan evaporation from mean maximum temperature and vapor pressure. *J. Appl. Meteorol.*, 2: 780–792.

FITZPATRICK, E. A., 1964. Seasonal distribution of rainfall in Australia analysed by Fourier methods. *Archiv. Meteorol., Geophys. Bioklimat., Ser. B*, 13: 270–286.

FITZPATRICK, E. A. and NIX, H. A., 1970. The climatic factor in Australian grassland ecology. In: R. M. MOORE, (Editor), *Australian Grasslands*. Australian National University Press, Canberra, A.C.T., pp.3–26.

FLOHN, H., 1957. Zur Frage der Einteilung der Klimazonen. *Erdkunde*, 11: 161–175.

FOLEY, J. C., 1945. Frost in the Australian region. *Australia, Bur. Meteorol., Bull.*, 32: 142 pp.

FOLEY, J. C., 1957. Droughts in Australia. *Bur. Meteorol., Bull.*, 43: 281.

GENTILLI, J., 1947. *Australian Climates and Resources.* Whitcombe and Tombs, Melbourne, Vic., 333 pp.

GENTILLI, J., 1948. Two climatic systems applied to Australia. *Australian J. Sci.*, 11: 13–16.

GENTILLI, J., 1949. Air masses of the Southern Hemisphere. *Weather*, 4: 258–261, 292–297.

GENTILLI, J., 1953. Die Ermittlung der möglichen Oberflächen und Pflanzenverdunstung, dargelegt am Beispiel von Australien. *Erdkunde*, 7: 81–93.

GENTILLI, J., 1955. Die Klimate Australiens. *Die Erde*, 1955 (3–4): 206–238.

GENTILLI, J., 1960. Il fattore termico nell'ecologia degli eucalitti. *Pubbl. Centro Sper. Agr. For.*, 4: 53–119.

GIBBS, W. J., 1946. Recent developments in weather analysis in Australia. *Australian Geographer*, 5: 47–51.

GIBBS, W. J. and MAHER, J. V., 1967. Rainfall deciles as drought indicators. *Bur. Meteorol., Bull.*, 48: 33 pp.

HOUNAM, C. E., 1955. Determination of the end of the growing season. *Australia, Bur. Meteorol., Meteorol. Studies*, 4: 16 pp.

HOUNAM, C. E., 1956. Evaporation pan coefficients in Australia. *Australia-UNESCO Symposium on Arid Zone Climatology, Canberra, 1956, Papers from Australia and New Zealand*, pp.6a–6o.

HOUNAM, C. E., 1961. Evaporation in Australia. *Australia, Bur. Meteorol., Bull.*, 44: 147 pp.

HOUNAM, C. E., 1964. The variability of Australian tank evaporation. *Australian Meteorol. Mag.*, 47: 26–39.

HUNT, H. A., 1914. Climate of Australia. In: *Federal Handbook, Australian and New Zealand Assoc. Advan. Sci.*, Government Printer, Melbourne, pp. 122–162.

JENNINGS, J. N., 1967. Two maps of rainfall intensity in Australia. *Australian Geographer*, 10: 256–262.

JOHANSSON, O. V., 1931. Die Hauptcharakteristika des jährlichen Temperaturganges. *Gerlands Beitr. Geophys.*, 33: 406–428.

JOHANSSON, O. V., 1942. Den årliga temperaturperioden och dess typer. *Mitt. Meteorol. Inst. Univ. Helsinki*, 49: 116 pp.

KARELSKY, S., 1960. Probable maximum rainfall at a geographical point—Climatological aspect. *Australian Meteorol. Mag.*, 29: 1–24.

KIDSON, E., 1925. Some periods in Australian weather. *Australia, Bur. Meteorol., Bull.*, 17: 5–33.

KUPFER, E., 1954. Entwurf einer Klimakarte auf genetischer Grundlage. *Z. Erdkundeunterricht*, 6: 5–13.

LAUER, W., 1952. Humide und aride Jahreszeiten in Afrika und Südamerika und ihre Beziehung zu den Vegetationsgürteln. *Bonner Geograph. Abh.*, 9: 15–98.

LAWRENCE, E. F., 1941. Climatic regions of tropical Australia. *Australian Geographer* 4 (1): 20–26.

LAWSON, G. I., 1964. Diurnal variation of pressure. *Weather*, 19: 390.

LEEPER, G. K., 1950. Thornthwaite's climatic formula. *J. Australian Inst. Agr. Sci.*, 16: 2–6.

LEEPER, G. K. (Editor), 1960. *The Australian Environment*. Melbourne Univ. Press, Melbourne, 151 pp.

LOEWE, F., 1944. Coastal fogs in Australia. *Australia, Bur. Meteorol., Bull.*, 31: 19 pp.

LOEWE, F., 1948. Variability of annual rainfall in Australia. *Australia, Bur. Meteorol., Bull.*, 39: 1–13.

MAHER, J. V., 1966. Climatological normals. *Australian Meteorol. Mag.*, 14:30–38.

MAINE, R., 1968. Restricted visibility statistics—selected Western Australian stations. *Meteorol. Summary, Bur. Meteorol., Melbourne*, 1968: 66 pp.

MCRAE, J. N. and MCGANN, J. J., 1965. Periodic fluctuations in wind, pressure and temperature at Port Hedland. *Australian Meteorol. Mag.*, 48: 1–7.

MEYER, A., 1926. Über einige Zusammenhänge zwischen Klima und Boden in Europa. *Chem. Erde*, 2: 209–347.

MILES, J. F., 1947. The pastoral and agricultural growing season in northeastern Australia. *J. Australian Inst. Agr. Sci.*, 13: 41–49.

MOLNAR, L., 1948. The peanut production scheme in British East Africa—with a comparison between the climates of certain localities in British East Africa and tropical Australia. *J. Australian Inst. Agr. Sci.*, 14: 125–137.

NIMMO, W. H. R., 1964. Measurement of evaporation by pans and tanks. *Australian Meteorol. Mag.*, 46: 17–53.

PATTON, R. T., 1930. The factors controlling the distribution of trees in Victoria. *Proc. Roy. Soc. Vic.*, 42: 154–210.

POLLAK, L. W., 1931. Korrelationen der monatlichen Anomalien der Lufttemperatur ausgewählter Pole mit jenen anderer Orte. *Gerlands Beitr. Geophys.*, 33: 70–111.

PRESCOTT, J. A., 1931a. The soils of Australia in relation to vegetation and climate. *C.S.I.R.O., Bull.*, 52: 82 pp.

PRESCOTT, J. A., 1931b. Atmospheric saturation deficit in Australia. *Trans Roy. Soc. S. Australia*, 55: 65–66.

PRESCOTT, J. A., 1934. Single value climatic factors. *Trans. Roy. Soc. S. Australia*, 58: 48–61.

PRESCOTT, J. A., 1936. The climatic control of the Australian deserts. *Trans. Roy. Soc. S. Australia*, 60: 93–95.

PRESCOTT, J. A., 1938a. Indices in agricultural climatology. *J. Australian Inst. Agr. Sci.*, 4 (1): 33–40.

PRESCOTT, J. A., 1938b. The climate of tropical Australia in relation to possible agricultural occupation. *Trans. Roy. Soc. S. Australia*, 62: 229–240.

PRESCOTT, J. A., 1940. Evaporation from a water surface in relation to solar radiation. *Trans. Roy. Soc. S. Australia*, 64: 114–118.

PRESCOTT, J. A., 1941. Papers on tropical Australia—the soils. *Australian Geographer*, 4 (1): 16–20.

PRESCOTT, J. A., 1942. The phase and amplitude of Australian mean monthly temperatures. *Trans. Roy. Soc. S. Australia*, 66: 46–49.

PRESCOTT, J. A., 1943a. A relationship between evaporation and temperature. *Trans. Roy. Soc. S. Australia*, 67: 1–6.

PRESCOTT, J. A., 1943b. The value of harmonic analysis in climatic studies. *Australian J. Sci.*, 5: 117–119.

PRESCOTT, J. A., 1946. A climatic index. *Nature*, 157:555.

PRESCOTT, J. A., 1949. A climatic index for the leaching factor in soil formation. *J. Soil. Sci.*, 1: 9–10.

PRESCOTT, J. A., 1951. Climatic expressions and generalized climatic zones in regard to soil and vegetation. *Spec. Conf. Agr. Australia, 1949, Proc.*, pp.27–33.

PRESCOTT, J. A., 1956. Climatic indices in relation to the water balance. Australia–*UNESCO Symposium on Arid Zone Climatology*, Canberra 1956, Papers from Australia and New Zealand, pp.5a–5g.

PRESCOTT, J. A. and LANE POOLE, C. E., 1947. The climatology of the introduction of pines of the Mediterranean environment to Australia. *Trans. Roy. Soc. S. Australia*, 71: 67–90.

PRESCOTT, J. A. and THOMAS, J. A., 1949. The length of the growing season in Australia as determined by the effectiveness of the rainfall: a revision. *Proc. Roy. Geograph. Soc. Australia, S. Australian Branch*, 50: 42–46.

PRESCOTT, J. A. and COLLINS, J. A., 1951. The lag of temperature behind solar radiation. *Quart. J. Roy. Meteorol. Soc.*, 77: 121–126.

PRESCOTT, J. A., COLLINS, J. A. and SHIRPURKAR, G. R., 1952. The comparative climatology of Australia and Argentina. *Geograph. Rev.*, 42: 118–133.

QUAYLE, E. T., 1918. On the possibility of forecasting the approximate winter rainfall for northern Victoria. *Australia, Bur. Meteorol., Bull.*, 5: 24 pp.

QUAYLE, E. T., 1938. Tropical control of Australian rainfall. *Australia, Bur. Meteorol., Bull.*, 15: 56 pp.

RADOK, U., 1948. On the nature of Queensland rainfall. *Australia, Bur. Meteorol., Bull.*, 39: 15–35.

REBER, G., 1967. Atmospheric pressure oscillations in Tasmania. *Australian Meteorol. Mag.*, 15: 156–160.

ROYAL AUSTRALIAN AIR FORCE, 1942. *Weather on the Australia Station*. Air Force Headquarters, Melbourne, Vic., Vol. I: 641 pp.; Vol. II: 470 pp.

ROYAL SOCIETY OF WESTERN AUSTRALIA, 1929. Salinity of rain in Western Australia. *J. Roy. Soc. W. Australia*, 15: 22–30.

SMITH, E. J., 1966. Cloud seeding experiments in Australia. *Proc. Symp. Math. Statistics Probability, 5th, Berkeley*, University of California Press, Berkeley, 5: 161–176.

SOUTHERN, R. L., 1964. Application of satellite data in the North Australian tropical region during the summer monsoon. In: J. W. HUTCHINGS (Editor), *Proceedings of the Symposium on Tropical Meteorology, Rotorua, New Zealand*. New Zealand Meteorological Service, Wellington, pp.572–531.

STERN, W. R. and FITZPATRICK, E. A., 1965. Calculated and observed evaporation in a dry monsoonal environment. *J. Hydrol.*, 3: 297–311.

STORMWATER STANDARDS COMMITTEE, 1958. *First Report on Australian Rainfall and Runoff*. Institution of Engineers, Australia, Sydney, 2 vol., pp.170.

SWAIN, E. H. F., 1928. *The Forest Conditions of Queensland*. Queensland State Forest Service, Brisbane, pp.15–27.

TAYLOR, G., 1920. *Australian Meteorology*. Oxford Univ. Press, Oxford, 312 pp.

TAYLOR, G., 1928. Climatic relations between Antarctica and Australia. In: (W. L. G. VOERG Editor) *Problems of Polar Research—Am. Geograph. Soc., Spec. Publ.*, 7: 284–299.

THORNTHWAITE, C. W., 1948. An approach toward a rational classification of climate. *Geograph. Rev.*, 38: 55–94.

TRANSEAU, E. N., 1905. Forest centres of eastern North America. *Am. Naturalist*, 39: 875–889.

TRELOAR, H. M., 1934. Foreshadowing monsoonal rains in Australia. *Australia, Bur. Meteorol., Bull.*, 18: 29 pp.

TRELOAR, H. M. and GRANT, A. M., 1953. Some correlation studies of Australian rainfall. *Australian J. Agr. Res.*, 4: 424–429.

TROLL, C., 1964. Karte der Jahreszeiten-Klimate der Erde. *Erdkunde*, 18: 5–28.

TROUP, A. J., 1967. Opposition of anomalies of upper tropospheric winds at Singapore and Canton Island. *Australian Meteorol. Mag.*, 15: 32–37.

TROUP, A. J. and CLARKE, R. H., 1965. A closer examination of Hobart rainfall fluctuations. *Australian Meteorol. Mag.*, 50: 35–43.

TRUMBLE, H. C., 1937. The climatic control of agriculture in S. Australia. *Trans. Roy. Soc. S. Australia*, 61: 41–62.

TRUMBLE, H. C., 1939. Climatic factors in relation to the agricultural regions of southern Australia. *Trans. Roy. Soc. S. Australia*, 63: 36–43.

TRUMBLE, H. C., 1945. Agricultural climatology in Australia. *J. Australian Inst. Agr. Sci.*, 11: 115–119.

TRUMBLE, H. C., 1948. Rainfall, evaporation and drought-frequency in S. Australia. *J. Agr. S. Australia*, 52: 56–64.

TUCKER, B. M., 1954. Some methods for climatic analysis. *C.S.I.R.O. Div. Soils, Tech. Mem.*, 13: 54.

VOLLPRECHT, R. and WALKER, D. R., 1957. Determination of the length of the growing season. *Australian Meteorol. Mag.*, 17: 37–46.

WADHAM, S. M. and WOOD, G. L., 1939. *Land Utilization in Australia*. Melbourne Univ. Press, Melbourne, 360 pp.

WALLACE, A. R., 1893. *Australia and New Zealand*. Stanford, London, 505 pp.

WALLIS, B. C., 1914. The rainfall regime of Australia. *Scot. Geograph. Mag.*, 30: 527–532.

WALPOLE, J., 1958. Maximum possible rainfall over Australia. In: *Conference on Estimation of Extreme Precipitation*, Bur. Meteorol., Melbourne, Vic., 245–271.

WATTERSON, G. A. and LEGG, M. P. C., 1967. Daily rainfall patterns at Melbourne. *Australian Meteorol. Mag.*, 15: 1–12.

WEBB, J. S., 1870. On the moon and the weather. *Trans. New Zealand Inst.*, 3: 62–65.

WHIGHT, J. M., 1964. The rainfall patterns associated with cyclone Audrey in Queensland, January, 1964. *Capricornia*, 1: 52–55.

WHITE, R. C. L., 1955. Drought and effective rainfall frequency in pastoral New South Wales, west of the wheat belt. *Australia, Bur. Meteorol., Meteorol. Studies*, 5: 13 pp.

WIESNER, C. J., 1970. *Climate. Irrigation and Agriculture*. Angus and Robertson, Sydney, 246 pp.

WILLS, J. T., 1887. Rainfall in Australia. *Scot. Geograph. Mag.*, 3: 161–173.

WILLIAMS, R. J., 1955. Vegetation regions. In: *Atlas of Australian Resources*. Dept. Natl. Develop., Canberra, A.C.T., 1 map.

WINTERBOTTOM, D. C., 1945. Water, Australia's problem. *Australian Geographer*, 5 (1): 20–28.

YAMAGUCHI, K., 1963. Coalescence rain mechanisms in Australia during 1960. *Australian Meteorol. Mag.*, 40: 21–34.

Chapter 7

Climatic Fluctuations

J. GENTILLI

The search for a cycle

The first colonists of New South Wales, arriving there towards the end of the 18th century, found among the aborigines tales of a terrible drought that had taken place many years earlier, but not long enough to be forgotten. There is no exact account of the conditions in the early years of settlement, but there are general reports on dry or wet years (RUSSELL, 1877) which do not show any set pattern. The very large and shallow Lake George, near Canberra, was full in 1824, but by 1836 it had completely dried up, and remained dry until 1849. By 1852 it again held much water. More or less concomitant periods of wet or dry weather were also experienced, and it was only natural that keen observers like Governor Brisbane in the 1820's and Jevons in the 1850's should mention the possibility, or perhaps the hope, that the weather might be repeated in cycles, which could be studied and eventually forecast.

Lunar cycles

JEVONS (1859) stated that at Sydney a dry period ended in 1798, followed by a wet period, 1799–1821; a dry period, 1822–1841 and a wet period from 1842 onwards. This gave periods of about 19 years, the well-known *Saros lunar cycle* of the ancients, revived by Toaldo in the 1770's and never wholly abandoned by some writers.

RUSSELL (1876, 1877), who as Government Astronomer was also in charge of meteorological observations in Sydney, reviewed the evidence for and against weather cycles of 1, 2, 3, 5–6, 6–7, 9, 10 years and also the sunspot cycle. He also considered periods of 12, 13, 19, 22, 30 and 50 years' duration, but could not find any strong support from the data, although he personally favoured the *19-year* period. Only a few years later, WALL (1883) could write that "it has been a favourite speculation to endeavour to draw from statistics a theory of periodicity; but nearly all theories ... have been given up as contrary to a more extended experience."

Early this century, however, Russell was still in favour of his 19-year cycle, stating that when the moon's course is to the southward, much more rain falls than when the moon moves to the northward, thus giving a cycle like 1867–1875 wet, 1876–1885 dry, 1886–1894 wet, 1895–1902 dry (the beginning of this sequence does not agree with the end of the sequence suggested by Jevons and quoted above). Wragge, then Government Meteorologist of Queensland, replied in his *Almanac* that heavy Queensland rains were coincident not only with a wide range of *lunar declinations* exceeding 28° north and south, but also with the maxima periods of *sunspots*. Dry seasons came on as the swing of the "lunar pendulum" became less, within 20°, and "solar energy apparently decreased".

It was recently claimed by BOWEN and ADDERLEY (1962) that there is a significant correlation between the *phases of the moon* and the frequency of heavy rainfall, and the evidence presented seemed to support the statement, especially in the negative form i.e., that there were fewer heavy falls of rain at the time of the full moon. O'MAHONY (1965) tested rainfall data from Sydney, for the two periods 1861–1910 and 1911–1960, for intercorrelation when arranged in moon-phase units, and found no significant statistical correlation. A detailed statistical test of any correlation among heavy rains (25, 50, 75 mm in 24 h) reduced the significance of the full-moon minimum, but did not eliminate it altogether. O'Mahony then tested the series of heavy falls of rain (over 25 mm in 24 h) at Sydney, Perth, Adelaide and Rockhampton, which all showed a definite minimum in correspondence with the full moon; he found that the overall relationship is not statistically significant, although it remains worthy of further investigation. Summing up, O'Mahony stressed "the inherent danger of attempting to assess visually and, therefore, qualitatively, the degree of relationship between two or more sets of curves, especially when the raw data have been subjected to a smoothing process," which can induce "correlations which are apparently significant".

BOWEN (1953, 1956) believes also that there is some relationship between meteoric dust and rainfall. Lunar and meteoric dust would periodically provide large amounts of the condensation nuclei needed for the heavier showers of rain, provided, of course, the other requirements for rain were also available. The hypotheses are being tested and it is not possible to express an opinion at this stage.

Solar cycles

KIDSON (1925) found significant negative correlations (–0.30, –0.35, –0.49, –0.35) between sunspot numbers and the mean latitude of anticyclones crossing Australia (smoothed to eliminate the three year cycle; see below); i.e., in years of sunspot maximum, the anticyclones tend to travel a little further north, but on the other hand, they are a little weaker and slower in their transit. Consequently, atmospheric pressure is also lower than usual (correlation coefficients of sunspot numbers with three-year smoothed mean pressure are: Darwin, –0.43; Thursday Island, –0.50; Brisbane, –0.39; Alice Springs, –0.52; Perth, –0.48; Melbourne, –0.36. The correlation of sunspot numbers with mid-latitude frontal rains (smoothed for the three-year period) is significant and positive (+0.87, +0.67, +0.77 in the wet southwestern, southern and Victorian coastal areas, respectively; Fig.1), that with monsoonal rains less so (Wyndham, +0.52; Darwin, +0.54; Thursday Island, +0.35). In other areas the agreement is disturbed by tropical cyclones (not segregated in Kidson's analysis) which seem to show a negative correlation with sunspot numbers, strong enough to alter (in eastern Australia) or even reverse (in Western Australia) the generally positive pattern (Fig. 1). Inter-anticyclonic rains also show a negative correlation, because they "were found to be heaviest when the anticyclones were farthest south. In each case, the essential condition is the development of pressure gradients favorable to easterly winds over a considerable range in altitude" (KIDSON, 1925). Fig.1 also shows that the pattern is complicated by a lag of about three years in the monsoonal regions.

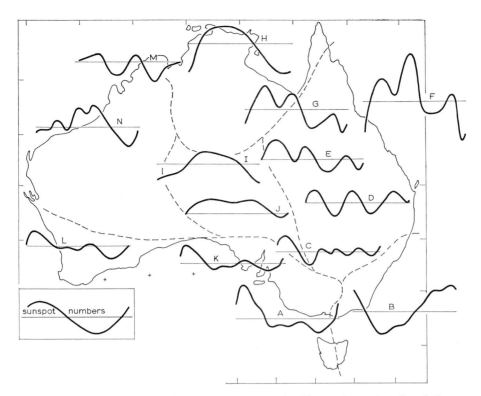

Fig.1. Mean district rainfall in the average sunspot cycle. The graphs are based on 3–4 sunspot cycles only, and might change considerably if the analysis were made for longer periods. (After KIDSON, 1925.)

"The drought which reached its maximum in 1902 was only less pronounced than that of 1914. Pressure shows a marked maximum in 1902 also. The year 1914 was the culmination of what was in all probability the worst drought in Australian history. It stands out as the year of record high pressure. The contrast with the following years is equally marked; from the latter half of 1915–1917 pressures were unusually low... This period was one, also, of extraordinarily good rains" (KIDSON, 1925).

Kidson's work had to be based on 3–4 solar cycles for most stations; the research should be taken up again now that longer records are available.

TAYLOR (1928) showed a very close negative relationship between sunspots and the number of dry months at Bourke, New South Wales, from 1870 to about 1905, but in later years the relationship was much less obvious.

QUAYLE (1938) studied the fluctuations of the rainfall in relation to the *sunspot cycle*, taken as 12 years, with the first and last years of each cycle overlapping. The rain vaguely tended to be less than normal in the years 2, 5–6, 9 and 12 (in the case of summer rain, this applied to the summer at the end of these years), but this was probably due to the southern oscillation.

CORNISH (1936) studied the 95-year long record of the rainfall at Adelaide and found that the respective dates by which the median and the quartiles of the year's rainfall have been received showed a remarkable cyclic variation, verified in a later study (CORNISH, 1954) for the median and most octile rainfall dates (Fig.2). The cyclic oscillation has a period of *23 years* and an amplitude of 30 days in the dates of incidence of the winter rains and of attainment of the yearly rainfall peak. The period is that of the

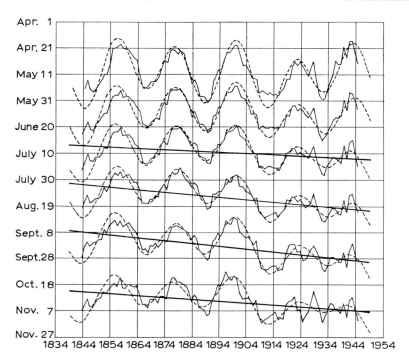

Fig. 2. 10–year means of the dates by which successive eighths of the year's total rainfall had been received at Adelaide (thin lines) with their computed harmonic curves (dashed lines) and their linear trend component (thick straight lines). (After CORNISH, 1954.)

cycle of magnetic polarity of sunspots: during a *P* sunspot cycle (polarity of the leading spot of a pair in the *Northern* Hemisphere of the sun being the same as that of the magnetic pole in the *Southern* Hemisphere of the earth), the onset of the winter rains is progressively earlier. *P* cycles occurred in the periods 1843–1856, 1867–1878, 1889–1901, 1913–1923, 1933–1945, 1954–1965.

During the *E* sunspot cycles (with polarity of the sunspots reversed), the onset of the winter rains is progressively later; these cycles occurred in 1856–1867, 1878–1889, 1901– 1913, 1923–1933, 1945–1954, 1965–.

Cornish points out that the cycle in the *time* of the rainfall does not bring about any change in its amount; no cyclic change in the quantity of rain received at various times of the year has been detected. LEEPER (1957a) stresses also that the data as shown, had been subject to severe smoothing by the running-average method, so that for instance, "the year 1889 had the heaviest January–April rain on record and had had nearly one-third of its rain by the end of April, but it appears on the graph as having a late break, because neighbouring years had dry summers and autumns, notably the preceding year, 1888, with the smallest January–April rain on record." O'MAHONY (1961) confirmed this 23-year cycle for Adelaide's rainfall only, with very faint traces of it in Melbourne's rainfall.

TROUP (1962) tested the difference between various meteorological elements in northern Australia in three years of sunspot minimum and three years of sunspot maximum. The values at sunspot minimum were obtained by averaging the three years centred on the preceding minimum with the three years centred on the following minimum, while the averages for the three years of maximum were centred on the intervening maximum of sunspots. For the period of sunspot maximum centred on 1917 temperature and pressure were lower and rainfall higher than at the preceding and following periods of sunspot minimum. By the period of sunspot maximum centred on 1928, the trend had been reversed, it was slightly warmer and drier than at the times of sunspot minimum, while changes in pressure were less noticeable. Troup suggests that the southern oscillation (taken as the pressure difference between the Pacific high and the Australian low) is positively correlated with the sunspot number, but there has been a slight secular change at some stations. "It is for instance possible that the development of the main pressure-fall area at sunspot maximum in the western Pacific in recent years, results in an increased southerly component of wind over Australia, with increased advection of warmer and drier air northward and a tendency for higher temperatures over North Australia and regions to the north." (TROUP, 1962.)

The southern oscillation

KIDSON (1925) investigated the existence of a three year cycle in atmospheric pressure at Darwin, and in rainfall over most of Australia. "The very severe droughts are usually the culmination of dry periods of several years, during which Darwin pressure is, at least mainly, above normal. The periods 1884–1885, 1894–1896, 1900–1902, 1911–1914, are illustrations. Following the year of maximum pressure and drought culmination is a year of rapid fall of pressure in which the drought is broken. The best examples of this are the years 1878, 1886, 1889, 1897, 1903, and 1915."

The pressure departures from normal show a tendency to a *three-year* period agreeing with the southern oscillation, with a slow increase over two years, and a rapid fall in the third; the driest year is usually the second one, which shows high pressure and anticyclonicity. Since in years of sunspot minima, most of Australia tends to receive less rain, it is very likely that the coincidence of minimum sunspots with the second year of the three year period, will produce very strong anticyclonicity and drought. In tropical Australia there is a lag of six months, the following summer being the one affected. On the other hand, tropical cyclones seem to be more frequent in anticyclonic than in cyclonic years.

O'MAHONY (1961) subjected rainfall data from the capital cities to very close statistical scrutiny, using correlograms and spectral curves. Many data revealed oscillations with a 2–3 year period, especially those from Perth (2.5–3 years for March, June, August, October), Melbourne (2.5 years for October, 3 years for August), Darwin (3 years for October; 2.5 years for December). A four-year period was found in the Adelaide rainfall; at Perth it only appeared in the March rainfall. A seven-year period appeared to be common to all localities in March and almost equally in June, and was very clear in the Hobart (March and June) and Darwin (March) graphs. A study of the fluctuations in pressure at Darwin and Daly Waters also revealed periods of 2–3 and 5–7 years' duration, the former confirming KIDSON's 1925 analysis, the second being possibly derived from it.

O'MAHONY (1961) also correlated the magnitude of the southern oscillation (as shown by anomalies of the Santiago–Djakarta pressure differences) with district rainfalls of the areas near Darwin, Sydney and Hobart, respectively: "These particular districts were chosen to provide a contrasting effect in results, and so, in a qualitative manner to improve the power of the test... In a particular year, when the anticyclonic belt remains

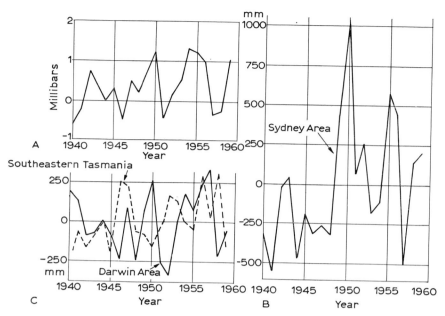

Fig.3. Graphs of: A. the annual anomalies of the pressure gradient between Santiago and Djakarta (indicating the strength of the southern oscillation); and the district rainfalls of: B. the Sydney area, and C. southeastern Tasmania (dashed line) and Darwin area. (After O'MAHONY, 1961.)

fairly uniform along the latitude of Sydney, rainfall (there) would be considerably below normal, while the travelling depressions further south would bring above-average rainfall to the Hobart area, and the more persistent southeasterly winds would keep the Darwin rainfall below average through the resulting weak penetration of the northwestern monsoon."

The similarity of the patterns observed proved quite striking (Fig.3) and can be summarized as shown in Table I.

TABLE I

CORRELATION BETWEEN SOUTHERN OSCILLATION AND RAINFALL OF AREAS NEAR DARWIN, SYDNEY AND HOBART

Southern oscillation ampl. and period	Press. gradient Santiago-Djakarta	Rainfall in Australian areas		
		Darwin	Sydney	Hobart
Large	weak	low	very low	high
Small	strong	high	very high	low
Correlation		+0.31	+0.66	−0.16

PRIESTLEY (1962) after studying the "sympathies" between monthly pressures and rainfall at Darwin confirmed the persistence of monthly pressure anomalies at Darwin, and ascribed it to the southern oscillation. The associations of pressure with subsequent rainfall are generally negative, i.e., the higher the pressure in the preceding months, the less it is likely to rain during the wet season. Extending the investigation, Priestley found a similar but generally attenuated negative association ("antipathy"?) between Darwin pressure and subsequent rainfall at Cairns and Cloncurry, Queensland. There is an exception: January rains at Darwin and February rains at Cairns show a weak positive association with prior Darwin pressure. One may suggest that higher than normal pressure at Darwin may be associated with the formation of tropical cyclones, more frequent in these months. Extending the induction, one might also suggest some association between southern oscillation and tropical cyclones in this region.

A powerful southern oscillation corresponds to a strengthening of the anticyclonic belt and probably to its keeping to a more steady course, with weaker and less frequent inter-anticyclonic troughs and other far-reaching consequences (see Chapters 5 and 6).

Other relationships may affect the two year cycle. FUNK and GARNHAM (1962) found a possible two year cycle in the ozone-content of the stratosphere above Aspendale, Victoria, and Brisbane, for the period 1955–1956 to 1961. The cycle affects only the spring maximum ozone-content, and not the autumn minimum; the authors suggest that it may be due to changes in the general subsidence pattern of ozone-rich stratospheric air, rather than to advection. This cycle is not as noticeable in the Northern Hemisphere, and the period of record in Australia is too short for any further hypotheses to be attempted.

Aspendale and Brisbane are in phase; the high spring maxima occur in the even years. The authors also pointed out that "in addition there seems to be an indication of a two year cycle in the sudden warming in the antarctic stratosphere." KULKARNI (1966) extended the analysis of Aspendale and Brisbane data to later years, and reported a fading of the cyclic pattern; SHAH (1967) interprets the fact that after 1962 the spring maximum is higher in odd years (1963, 1965) as an indication that the ozone cycle was reversed in 1963. BERSON (1966) gives a survey of the whole field, beginning with the recall of the fact that the wind oscillation in the tropical stratosphere, possessing a quasi-biennial rhythm in the zonal component, was only first reported in 1960. There is an "equatorial oscillation of about 26 months average period and 20-knot maximum amplitude (at ± 25 km height)... The basic features... are: (*1*) a vertical downward propagation of the (temporal) maximum amplitude at a rate of about 1 km per month and disappearance at the tropopause level; (*2*) maximum amplitude at the equator and rapid amplitude attenuation to 15°N and S. ... soundings in the Australasian and South American sectors, and over Antarctica, have yielded evidence of a more pronounced extratropical oscillation. There is ... a sharp increase in the temperature amplitude at the 50–60-mbar levels (± 20 km) from middle latitudes to Macquarie Island... Furthermore, Sparrow and Unthank have reported a comparatively strong quasi-biennial zonal wind oscillation at 60,000 and 70,000 ft. over Hobart."

The equatorial oscillation in zonal wind is still conspicuous at Darwin, but while still in phase it has less than half the amplitude at Alice Springs, where it shows some pattern of phase interference with oscillations of similar periodicity but with a phase lag, as have been observed in the Antarctic stratosphere. BERSON (1966) writes that "a detailed correlation analysis is being carried out of the coupling mechanism between quasi-biennial oscillations in polar and tropical latitudes of the stratosphere." TUCKER (1966) isolated the non-seasonal zonal component of the wind by taking 12-monthly running averages, at 21,000 m, for Singapore, Lae, Darwin, Townsville, Alice Springs, Laverton (Melbourne) and Hobart, ranging from 3°N to 43°S. The westerly component is strongest at the end of the odd years, weakest at the end of the even years. Between 12° and 30°S (approximately) no actual westerlies appear, but the easterlies become much weaker at the corresponding times. Beyond latitude 40°S no actual easterlies appear, but the westerlies become weaker in the even years. However, the clearness of the phase decreases with increasing latitude, as noted above, showing the influence of the Antarctic oscillation. HOPWOOD (1968) tested more recent Darwin wind data at 21,000 and 30,000 m to find what happened after 1963, a year in which the easterly component was stronger and more persistent than usual. The oscillation lasted 22 months in 1960–1961, and 27 months from the end of 1964 to the end of 1966, while the intervening phase lasted 34 months, from the end of 1961 to late 1964. It is this persistent easterly phase which has led to the possible reversal of the cycle, so that the peak of westerly flow occurred at the end of the even year 1966, instead of about one year earlier as expected. TUCKER and HOPWOOD (1968) gave a full analysis of the data from Lae to Hobart, revealing that the phase of the 26-month oscillation is propagated downwards in the low latitudes but is independent of height in the middle latitudes. The amplitude reaches a maximum at about 25 km in the low latitudes, while in the middle latitudes it continues to increase to the present effective limit of observations, about 30 km; however, the amplitude decreases very rapidly between 12° and 23°S, while farther south it increases very gradually, and does not reach much more than half of its amount north of the tropic.

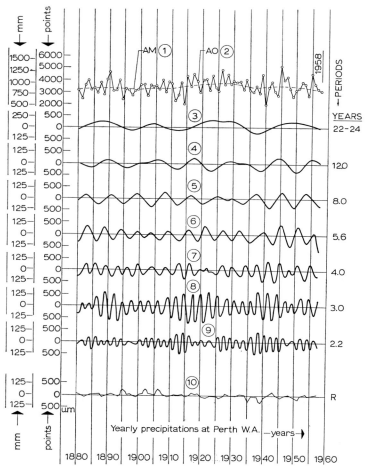

Fig.4. Period analysis of the yearly rainfall of Perth, showing (AM1, dashed line) the long term variations, and (AO2, thin line) the actual data above, and (below, 3–9) the amplitude of each period's wave in mm and points (0.01 inches), with (10) the residual. (After A. Rima, personal communication, 1966.)

The fact that the critical minimum of the 26-month wave amplitude occurs at the tropic may be a pointer to its close causal derivation from insolation.

Further work may be possible after the chief periods in Australian climate have been isolated. We owe to RIMA (1961, and personal communication, 1966) the filter analysis of the yearly rains at Perth, according to VERCELLI's (1940) method (Fig.4). As KIDSON (1925) and O'MAHONY (1961) had also found using other methods, the three-year period stands out very significantly; the 2.1 or 2.2 year period is also conspicuous. Longer periodicities appear less regularly, e.g., the 5.6 year period which at Sydney may be associated with exceptionally heavy rains which, however, are very erratic. Much further work is needed to test correlations, associations and causalities.

Climatic trends and fluctuations

It was unfortunate that the search for a regular cycle should have diverted earlier research away from the general study of climatic trends and fluctuations.

Climatic fluctuations may be classified according to their nature, their amplitude, their duration. As to *nature* they may be rhythmic or arhythmic; pulsating, intermittent or sporadic; recurring or nonrecurring. As to *amplitude*, they may be micro, meso, macro or mega-sized, regular or aberrant. As to *duration*, they may vary from instant to secular to millennial and beyond. Cycles come within the rhythmic class, and their waves may be of various forms; they may also be symmetrical or asymmetrical, and relatively to other waves, they may be concurrently in phase, or show a lag in time or other relationships, if any. All these characteristics are important. An *aberrant mega-sized* sporadic drought or frost may have a catastrophic effect on some plant or animal species and not on others, according to the particular phase in the life-cycle affected by it: we owe to E.P. Hodgkin (personal communication, 1959) the account of the catastrophic coincidence during two days of an exceptional heat wave with an exceptionally low tide along the southwestern shore, circumstances which caused the death of millions of littoral mollusks. A *prolonged macro-sized*, arhythmic drought or frost may prove just as catastrophic: hundreds of thousands of mulga trees (*Acacia aneura*) have died in those parts of southeastern and northwestern Australia where the rainfall has been less than normal many times and for prolonged periods in the last half century, even in areas where sheep and cattle cannot be blamed for it. (For a review of droughts, see FOLEY, 1957.)

Long-term variations, whether rhythmic or not, and even if quite small in magnitude, may affect areal distributions of organisms, especially if their wave-length or duration as the case may be, exceeds the life span of the individuals affected, because the species would tend gradually to follow their usual climatic environment in its spatial advance or retreat. There would thus be slow migrations of species and communities, with consequent alterations in the landscape.

Australian climatic records provide evidence of changes, which may perhaps not be part of any long-term trend or any cycle, but are very real and have noticeable consequences in the day-to-day weather and in the long-term changes in the biotic environment.

The climate of the southwestern quarter

Records of rainfall at Perth are available from 1877 onwards, but there is a break in the series due to the opening of the new Observatory in 1897. Before that year the observations had been made in the Government Gardens, very near the Swan River. The Observatory, built on a hill, is not only higher above sea level, but also differently exposed to the weather. A statistical analysis revealed that observations of the *quantity* of the rainfall in the two localities hardly showed any significant difference, whereas observations of the *number of wet days* disclosed a substantial difference, so that Government Garden records and Observatory records should be treated separately for that purpose. The treatment of data on the quantity of the rainfall as a continuous sequence, i.e., using the data from Government Gardens from 1877–1896 and the data from the Observatory from 1897 onwards, seems, therefore, justified.

The mean annual rainfall for the period 1877–1945 was found to be 885 mm. The least-square straight line passes through 827 mm in 1877 and 943 mm in 1945, showing an overall increase of 116 mm (14.1%) during the 69-year period (GENTILLI, 1952).

The same statistical treatment gives the values for the monthly rainfall shown in Table II.

TABLE II

COMPUTED LEAST-SQUARE VALUES OF MONTHLY RAINFALL, PERTH

Month	1877 (mm)	Mean (mm)	1947 (mm)	Change 1877–1947	
				(mm)	(%)
January	10	8	7	−3	−35
February	12	10	7	−5	−41
March	16	21	26	+10	+60
April	40	44	48	+8	+19
May	123	129	135	+12	+9
June	149	180	211	+62	+42
July	152	170	189	+37	+25
August	143	145	147	+4	+3
September	81	86	91	+10	+11
October	52	55	59	+7	+12
November	19	19	19	0	0
December	15	14	13	−2	−12

The table discloses that summer rainfall (December–February) has slightly decreased, whereas the rainfall of winter type (April–October) has increased considerably. The increase has reached the highest value in June, with 62 mm for the 71 years, i.e., 42% of the initial amount. The increase in the July rainfall, 37 mm (25%), comes next. The increase in the May rainfall, 12 mm (9%), is much smaller, and the increase in the August rainfall, 4 mm (3%), is the smallest of all those experienced in the cooler months. The very high percentage increase in the March rainfall (60%) is probably due to a slight increase in the number of tropical cyclones, or at least to the fact that the anticyclones travelling further south allow more tropical cyclones to affect Perth.

A change in the regime of the rainfall is also evident. In the computed values for 1877, July was the wettest month, and the decrease in the amount of rain from July to August

TABLE III

NUMBER OF DRY MONTHS PER DECADE, PERTH

Decade	Months with less than 63.5 mm	Months with 63.5–127 mm	Months with 127–190 mm	Months with 190–254 mm	Months with less than 127 mm	Months with 127–254 mm	Total dry months (mm)
1877–1886	19	2	14	9	21	23	44
1887–1896	25	7	7	8	32	15	47
1897–1906	25	14	7	3	39	10	49
1907–1916	23	14	4	7	37	11	48
1917–1926	20	7	6	8	27	14	41
1927–1936	24	14	3	8	38	11	49
1937–1946	26	7	5	3	33	8	41
1947–1956	24	8	5	6	32	11	43

was not very great. Now June is the wettest month, while the rainfall in August has hardly changed.

Even though the amount of summer rain has decreased slightly, there is no evidence of any greater severity of summers. Some data have been collated and are shown in Table III. A partial analysis of trends in the rainfall of Albany, Western Australia, was made for the 73 years from 1877 to 1949. The total annual rainfall reaches an average of 954 mm, and the least-square values found were 871 mm for 1877 and 1,036 mm for 1939, i.e., an increase of 165 mm (19%) over the 73 years; the winter months were chosen for detailed study. The values found are shown in Table IV.

TABLE IV

COMPUTED LEAST-SQUARE VALUES OF MONTHLY RAINFALL, ALBANY

Month	1877 (mm)	Mean (mm)	1949 (mm)	Change 1877–1949 (mm)	(%)
June	119	139	159	+40	34
July	119	146	172	+53	44
August	130	132	135	+5	4

Remarkable changes are revealed by this table. August was the wettest month at the beginning of the period, but the conspicuous increase in the rainfall of June and July (amounting to nearly 57% of the increase for the annual rainfall) has now made July the wettest month, and the peak of the annual rainfall has clearly shifted to an earlier time of the year (GENTILLI, 1952).

The shift in the position of the pressure belts detected by LAMB (1964) seems to point to a strengthening of the anticyclones, rather than merely to their migration to different latitudes. If the latter were the case, the mid-latitude fronts should reach further north; if the former hypothesis is correct, the northward shift of the anticyclones should produce higher rainfall only in the southern areas, while the northern reaches would be made drier by the greater prevalence of anticyclonic and continental air. The climatic record of Onslow, Western Australia (Table V), at the extreme fringe of the frontal-rain area, over

TABLE V

COMPUTED LEAST SQUARE VALUES OF MONTHLY RAINFALL, ONSLOW

Month	1883 (mm)	Mean (mm)	1948 (mm)	Change 1883–1948 (mm)	(%)
January	21	24	28	+7	+31
February	(−8)	27	(62)	+70	(+852)
March	15	43	71	+56	+363
April	25	26	27	+2	+7
May	44	41	39	−5	−12
June	46	40	33	−13	−28
July	20	19	18	−2	−8
August	12	11	10	−2	−17
September	1	1	1	0	0
October	>0	<1	1	+<1	(+200)
November	>0	1	2	+<2	(+600)
December	2	4	6	+4	(+229)

the 63 years from 1886 to 1948, shows a decrease in the winter rains which are of frontal origin; this may be taken to support the view that the anticyclones had become more pronounced during the period examined. LAMB's (1964) finding that the belt of low pressure had moved further south is also confirmed by the general increase of the summer rains, although the overall increase in the frequency of tropical cyclones (Fig.5) accounts for most of the spectacular increase in the rainfall of February and March.

While the rainfall of Onslow is too erratic to be taken as a reliable climatic indicator, its changes seem to conform to the general interpretation suggested by the changes found in the rainfall of Perth and Albany, combined with the shift in the pressure belts.

It is quite clear that some change has taken place over the period under consideration, but it would not be justified to say that it is part of a definite trend. The fact remains that during June there were over 38 mm of rain (on the average) during the first 40 years on record and less than 30 during the last 20 years; and in July the average of more than 22 mm for the first 30 years has fallen no less than 18 mm for the last 30 years, so that,

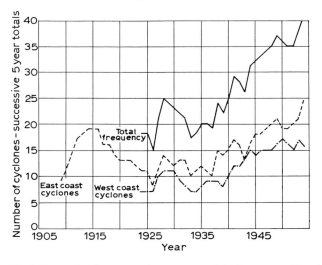

Fig.5. Successive 5-year running averages of the frequency of tropical cyclones. (After Brunt and Hogan.)

for instance, the three winter months had an average total rainfall of 42 mm during the 10 years ending 1945, against an average of 80 mm during the 10 years ending 1895.

An analysis of temperature records over the period from 1876 onwards at Perth, allowing for the change of site from the Government Gardens (1876–1896) to the new Observatory (1897 onwards, with an overlap between 1897 and 1926 that allows a very thorough comparison to be made) disclosed very minor changes, with the summers slightly cooler (February down 0.6°C) and the late autumn and winter slightly warmer (May up 0.7°C; June up 1°C). If the expansion of the city normally induces an increase in temperature, then the slight cooling of the summer (which extends to March) becomes more significant, while the winter warming may be entirely due to the great increase in the number, size, and mass of buildings.

The climate of the southeastern quarter

In southeastern Australia the rainfall, which had been slowly and very irregularly increasing during the second half of the last century, began a period of decline from about 1890 (Fig.6). In the semiarid belt this decline proved critical and in some areas actually disastrous.

DEACON (1953) took 10 year running averages of the mean daily maximum temperature for some typical Australian inland stations (Alice Springs, Northern Territory; Bourke, Narrabri, and Hay, New South Wales), and from the graphs divided the period of study

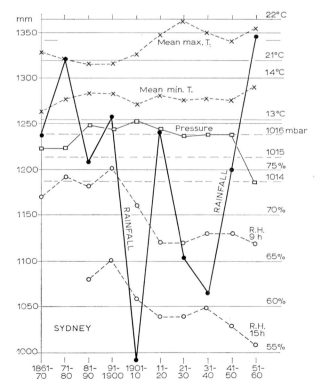

Fig.6. 10-year stationary means of daily temperature, pressure and humidity and yearly rainfall at Sydney; notice how conflicting trends could be obtained by considering limited periods only. Rainfall scale on the left, other scales on the right.

into the 30 year sections, 1881–1910 and 1911–1940, leaving aside the years after 1940 because they seemed to show a tendency to reverse the earlier trend. He then computed the differences between the mean daily maximum temperatures for the two periods at 14 stations, finding a general decrease ranging from 2.6°C at Cooma, New South Wales to 0.3°C at Albury, New South Wales. The temperatures at other times of the year were also tested but failed to show any significant change.

LEEPER (1957b) pointed out that the actual history of each weather station (he chose the examples of Ballarat and Bendigo, Victoria) may contain facts such as changes of site or observer, which can explain certain variations in the records. Data from earlier years, before 1880, may be suspect because so many observers lacked training. Any apparent "climatic changes" in the years immediately after 1908 may be due to the fact that the local observers had been changed by the newly formed Commonwealth Bureau of Meteorology, after it had taken over from the state organizations.

An analysis of changes in the rainfall, made by DEACON (1953) over the same period for the southeastern quarter of the mainland, showed a very consistent geographical pattern in the changes in summer rainfall, with the zero line corresponding closely to the line dividing predominantly summer rains from predominantly winter rains. The summer rains had decreased by up to 19% in the area around 30°S, 150°E, while they had increased considerably in the regions where they were most insignificant and irregular, with a maximum relative increase of 44% at Adelaide.

The winter rainfall seemed to change much less during the same periods, except in the vicinity of the meridian of 141°E, where changes of over 20% were reached. In central and eastern New South Wales, the change in the amount of winter rains is practically insignificant. In South Australia and western Victoria, there was a distinct tendency for the winter rains to come a little later in recent years, with a lag which could be estimated at about two weeks as from the earlier 30 years to the later 30 years, as already found by CORNISH (1936) for Adelaide.

Deacon suggests that "the increased summer rainfall in South Australia and Victoria in 1911–1940, as compared with 1881–1910, is most probably a result of increased activity in the troughs between the migratory anticyclones, and it points to an increase in meridional interchange, which, together with the resultant increase in cloud amount, would account for the decreased inland maximum temperatures at this season." He also thinks that the analysis of pressure data for the same periods suggests that the belt of high pressure did not move so far north in 1911–1940 as it used to do in the earlier period.

LEEPER (1957) pointed out that Deacon's choice of periods of reference and of seasons of rain was unfortunate, because "there were several unusually dry summers in the 1880's and there had been several wet summers in the 1870's, so that this choice gives too low a base-line for reference. Secondly, in western Victoria (though not in Adelaide) February, 1911, was extraordinarily wet." As a result, the effect "is shown far more by February than by the other two months: and if a line is drawn after 1911 instead of 1910 the percentage increase of the Februaries drops to a low figure. But a major objection is to the choice of December, January and February for the calculation. The first half of March belongs to summer more than does the first half of December; it matters little whether an erratic monsoonal storm comes in late February or early March."

Had the 1871–1889 decade been also included in the periods tested, the 1911–1940 increase would have dropped to an insignificant 6%. The various combinations of

TABLE VI

INCREASE IN ADELAIDE RAINFALL 1911–1940
(After LEEPER, 1957a)

Base period	Increase in corresponding months, 1911–1940 (%)
December–February, 1881–1910	44
January–March, 1881–1910	17
January–March, 1871–1910	6
December–March, 1881–1910	28
December–March, 1871–1910	15

decades and the respective increases in rainfall during 1911–1940 are shown in Table VI. In western Victoria, some of the supposed increase during 1911–1940 turns out to be a decrease if February, 1911, is attached to the base period, i.e., if the base period is made 1881–1911.

ALEXANDER (1957) applied the suggestion made by LEEPER (1953) that *runs of successive wet or dry years* be tested, instead of their average amounts of rain. The study of the rainfall of Melbourne, over 101 years, shows that runs of 1, 2, 3 . . . 8 successive years above or below the median (655 mm, against a mean of 660) did occur very much according to random expectation. Closer examination of the results seems to show, however, that while the combined numbers of runs above and below average did conform to random expectation, thereby excluding the likelihood of any cycle or pronounced trend, *two-year runs of wetter years* exceeded random expectation and *four-year runs of drier years* did likewise. This difference may be due to chance in the sample period (101 years) chosen, but on the other hand it may be significant. It certainly appears throughout Australia (Fig.7). These results are confirmed for later years (MAUNDER, 1970).

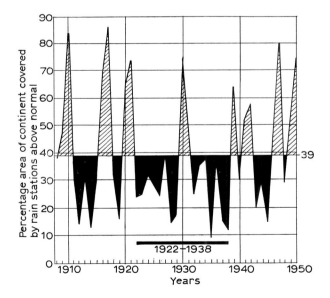

Fig.7. Percentage of Australia's area with yearly rainfall above normal; notice the short runs of wet conditions and the longer runs of dry conditions. (Bureau of Meteorology.)

The analysis of the rainfall at Adelaide carried out by CORNISH (1936, 1954) and already discussed above, has also revealed a slight but progressive *retardation of the winter rains*, so that while their computed midpoint should have fallen about July 4 in 1844, it was expected by about July 14 in 1944. The rains of later winter seem to have decreased slightly in their concentration after the peak, because the next octile of the annual rainfall, due about August 5 in 1844, was delayed until about August 23 in 1944 (Fig.1). From the median date onwards, the retardation has been quite regular, while it is absent from the rainfall of the first half of the year.

It should be mentioned that, at any station subject to the 23-year cycle found by Cornish, the 30-year period, 1881–1910, has 19 years in the retarding rainfall phase and 11 in the advancing-rainfall phase, while the reverse applies to the 1911–1940 period. Cornish has stressed that in Adelaide this cycle affects the timing and not the amount of the rain, but this may not be so at other stations.

The overall pattern of change

In a land with such enormous areas of arid or semiarid climates, even the slightest spread of desiccation has important effects. The cattle population around Alice Springs has decreased from over 350,000 in 1958 to 165,000 by 1964, only because of shortages of pastures and water; this is one of many occurrences of this kind, so many in fact throughout the years, that the public tends to dismiss them as local events due to overstocking and overgrazing. In the same year, however, the Soil Conservation Commissioner of Western Australia stated officially that large parts of the pastoral areas may become man-made deserts within 50 years, having become very vulnerable because of frequent droughts, variable rainfall and high temperatures; human action only accelerates and worsens these processes which have been started and established by climatic conditions. It is, therefore, necessary to attempt a general evaluation of climatic conditions and changes over the whole of Australia in order to appraise the situation.

From the more detailed studies of the southern parts of Australia, it is evident that temperatures have changed very slightly indeed (Fig.6); it was, therefore, decided to assume for the first general evaluation that temperatures had remained unchanged, also in view of the great errors that changes of site or observer may have introduced into a temperature record, while rainfall records are less sensitive, and to some extent self-checking, being so numerous at any given time.

Since only a few stations had been in operation before 1880, it was also decided to use the periods 1880–1910 and 1911–1940 as previously done by DEACON (1953). A complete tabulation was made of all stations with rainfall data from these periods, including also those which began recording in the early 1880's, but keeping their results separate. The map (Fig.8) shows the isopleths of the yearly rainfall difference (1881–1910 *minus* 1911–1940) as continuous lines where they are based on data for the whole period and dashed lines where a few of the earlier years were missing. Apart from localized exceptions, the summer rains have greatly decreased and the winter rains increased; but the area with significantly increased rains is about 250,000 km² against 2,500,000 km² with significantly decreased rains. It should also be noted that a decrease of 75 mm in the yearly total is far more serious at Bourke, New South Wales, than, say, at Armidale, with cooler summers and a much higher rainfall.

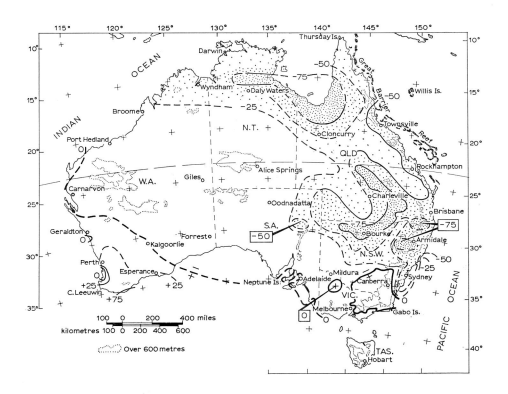

Fig.8. Mean yearly rainfall variation from the period 1881–1910 to the period 1911–1940; notice the large areas of decreased rainfall.

An evaluation of the general implications could be made by classifying the climate of the various stations and producing climatic maps for the two periods. It is not possible to use the system of classification outlined in Chapter 6, because evaporation data are not available for the earlier years; climatic moisture has been assessed by the WANG (1941) formula $r[12r — 20(t + 7)]$ where r is the total precipitation in mm and t the mean temperature in °C, with a value of 3,000 for the boundary between the dry and the wet climates. As LAUER (1952) pointed out, the result differs but little from that obtained by using De Martonne's $12r/(t + 10) = 20$. The Wang index allows the computation to be extended to the colder climates, excluded from De Martonne's index by the "+10" intercept; however, use of the latter index would give very similar results for Australia. The innovation is the introduction of subdivisions, respectively *perarid, arid, semiarid* for the dry climates, and *subhumid, humid, perhumid,* and *hyperhumid* for the wet climates.

The different boundaries of the corresponding climatic regions obtained for 1881–1910 and 1911–1940 are shown in Fig.9. The belts with hydric regression have been stippled. Near Mitchell, Queensland, Bourke, New South Wales and Farina, South Australia, the regression of hydric values has reached a depth of some 100 km; even though it is agreed that these values are arbitrary and the 1881–1910 data less detailed than the recent ones, the regression is a very real and significant one. It may be said also that at some places, more recent years have shown a reversal of the previous trend, but the fact remains that, for instance, a strip of country west of Balranald and Bourke, New South Wales, which, on the average, had been semi-arid for 30 years, through the desiccation of the

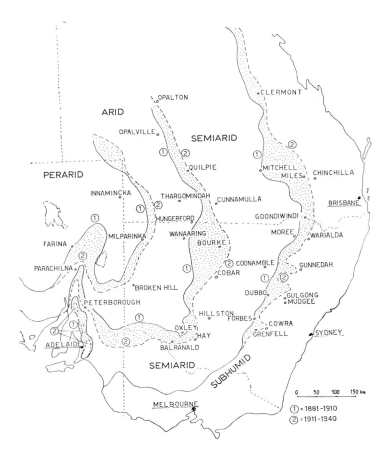

Fig.9. Shifts in the climatic belts over 60 years: position of boundaries in 1881–1910 shown by line *1*, position in 1911–1940 shown by dashed line *2* in each case.

following 30 years became arid, a change which because of its prolonged duration must have had a profound effect on plant and animal life.

The secular variation of aridity was studied by AMBE (1967). He considered "arid" any region which did not have at least one humid month within the year. The definition of a humid month is one in which:

$$11.5 \left(\frac{P}{T-10} \right)^{\frac{10}{9}} > 1$$

this being the formula for the estimate of precipitation effectiveness given by THORN-THWAITE (1931). AMBE (1967) found that the arid region of Australia *extended to the east coast* in the years 1932, 1935, 1936, 1941, 1944, 1952, 1957, 1959, 1960, this being the last year included in the survey. It was smaller, not reaching the east coast and conforming more closely to the *average extent*, in the years 1931, 1933, 1937, 1938, 1939, 1940, 1945, 1946, 1947, 1950, 1953, 1954, 1956. It was much *smaller than average* in 1934, 1942, 1943, 1948, 1949. It did not fall into a definite class in 1951, 1955, 1958. Between 1931 and 1960 the mean position of the eastern boundary of the arid region between 25° and 35°S advanced eastwards at the rate of 1° of longitude every 6 years. The northern boundary of the arid region moved southwards by about 4° in the first 15 years, then northwards by some 5° in the next 15 years. The arid region varied from less than 40% of the continent in 1949 to over 80% in 1935, 1952, 1958, 1960. There was practically no trend in the 1940; the arid region expanded considerably during the 1950's, but this expansion only restored it to the extent it had in the later 1930's. Ambe found a coefficient of correlation of 0.48 between the extent of the arid regions in Australia and South America, the only intercontinental value of any significance.

KRAUS (1954) studied the rainfall of southern New South Wales and Victoria, taking data from Bukalong (since 1858) and Deniliquin (since 1859), and found that the autumn and spring rains were much less in the later period. The rain for Bukalong was more than one-third down, and "something like a discontinuous break in the rainfall regime occurred about 1893–1894."

Additional evidence may be provided by frequency distributions, such as are shown in records for Clarence Heads, New South Wales. Rains of more than 200 mm per month have occurred 57 times in 1881–1910 and 55 times in 1911–1940, but this included eight months with over 400 mm in the first period and four in the second one; and sixteen months with 300–400 mm in the first period and ten in the second one: rains of over 300 mm per month were much more frequent in the earlier period. This might be due at

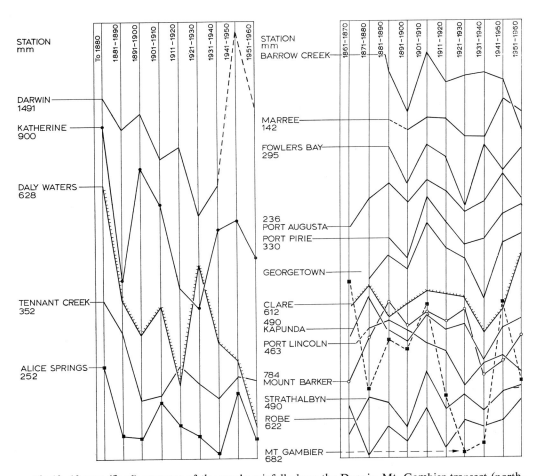

Fig.10. 10-year (fixed) averages of the yearly rainfall along the Darwin–Mt. Gambier transect (north-northwest–south-southeast) to show the general pattern of variation in the interior. The scale is the same for all graphs, but no base-line has been set in order to save space; the mean yearly rainfall of each locality (mostly for the period 1911–1940) is also shown.

least for the summer and autumn months, to a change in the frequency of tropical cyclones (Fig.5), or to a slightly different track followed by a few of them, but it is undoubtedly a significant difference between the two 30-year periods. As to the winter months, the same difference is evident, and it may be necessary to assume a change in the conditions of the ocean surface in order to explain it.

A short distance further north, O'MAHONY (1961) found a distinct break in the series of rainfall records at Brisbane and Ipswich; the period of abundant rains ends abruptly with 1893, and from 1894 onwards no more yearly totals above 1,600 mm have been recorded at Brisbane, or above 1,150 mm at Ipswich, while there had been one every three years before 1894.

An analysis of the rainfall at Townsville and Darwin (KRAUS, 1955) also confirms the change in the 1890's: both localities had more rain up to 1897–1898 and less afterwards (see Darwin's graph, Fig.10). Along the arid fringe, e.g., at Georgetown, Queensland, this decrease in the summer rains must have been of much greater consequence than in the wetter areas.

Kraus expects the decrease in spring and autumn rainfall at the end of the last century to be due to the temporary absence of east-coast depressions. A combination of dry springs and autumns in the southeast, with a monsoon failure or an exceptionally short monsoonal season in the tropical belt, occurred in 1922–1923 and 1933–1935. Kraus assumes that a strong equatorial development of the upper westerlies weakens or shortens the monsoon and at the same time inhibits the development of the "wet east-coast cyclones" (the situation which we have termed the inter-anticyclonic front or trough). Such development was apparently rare between the 1860's and the 1890's, periods of high average monsoon rains in the north, wet springs and autumns in the south, and extensions of coastal types of climate comparatively deep into inland New South Wales. In the middle of the 1890's, the mean strength of the westerlies seems to have increased rather suddenly, and the mean position of their maximum shifted towards the equator. This was associated with repeated monsoon failures or short monsoonal seasons in the north. In southern New South Wales and eastern Victoria this period was characterized by an almost catastrophic reduction of the spring and autumn rains, by comparatively dry summers and by fairly wet winters. The peculiarities of South Australian climates, especially the dry summer, extended eastwards into southwestern New South Wales.

Over the past 40–50 years, the equatorward development of the westerlies has decreased again and its extensions have become less frequent, so that summer rains have increased. Spring and autumn rains have increased to a lesser extent, and winter rains have in general decreased.

A general survey of secular changes, also with regard to the southern oscillation, is briefly given by TROUP (1962), who refers to the relevant literature and supplies a very useful bibliography.

Changes in pressure have been investigated by LAMB (1964) who thinks that, in Australia, in the period 1850–1889 the belt of subtropical high pressure in January usually was at about 37°30'S; it tended to shift slowly towards higher latitudes, until in the period 1900–1939, it was at about 39°30'S. Accordingly, the intertropical trough which had been at about 13°30'S in 1850–1889, had moved to about 17°S during 1900–1939. This would point to stronger monsoons in the later period, which seems to conflict with Kraus' view. It would support our findings that at the same time, the summer of the southwest became drier and probably also slightly hotter.

As to the winter conditions, LAMB (1964) believes that in July the belt of subtropical high pressure used to be at about 32°S in 1850–1889, but the trend had gradually been towards more northerly paths so that in 1900–1939, the average position of this belt was at about 29°30'S. This would make the southern winters wetter and the northern winters drier. It is interesting to note that while in 1850–1889 the belt of high pressure shifted about 5°30' in latitude between January and July, by 1900–1939 the seasonal shift had increased to some 10°. Lamb used 40-year running averages which can be quite treacherous, and it is quite likely that slightly different results would be obtained by using other methods, but the general nature of the change is obviously there, and is borne out by other research, as outlined above.

KRAUS (1962) found a return to wetter conditions in the years following 1940, and "the wetter regime appears to have set in only in the second half of the 1940's". Taking the whole period 1881–1940 as the normal, Kraus found that during 1941–1960, the yearly rainfall had increased in Queensland by 18% at Townsville and 3% at Brisbane, in New South Wales by 9% at Bourke, 13% at Charlton, 11% at Sydney and 4% at Murray Downs, and in Victoria by 1% at Gabo Island. It should be noted, however, that most of the increase is due to the extraordinarily rainy year 1950 and to the rainy biennium, 1955–1956 (Fig.3, 6), and there is no assurance that such events will be repeated in the immediate future.

Fig.10 shows stationary 10-year averages of the yearly rainfall at a number of stations along a north-northwest–south-southeast transect from Darwin in the north to Mount Gambier, South Australia, in the south. Darwin shows an increase but such increase is already proportionally less substantial at Katherine, and non-existent at Daly Waters. The downward trend is still weakly noticeable at Alice Springs. Further south, it is only the stations right at the western foot of the hills or on the slopes facing west, which show an increase, e.g., Port Pirie, Georgetown, Clare, or those exposed to the south, such as Strathalbyn. At Adelaide, however, the rainfall has not increased.

That even slight changes in the amount of water available can have far-reaching consequences is shown by the biological events that have been noticed in Western Australia. In recent years there has been a die-back of jarrah (*Eucalyptus marginata*) which is the chief economic tree in southwestern forests. The FORESTRY AND TIMBER BUREAU (1964) reports that die-back occurs on all topographic situations except slopes in excess of 10°; there is a significant concentration at the heads of local drainage lines, with no relation to aspect. In less affected areas, the die-back is restricted to gullies and low saddles. All told, the disease affects long-lived trees, it was not known in the earlier years of forestry, and it has become serious and widespread since the increase in rainfall; it also tends to occur where excess water concentrates.

The writer has observed over the last 25 years the growing diffusion and frequency of young trees and saplings of the western flooded gum (*E. rudis*) which needs plentiful water near its roots during part of the year, and is, therefore, a tree of river banks and creek beds. This increase is an indication of greater run-off, and it has been noticed also where no clearing of the vegetation has taken place.

The rising water-table has resulted in waterlogging or flooding of low-lying areas which had been dry for many years, so much so that roads and houses had been built on them.

What increases in rainfall would be needed to produce wholesale changes in the landscape?

In a land of flat climatic gradients and broad ecotones such as is much of Australia, even a slight change can produce great consequences. An increase of 25–30 mm in the yearly rainfall of the northwestern coast would be enough to create a semi-arid corridor on the western edge of the desert country, allowing plant and animal migration far more freely than at present. The arid edge would recede by 150–300 km, probably enabling eucalypts to grow much further inland than at present, and increasing the semi-arid belt by some 300,000 km².

Conditions for optimal forest growth, i.e., mesophytic or laurisilvan instead of sclerophyllous forest environment in the southwest, would require an additional 225 mm of rain during the summer, 80 mm during the autumn and another 80 mm during the spring, while the winter rains could still decrease by a substantial amount. This would correspond to a poleward shift in latitude of about 10° for Perth, 7° for Albany, or better, require a corresponding equatorward shift in the westerly winds, or a much less substantial one if the anticyclones became weaker.

Much more momentous and hypothetically plausible is a decrease in the rainfall, caused by the strengthening of the anticyclones or by their following a more southerly path. A decrease of only 125 mm in the yearly rainfall, if suitably spread throughout the wetter parts of the year, would bring the desert margin to where the wandoo woodland is found at present. The enormous semi-arid belt would become arid. The southwest would remain as a humid refuge, because even if the yearly rainfall decreased by 750 mm the edge of the arid zone could not reach the southwestern edge of the plateau, where the denser forests grow at present. In the extreme north, on the other hand, a decrease of only 125–150 mm in the annual rainfall would bring the arid zone right to the coast in the Kimberleys.

References

ALEXANDER, G. N., 1957. Are there climatic cycles?—Melbourne's rainfall examined. *J. Australian Inst. Agri. Sci.*, 23: 235–237.

AMBE, Y., 1967. Secular variation of aridity in the world. *Japan. J. Geol. Geography, Trans.*, 38: 43–61.

ARCTOWSKI, H., 1909. *L'enchaînement des Variations Climatiques.* Soc. Belge d'Astronomie, Bruxelles, 135 pp.

BERSON, F. A., 1966. Polar lobe of the quasi-biennial stratospheric wind oscillation. *Nature*, 210: 1243–1244.

BOWEN, E. G., 1953. The influence of meteoritic dust on rainfall. *Australian J. Phys.*, 6: 490–497.

BOWEN, E. G., 1956. A relation between meteoric showers and the rainfall of November and December. *Tellus*, 8: 393.

BOWEN, E. G. and ADDERLEY, E. E., 1962. A lunar component in precipitation data. *Science*, 7: 749.

CORNISH, E. A., 1936. On the secular variation of the rainfall at Adelaide, South Australia. *Quart. J. Roy. Meteorol. Soc.*, 62: 481–490.

CORNISH, E. A., 1954. On the secular variation of rainfall at Adelaide. *Australian J. Phys.*, 7: 334–346.

DEACON, E. L., 1953. Climatic change in Australia since 1890. *Australian J. Phys.*, 6: 209–218.

FOLEY, J. C., 1957. Droughts in Australia. *Bur. Meteorol., Bull.*, 43: 281 pp., 20 maps.

FORESTRY AND TIMBER BUREAU, 1964. *Annual Report for 1963.* Forestry Timber Bur., Canberra, A.C.T., 80 pp.

FUNK, J. P. and GARNHAM, G. L., 1962. Australian ozone observations and a suggested 24 month cycle. *Tellus*, 14: 378–382.

GENTILLI, J., 1952. Present climatic fluctuations in Western Australia. *West Australian Naturalist*, 3: 155–165.

HOPWOOD, J., 1968. A note on the quasi-biennial oscillation at Darwin. *Australian Meteorol. Mag.*, 16: 114–117.

JEVONS, N. S., 1859. Some data concerning the climate of Australia and New Zealand. In: *Waugh's Almanac*, Waugh, Sydney, pp.XV–XVI and 47–98.

KIDSON, E., 1925. Some periods in Australian weather. *Australia, Bur. Meteorol., Bull.*, 17: 5–33.

KRAUS, E. B., 1954. Secular changes in the rainfall regime of southeastern Australia. *Quart. J. Roy. Meteorol. Soc.*, 80: 591–601.

KRAUS, E. B., 1955. Secular changes of east coast rainfall regimes. *Quart J. Roy. Meteorol. Soc.*, 81: 430–439.

KRAUS, E. B., 1962. Recent changes of east coast rainfall regimes. *Quart J. Roy. Meteorol. Soc.*, 89: 145–146.

KULKARNI, R. N., 1966. The vertical distribution of atmospheric ozone and possible transport mechanisms in the stratosphere of the Southern Hemisphere. *Quart. J. Roy. Meteorol. Soc.*, 92: 363–373.

LAMB, H. H., 1964. Neue Forschungen über die Entwicklung der Klimaänderungen. *Meteorol. Rundschau*, 17: 65–74.

LAUER, W., 1952. Humide und aride Jahreszeiten in Afrika und Südamerika und ihre Beziehung zu den Vegetationsgürteln. *Bonner Geograph. Abh.*, Heft 9: 9–10; 12–13; 15–98.

LEEPER, G. W., 1953. A note on alleged runs of dry and wet years. *J. Australian Inst. Agri. Sci.*, 19: 48–49.

LEEPER, G. W., 1957a. Changing climates and agriculture. *J. Australian Inst. Agri. Sci.*, 23: 232–235.

LEEPER, G. W., 1957b. A note on alleged cooler summers. *J. Australian Inst. Agri. Sci.*, 23: 342–344.

MAUNDER, W. J., 1970. *The Value of the Weather*. Methuen, London.

OLSEN, A. M., 1964. Tasmanian coastal waters. *Fisheries Newsletter*, 23(3): 9.

O'MAHONY, G., 1961. Time series analysis of some Australian rainfall data. *Australia, Bur. Meteorol., Meteorol. Studies*, 14, 65 pp.

O'MAHONY, G., 1965. Rainfall and moon phase. *Quart J. Roy. Meteorol. Soc.*, 91: 196–208.

PRIESTLEY, C. H. B., 1962. Some lag associations in Darwin pressure and rainfall. *Australian Meteorol. Mag.*, 38: 32–41.

PRIESTLEY, C. H. B., 1963. Some associations in Australian monthly rainfalls. *Australian Meteorol. Mag.*, 41: 12–21.

QUAYLE, E. T., 1938. Australian rainfall in sunspot cycles. *Australia, Bur. Meteorol., Bull.*, 22, 56 pp.

RIMA, A., 1961. Sulla correlazione tra osservazioni mensili solari e terrestri. *Boll. Soc. Ticin. Sci. Nat.*, 1960–1961: 1–15.

RUSSELL, H. C., 1876. Meteorological periodicity. *J. Roy. Soc. New South Wales*, 10: 160–176.

RUSSELL, H. C., 1877. *The Climate of New South Wales*. Potter, Sydney, 254 pp.

SHAH, G. M., 1967. Quasi-biennial oscillation in ozone. *J. Atmospheric Sci.*, 24: 396–401.

SPARROW, J. G. and UNTHANK, E. L., 1964. Biennial stratospheric oscillations in the Southern Hemisphere. *J. Atmospheric Sci.*, 21: 592–596.

TAYLOR, G., 1920. *Australian Meteorology*. Oxford Univ. Press, Oxford, 312 pp.

TAYLOR, G., 1928. Climatic relations between Antarctica and Australia. In: W. L. G. JOERG (Editor), *Problems of Polar Research—Am. Geograph. Soc., Spec. Publ.*, 7: 284–299.

THORNTHWAITE, C. W., 1931. The climates of North America according to a new classification. *Geograph. Rev.*, 21: 633–655.

TROUP, A. J., 1962. A secular change in the relation between the sunspot cycle and temperature in the Tropics. *Geofis. Pura Appl.*, 51: 184–198.

TUCKER, G. B., 1966. A preliminary report of work on aspects of the lower stratosphere. *Australian Meteorol. Mag.*, 14: 22–29.

TUCKER, G. B. and HOPWOOD, J. M., 1968. The 26-month zonal wind oscillation in the lower stratosphere of the Southern Hemisphere. *J. Atmospheric Sci.*, 25: 293–298.

VERCELLI, F., 1940. Guida per l'analisi delle periodicità nei diagrammi oscillanti. *Comit. Talassogr. Ital., Centro Nazion. Ric., Mem.*, 285: 1–12.

WALL, H. B. DE LA P., 1883. *Manual of Physical Geography of Australia*. Robertson, Melbourne, 194 pp.

WANG, T., 1941. Die Dauer der ariden, humiden und nivalen Zeiten des Jahres in China. *Tübinger Geol. Geograph. Abh., Reihe II*, 33 pp.

Chapter 8

The Climate of New Zealand – Physical and Dynamic Features

W. J. MAUNDER

Situation—location and physiography

New Zealand (267,000 km²), situated between latitudes 34°S and 47°S, is narrow and mountainous, with a major range running northeast–southwest, and subsidiary ranges generally paralleling the major one (Fig.1).

The distance from the extreme north to the extreme south is about 1,930 km, while from west to east it is 400 km at the widest point. No part of New Zealand lies further than 130 km from the sea; the whole country is well exposed to oceanic influences.

The high mountain ranges cause vertical movements in the atmosphere and much mixing of the various layers up to considerable heights. These processes in turn cause profound modification in the distribution of water vapour and temperature in these layers. In consequence the distribution of rainfall is highly irregular, low clouds can seldom form continuous sheets of any great extent, there is a high percentage of bright sunshine and the diurnal variation of temperature is surprisingly large (KIDSON, 1950).

An important physical aspect is the *orientation* of the main mountain divide. The divide (Southern Alps in the South Island) runs along a line of 220°, with the result that winds from a direction west of 220° are effectively blocked from the east of the divide, while winds from east of 220° (which includes southerlies) are effectively blocked from reaching the western parts of the South Island. This blocking effect is particularly important in southwesterly conditions, where a small deviation in the wind flow will make one coast exposed to winds and the other coast sheltered and vice-versa, with consequent rapid changes of weather.

The country is very isolated: in a hemisphere centered on New Zealand, the only continental masses are Australia (1,600 km to the west) and Antarctica (2,200 km to the south). The rest—apart from Indonesia and the islands of the Pacific—is water. The vast area of water considerably modifies most air masses reaching the country, the long sea passage effectively changing "tropical dry" and "tropical moist" to "temperate maritime," and "polar" to "cool maritime". Thus, there is a limited range of temperatures, the highest and lowest air temperature recorded in New Zealand (to 1967) giving an absolute range of 58°C.

The presence of extensive water surfaces ensures an abundant supply of moisture, the average precipitation in several mountain areas being in excess of 5,000 mm per year. The chain of high mountains, however, results in sharp west-east rainfall gradients, yearly precipitation in Fiordland averaging over 6,000 mm compared with 330 mm in parts of Central Otago.

213

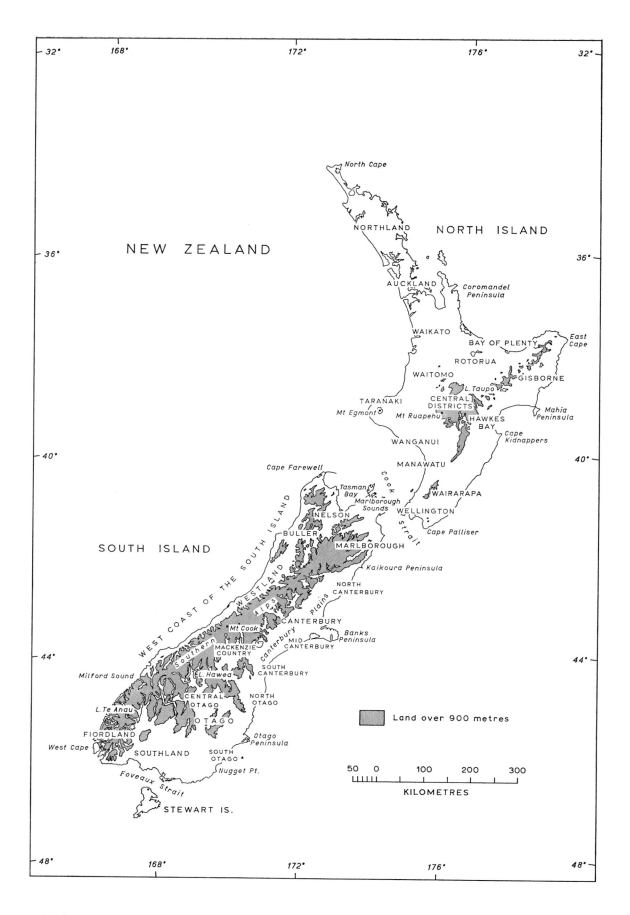

NEW ZEALAND

NORTH ISLAND

North Cape

NORTHLAND

AUCKLAND

Coromandel Peninsula

WAIKATO

BAY OF PLENTY

East Cape

ROTORUA

WAITOMO

GISBORNE

L. Taupo

TARANAKI

CENTRAL DISTRICTS

Mt Egmont

Mt Ruapehu

HAWKES BAY

Mahia Peninsula

WANGANUI

Cape Kidnappers

MANAWATU

Cape Farewell

Tasman Bay

WAIRARAPA

Marlborough Sounds

WELLINGTON

NELSON

BULLER

Cape Palliser

SOUTH ISLAND

MARLBOROUGH

Kaikoura Peninsula

NORTH CANTERBURY

Mt Cook

CANTERBURY

Banks Peninsula

MACKENZIE COUNTRY

MID CANTERBURY

Milford Sound

L. Hawea

SOUTH CANTERBURY

CENTRAL OTAGO

NORTH OTAGO

L. Te Anau

OTAGO

FIORDLAND

Otago Peninsula

West Cape

SOUTHLAND

SOUTH OTAGO

Nugget Pt.

Foveaux Strait

STEWART IS.

WEST COAST OF THE SOUTH ISLAND

WESTLAND

ALPS

Southern

Canterbury Plains

C O O K   Strait

☐ Land over 900 metres

| 50 | 0 | | 100 | 200 | 300 |
KILOMETRES

Surface dynamic features

The basic weather pattern is a succession of anticyclones (or ridges) and troughs. The troughs are usually associated with cold fronts, and often become deformed, resulting in the formation of wave depressions. Typical weather maps are shown and discussed in ROBERTSON (1966) and WATTS (1945, 1947).

Seasonal movement of pressure systems and frequency of patterns

The north–south pressure gradients are steeper in the half year from August through October to February when the pressure systems are moving southwards, than in the other half year from March to July when the system is moving northwards (KIDSON, 1932a). The gradient shows a minimum in February followed by a maximum in April; after this comes the principal minimum with the lowest point in June, then a rapid rise from September–October, the time of principal maximum. This is followed by a slight fall to December, then a more rapid fall to the minimum of February (KIDSON, 1932a).

Frequency of weather patterns

Two investigations have been made of the frequency of the various weather patterns in the New Zealand area. The first (KERR, 1944), classified weather patterns into five types, for each of the four seasons (Table I).

TABLE I

FREQUENCY OF OCCURRENCE (%) OF WEATHER PATTERNS IN THE AREA 130°–180°E AND 20°–50°S (From KERR, 1944)

Type	Winter	Spring	Summer	Autumn	Year
Anticyclones	8	6	8	22	11
Undeformed troughs	42	49	41	34	42
Deformed troughs	50	43	41	35	42
Tropical cyclones	0	0	5	2	2
Uncertain	0	2	5	7	4

It is clear that the troughs of low pressure provide most of the weather patterns in the area considered, which it should be noted extends far beyond the immediate New Zealand area. The frequency of the various types (Table I) shows little seasonal variation except for an increase in anticyclonic activity in autumn, allied to a corresponding decrease in the frequency of troughs, and an absence of tropical cyclones in winter and spring.

A more detailed analysis by WATTS (1947) shows the percentage of days in which various weather patterns are in control for the three-fold division of New Zealand, tabulated from the 6 a.m. weather charts for the period of 1,096 days from July, 1943 to June, 1946 (Table II).

Anticyclonic weather with light winds is the dominant single weather type, anticyclones "controlling" the weather on 28% of all days in the north, 19% in the central area, and

Fig.1. New Zealand—principal physical features and geographical areas.

TABLE II

FREQUENCY OF OCCURRENCE (%) OF CONTROLLING FACTORS IN NEW ZEALAND
(Adapted from WATTS, 1947)

Controlling factor	Northern[1]	Central[2]	Southern[3]
Anticyclones	28	19	20
Moderate or strong northwesterly or northerly flow:			
(a) moderate north to northwest	9	9	8
(b) strong northwest	3	11	7
	12	20	15
Moist easterly to northerly flow:			
(a) moist north	5	4	3
(b) moist northeast to east	6	4	5
	11	8	8
Cold front	5	6	10
After cold fronts:			
(a) northwest flow	1	1	1
(b) west flow	4	4	4
(c) southwest flow	11	9	11
(d) south flow	3	7	8
(e) east to southeast flow	2	8	7
	21	29	31
Warm front or stationary front:	3	3	3
Depressions	6	6	6
Factors not associated with regular series:			
(a) light to strong southwest to west flow	11	7	6
(b) light to strong south to east flow	3	2	1

[1] North of Napier and New Plymouth.
[2] Central Region.
[3] South of Farewell Spit and Cape Campbell.

20% in the south. More frequent than the anticyclones, are the group of cold fronts together with weather following a cold front. This group dominates the weather on 26% of the days in the north, 35% of the days in the central area, and 41% of the days in the south.

Direction of airflow

The direction of airflow into New Zealand is so important, that a further division into "wind flows" is warranted. This shows the predominance of westerly conditions—winds from the northwest, west, or southwest "controlling" the weather 39% of the time in the north, 41% in the central area, and 37% in the south. In general they bring cloudy and often rainy conditions in western areas, and partly cloudy but often clear conditions in eastern areas.

Of the other significant wind flows, moist northerly to easterly conditions (which generally bring rain and cloud to northern and eastern areas) are experienced 11% of the time in the north and 8% of the time in the other two areas. Colder easterly-southerly conditions (which usually bring cloud and light rain to eastern and southern areas) are experienced 8% of the time in the north, and 16–17% of the time in the centre and south.

Features of weather systems

In the Australasian region there is no semi-permanent anticyclone, such as that which exists in subtropical latitudes in the Indian and eastern Pacific Oceans. Instead, a continual eastward migration of anticyclones takes place, roughly at weekly intervals (ROBERTSON, 1967).

During the summer (October–March) the anticyclones cross the Tasman Sea to the east-northeast (relative to their Australian location), whereas the movement in the winter (April–August) is to the east-southeast. Table III shows also that over the New Zealand area (170°E), anticyclones reach their southernmost positions (38.5°S) in February, and their northernmost position (32.7°S) in October. The mean *velocity* of the anticyclones (over the area 120°–170°E) also shows a seasonal variation from an average velocity of 9.1° longitude per day in November–December, to 6.6° longitude per day in May. During the three "winter" months, May, June and July, the average velocity is 6.7° longitude per day compared with an average velocity of 9.1° longitude per day in the spring–summer months of October, November and December (KIDSON, 1947).

TABLE III

MEAN LATITUDINAL POSITIONS OF ANTICYCLONES
(Adapted from KIDSON, 1947)

Longitude	Jan.	Febr.	Mar.	Apr.	May	June	July	Aug.	Sept.	Oct.	Nov.	Dec.
150°E	38.6	39.5	38.7	36.2	34.9	33.6	33.4	33.1	32.8	34.3	35.7	37.5
170°E	36.3	38.5	38.4	37.2	36.7	35.8	35.9	34.5	32.8	32.7	33.4	35.3
170°–150°E	−2.3	−1.0	−0.3	+1.0	+1.8	+2.2	+2.5	+1.4	0.0	−1.6	−2.3	−2.2
Mean velocity[1]	8.6	8.5	8.5	7.6	6.6	6.7	7.0	7.8	8.1	9.0	9.1	9.1

[1] Degrees longitude/day 120°–170°E.

Anticyclonic weather

Anticyclones usually bring settled weather with light winds, although local sea breezes can be quite strong (up to 10–12 m/sec) at times. Fog is also a common feature of anticyclonic weather, especially in inland areas, and on the eastern coast of the South Island. The different nature of the air on the eastern and western sides of anticyclones has an important influence on the weather. On the eastern side of an anticyclone, the air flow is normally from the southwest (usually cool maritime air), although on occasions it is from the south or southeast with cool and usually relatively humid air which may give widespread drizzle to the eastern parts of New Zealand. On the western side of the anti-cyclone, however, the airflow is usually from the north or northwest (usually warm maritime air), bringing mild and relatively humid conditions with clouds in many areas exposed to the west and northwest.

Low-pressure troughs

The mild moist northwest air stream for each preceding anticyclone is replaced by a cool moist westerly to southerly airstream from the following one. The confluence of these two contrasting air-streams results in the formation of a *trough* of low pressure, usually but not always associated with a *cold front*, which is accompanied by the usual unstable conditions associated with such a system. The southern portion of the cold front is normally part of one of the major depressions which travel from west–east across the Antarctic ocean to the south of New Zealand. The northern section of the cold front, on the other hand, usually extends to the northwest and it is on these trailing fronts that most wave depressions develop, this type of wave formation occurring fairly frequently over the northwest Tasman Sea in the vicinity of Lord Howe Island.

The weather conditions associated with a low pressure trough are best discussed under the various airflow patterns described by WATTS (1945).

As an anticyclone moves across to the east of New Zealand, the airflow over the country becomes progressively more west or northwesterly, the northwest airflow developing more rapidly over the southwestern areas of the South Island and about Cook Strait than elsewhere. If the *northwest flow* is sufficiently moist (as is usually the case), clouds develop in exposed western areas, from north Taranaki southwards, while a stronger and very moist flow will produce overcast conditions over most of western New Zealand. In these conditions, Tasman Bay and most eastern districts remain relatively clear, although a very strong northwesterly flow will produce a high stratiform layer over these areas (especially south of Napier). In such conditions, a very warm föhn wind prevails in most eastern areas, especially in Canterbury, where it is known as the "nor'wester" (KIDSON, 1932b).

Cold fronts

The frequency of cold fronts varies from 5% in the north and 6% in the centre areas to 10% in the south (Table II). Generally, cold fronts move over New Zealand from the southwest, crossing Southland, Otago, Canterbury and the North Island in succession. However, on about half the occasions, the cold front which crosses Southland and Otago weakens or dissipates before reaching Cook Strait.

In both eastern and western districts, the weather accompanying a *cold front or trough passage* is generally of fresh to strong northerly-westerly winds and clouds, followed by fresh and occasionally strong winds from a southerly quarter. Usually the actual passage is marked by a short period of moderate, or occasionally heavy, rain followed by colder showery conditions. The exact nature of the cold frontal or trough weather varies; on occasions there is little or no rain with its passage, the nature of the weather depending chiefly on the locality and on the direction of the airflow which follows.

Five cases of *airflow behind cold fronts or troughs* can be distinguished: northwest, west, west-southwest, south-southwest–south, and southeast.

Northwesterly conditions occur when either one cold front or trough is followed by a second cold front or trough, or when a depression over the southwest Tasman Sea maintains a northwesterly flow over the central and eastern Tasman Sea. The weather in such airflow has been discussed above.

A *westerly* flow following a cold front or trough passage, results in western areas being exposed while eastern areas are sheltered. In such a case the rain band of the front or trough will give a period of moderate-heavy rain in most western areas, and in Southland and coastal South Otago. In the westerly airflow following the cold front, conditions improve to broken cumulus and occasional showers. In the Bay of Plenty, eastern districts of the North Island, Tasman Bay, Wairau Plain and most of Canterbury and Central Otago, conditions usually remain clear or partly cloudy throughout.

The *west-southwesterly* type of cold front or trough passage is similar and affects similar areas to that of the westerly type, except that generally there is little or no rain in Nelson, Marlborough, Cook Strait and southern Manawatu—these areas, together with most eastern areas being sheltered from this flow.

In the *south-southwesterly–southerly* airflow type, the southern areas of New Zealand are exposed, whereas most other areas are usually sheltered. Generally some frontal rain occurs on both coasts, followed by a usually rapid clearance, but in the Bay of Plenty and Canterbury little change occurs. Showery conditions usually prevail in Southland and coastal Otago.

With a following *southerly–southeasterly* airflow, there is a fairly rapid reversal of pre-frontal or pre-trough conditions. Western districts cease to be the windward coast, and, therefore, a frontal or trough passage in western areas (especially the South Island) is followed by a rapid clearance to clear conditions. By contrast, areas exposed to the southeast become the windward coast, and thus this type of frontal or trough passage is followed by overcast southeasterly conditions in most of Southland, Otago and Canterbury, and in many cases also in eastern areas of New Zealand north of Canterbury. If the following flow of air is between 150°–170°, these southeasterly conditions will penetrate also through Cook Strait to the south Taranaki area. The actual frontal passage or trough in eastern districts is usually marked by a rapid wind shift from a westerly to a southerly quarter, accompanied by a period of moderate to heavy precipitation.

Wave depressions

As has been stated, the northern section of a cold front usually extends to the northwest and frequently a wave develops on the front. WATTS (1945, 1947) has distinguished a number of types of wave formations depending on the position of the cold front over New Zealand when the wave develops.

A wave developing *south of Australia* will generally (if no further deformation occurs) move to the southeast, the northern part of the deformed front approaching New Zealand as a simple cold front, with a transition from moist northwesterlies to cooler southerly or southwesterly winds.

If a wave develops on a *cold front which has not reached New Zealand*, it will generally move along the frontal system towards the southeast, either crossing over southern New Zealand or passing to the south. In the former case, a cold front will cross New Zealand; in the latter case "depression weather" with a clockwise airflow and rain, will occur in the south of New Zealand, with a cold front passage over the rest of the country. The cold front in both cases is usually preceded by a northerly or northeasterly airflow, with the result that low cloud cover extends over most of northern New Zealand, Taranaki, Nelson, and Marlborough Sounds, often with drizzle and light rain in the Bay of Plenty and Tasman Bay areas, which are usually sheltered from most other wind flows. The cold front passage in this case, crosses most of western New Zealand simultaneously with an accompanying rain belt, but in eastern areas south of East Cape there is little evidence of a frontal passage.

If the southern portion of the *cold front has moved onto New Zealand* before the wave develops on the front to the northwest, the weather pattern becomes more complex, since in this case, a wave travelling southeast along the front will cross over some part of New Zealand. Irrespective however of the actual path of the depression, some and in a few cases most of New Zealand, will experience low clouds and rain, as each coast is exposed in turn to the clockwise circulation of wind around the depression centre. A depression of this type usually takes 36–48 h to cross the country, although it may on occasion take much longer, giving prolonged rain periods to some areas.

Occasionally *a cold front passes over the whole of New Zealand* before deformation occurs. If the deformation occurs west of the North Island, the depression may either move to the east passing to the north of the North Island, with the result that northern areas will be exposed to moist easterly winds; or the depression may move south or southeastwards, passing over or to the south of the country. In such a case the cold front which passed over the country, returns as a warm front, with warm humid air from the northwest being directed onto the exposed western coast. In eastern areas middle and high cloud associated with the warm front occurs—with little precipitation, if any. The warm front, with its sector of warm humid air, will in turn be followed by a "cold type" front and generally westerly to southerly conditions. It should be noted that in this case, the effect of a cold front crossing New Zealand is duplicated—acting initially as a cold front but also, if the wave develops, as a warm front.

The final case of wave formation is a *wave developing to the east of New Zealand* on a cold front that has passed over New Zealand. In such a case, a depression forming to the east of the North Island will bring warm humid easterly conditions to exposed eastern areas, with low clouds, rain and poor visibility. Depending on the movement of such a depression, winds to the south of the centre will be from the east or southeast, which are frequently intensified through Cook Strait. Occasionally the depression will move towards the coast, so that the accompanying warm air mass and warm front will move onto New Zealand. In almost all cases of a depression forming to the east of the country, moderate-heavy rain may occur in eastern areas, which may be prolonged for two or three days. Western districts in such conditions are usually clear.

Westerly wave series

A fairly frequent occurrence in the area south of New Zealand is the passage of a *series of wave depressions*. These usually occur when there is a strong zonal flow over New Zealand, with relatively high pressure over the North Island and low pressure to the south of the South Island. The passage of these waves, which move rapidly eastwards to the south of New Zealand, is accompanied by fluctuations in the wind flow from northwest to southwest. During this process western areas of the South Island, Southland and coastal Otago are subject to changeable showery weather especially in the southwesterly airflow. Through Cook Strait the airflow usually remains northwesterly, increasing as the north–south pressure gradient increases. Such a system usually lasts for some days.

Disturbed southwesterlies

A similar weather series is the disturbed southwesterlies with series of "cold fronts" passing over New Zealand from the southwest. These "cold fronts" are generally associated with a very deep depression passing well to the south of New Zealand, and move from the southwest to the northeast over the country, in particular the southern areas, at intervals of 12–36 h. The first of the series may result in the passage of a simple "cold front"—involving a change from a warm moist northerly or westerly airstream to a cooler westerly to southerly airstream. Then at intervals of between 12 and 36 h, a second, and sometimes a third and fourth, "cold front" crosses the area, each "cold front" being followed by colder air. Usually the last "cold front" is followed by (for New Zealand) very cold air, which in winter may bring snow to low altitudes in Southland and Otago and to higher altitudes in many other areas.

The disturbances in the southwesterly airflow, which occur when an anticyclone covers the northern and western parts of the Tasman Sea and a depression is relatively slow moving to the southeast of New Zealand, have been examined by HILL (1959). He suggests that there are involved (*a*) true fronts approximating the classical model; (*b*) clearly non-frontal troughs accompanied by wind shift nearly coincidental in time at all heights; (*c*) upper level troughs with little reaction on the winds in the lower level but nevertheless accompanied by significant clouds and precipitation. However, even such a variety of types does not completely define the situation as (doubtless owing to orographic effects) it is not uncommon for the apparently non-frontal troughs to appear as cold fronts near the ground, especially over the eastern parts of the South Island.

These disturbed southwesterlies are a fairly common feature of New Zealand weather patterns, and are usually associated with series of "cold fronts," but as mentioned by HILL (1959) these "cold fronts" may also be clearly non-frontal troughs in a disturbed southwesterly flow.

Tropical cyclones

During the summer months, cyclones of tropical origin occasionally affect the weather of New Zealand. VISHER (1922), HUTCHINGS (1953), and GABITES (1956) have compiled information on the frequency of tropical cyclones in the New Zealand area.

GABITES (1956) continuing the work of HUTCHINGS (1953) has surveyed tropical cyclones in the South Pacific for the years 1940–1956. A tropical cyclone for the purpose of this survey was defined as a depression in which surface winds of 21 m/sec or more were reported on the synoptic charts of the General Forecast Office, New Zealand Meteorological Service, in the area 150°E–150°W.

The cyclone season is the late summer or early autumn, the frequency in the South Pacific (for the 16–17 year period) being about one in two years in December, slightly over one per year in each of January, February and March, and one per four years in April. During this period of 16–17 years, however, the number of cyclones per season has varied considerably, 1955–1956 being a peak cyclone season with nine cyclones, compared with an average of four per season over the whole period. During 1944–1945, no cyclones were reported, the only "no cyclone" season in the period.

The direction of movement of the cyclones also shows considerable variety and irregularity. About half of them start with an eastward movement and half with a westward movement, the initial eastward movement being predominant in all months except February (GABITES, 1956).

These tropical cyclones bring heavy rains and strong winds to northern and eastern parts of the North Island if they pass over or sufficiently close to New Zealand; on occasions they cause a great deal of damage to property. BARNETT (1938) analyses the effect of a tropical cyclone which affected northern New Zealand in February, 1936, resulting in much damage and at least two deaths, while KERR (1962) mentions a violent cyclone which struck Northland in March, 1959, with a wind gust of 49 m/sec recorded at Kaitaia.

One of the most destructive storms, however, in terms of property damage and loss of life, occurred on April 10, 1968 when wind gusts of 55 m/sec were recorded in Wellington, the hurricane force winds being one of the causes of the sinking of the interisland steamer "Wahine" with the loss of 51 of the 600 people aboard.

Tornadoes

Tornadoes in New Zealand are discussed by SEELYE (1945) who says that a typical well-developed tornado in New Zealand would be one with a diameter of 18 m and a track of 3 km. They may occur at any time of the day, but most occur during the afternoon. Their average strength is greatest during the period May–October. The western districts, especially Westland and north Taranaki, are the most susceptible areas; on the average probably 25 tornadoes occur each year throughout the whole country.

Upper dynamic features

The upper flow

An analysis of the upper level flow in the New Zealand area has been made by GABITES, (1953a) using data from four radiosonde and two radar wind stations for periods between 1944 and 1952.

The mean flow between latitudes 20°–50°S in the New Zealand region at levels from 3–15 km, is from the west throughout the year, meridional components appearing negligible (Table IV).

The mean west–east airflow reaches its maximum strength in a "mean" jet stream near

TABLE IV

DIRECTION OF MEAN WIND (DEGREES): 1956–1960*

Pressure level (mbar)	Approx. height (1000 m)	Whenuapai		Ohakea		Wellington		Christchurch[1]		Invercargill	
		Jan.	July	Jan.	July	Jan.	July	Jan.	July	Jan.	July
100	16.1	263	266	253	268	263	271	268	268	271	270
200	11.9	271	271	267	267	266	265	267	265	272	266
300	9.1	267	273	266	265	264	268	268	263	276	264
400	7.3	265	267	267	259	262	262	267	260	275	261
500	5.5	261	265	266	257	260	259	265	255	277	261
600	4.3	257	264	265	254	260	257	263	252	277	263
700	3.0	253	265	268	255	264	253	263	250	279	267
800	2.1	252	263	277	258	276	245	273	254	286	272
900	0.9	251	257	297	282	320	258	307	258	285	284

* From New Zealand Meteorological Service (1962), Misc. Publ., 114 (1)(2)(4)(5)(6).
[1] 1957–1960

the 13.7 km level in latitudes 25°–30°S, with the maximum of about 60 m/sec occurring in the winter period, May–August, these winds decreasing to 26 m/sec in February.

The *tropical tropopause* is found between 15.2–16.8 km, usually appearing in summer as far south as Hokitika. The *polar tropopause* varies between 9–12 km. In general, GABITES (1953a) concluded that the strongest mean winds are found some 3 km below the tropical tropopause where it occurs, and near or just below the polar tropopause, except that in the winter over southern New Zealand, strong stratospheric cooling in high latitudes results in the stronger winds in this area in winter being located in the lower stratosphere. PORTER (1953) analysing the 300 mbar westerly wind flow over Australia and New Zealand for the period 1949–1952, said that the most predominant feature is the strong flow in winter between 20°–35°S, with a maximum exceeding 41 m/sec in the zone 25°–35°S. In summer, however, this maximum tends to disappear, and strong winds are found most often in the 45°–50°S latitude zone. Across the 170°E meridian, considerable variation occurred in the short period examined, westerly components of over 88 m/sec contrasting with easterly components of nearly 31 m/sec.

Temperature differences in the troposphere between equatorial and polar areas tend to be concentrated into narrow zones, which can often be linked with the frontal zones between different air masses. In general these *temperature gradients* are steeper in winter. GABITES (1953b) finds that in the New Zealand–Fiji area, the mean gradients during the summer are of the order of 0.5°C/degree of latitude, whereas in the winter the mean gradient exceeds 1.0°C/degree of latitude through most of the troposphere from 25°–35°S, this zone being, as would be expected, also the location of the strong wind belt.

Long period fluctuations of upper level winds and temperatures at Nandi (Fiji), Auckland and Invercargill were assessed by FARKAS (1964) who showed that 28–29 month stratospheric wind oscillations of small amplitudes exist in the middle and higher latitudes. In addition, larger amplitude wind and temperature oscillations of 22–23 months were present in the 1953–1963 period studied. Other studies of the upper atmosphere over the New Zealand area include an analysis of the vertical ozone distribution (BOJKOV and CHRISTIE, 1966), a meridional cross-section (HUTCHINGS, 1950), and an assessment of the magnitude of the torque exerted by the Southern Alps (HUTCHINGS and THOMPSON, 1962).

Correlation with surface dynamic features

As stated above, the mean wind flow over the New Zealand region at levels up to 15 km is generally westerly throughout the year, easterly conditions and "blocking" occurring infrequently. Little has been published on the relationship between the upper and surface dynamic features (see, however, typical surface and corresponding upper level maps in ROBERTSON, 1966), but as suggested by HILL (1959), the relationship at least in a southwesterly airflow is not a simple one, it being not at all clear what causative relationship exists between the dynamic and thermal aspects of disturbances in a southwesterly airflow over New Zealand.

Mean convergence and divergence zones in the South Pacific are discussed by CURRY (1960) who notes that ridges occur frequently over the Andes as well as in the Australia–New Zealand area, whereas troughs have their mean positions in the central Indian Ocean and the central South Pacific. LAMB (1959) suggests that the reasons for these locations are thermal, and HOFMEYR (1957) proposes a linkage of the central South Pacific trough with the low pressure of the tropical Pacific. CURRY (1960), however, indicate that it appears possible that the Andes, in deflecting the westerlies, produce divergent flow on and to the windward of the ranges, which could presumably set up steady waves around the latitude circles. He also attempts to show the zone of divergence and convergence associated with the troughs and ridges, with New Zealand centered in an area of divergence, while an area of convergence extends southwest of a line drawn from about Macquarie Island to the southwest corner of Australia, and a further convergence area is about 800 km to the northeast of Northland.

CURRY (1960) also suggests that in the Southern Hemisphere the cycle between meridional and zonal flow is not nearly so pronounced nor so regular as in the Northern Hemisphere, pointing out that high latitude cyclones in the area to the south of New Zealand occur in "families," followed by a lull, and that this suggests that there is a *cycle between weak and intense zonal flow*, with meridional flow occurring only occasionally and in restricted sectors.

Much of New Zealand's worst weather is associated with *upper level depressions*, or pools of relatively cold air. They have been studied by KERR (1953) for a 12 month period. They appeared on 231 days (153 in winter and spring, only 38 in summer). For all three types of cold pools considered, the average eastward movement varied from 4.0° longitude per day in summer to 5.8° longitude per day in autumn, with an average for the year of 5.5° longitude per day. The north–south motion was much smaller, there being a southward movement of 1.2° latitude per day in autumn and a northward movement of about 0.8° latitude per day in other seasons.

During the year studied, 25 cold pools crossed New Zealand, 14 of them associated with heavy rain. KERR (1953) further showed that in four of the remaining cold pool situations, the depression did not extend to the 300 mbar level, whereas of eight other cases where heavy rain did occur, six had a depression, or a very pronounced trough, present at the 300 mbar level.

It would appear, therefore, from KERR's (1953) work that heavy rain situations in New

Zealand occur either when a depression exists up to the 300 mbar level, or when a cold pool is present in the 1,000–500 mbar thickness pattern, the probability of heavy rain (and consequent flooding) being more likely if these two conditions occur together. Further aspects of the dynamic climatology of New Zealand are examined by RAYNER (1965), and in a further paper a cross spectral analysis of meteorological time series, using New Zealand data, is made as one possible way of bridging the gap between climate and the instantaneous features of the atmosphere (RAYNER, 1967).

Sources used in the New Zealand chapters

Unless otherwise stated, all data used in these chapters have been obtained from the New Zealand Meteorological Service. The actual data used are principally of two kinds: (*1*) Data supplied by the New Zealand Meteorological Service usually in the form of published summaries or averages (e.g., "*Mean monthly rainfall for selected stations*"). The source of such data is shown in the tables thus: "New Zealand Meteorol. Serv. data".

(*2*) Most of the other data, especially those relating to extremes, have been obtained from unpublished records and summaries held at the Meteorological Office, Wellington. All other data sources are shown on the respective tables or maps.

Official published data

The principal sources of meteorological data are as follows (the data in each case referring to actual months or years):

Data from climatological stations

(*1*) 1853–1927 Monthly and annual data published in *Statistics of New Zealand*, an annual Government publication.

(*2*) 1892–(cont.) A summary of the weather for the year, and a selection of data from a few of the climatological stations is published annually in the *New Zealand Official Year Book*. (Government Printer, Wellington.)

(*3*) 1904–(cont.) A selection of monthly data for all climatological stations is published in the *New Zealand Gazette*. (Government Printer, Wellington.) The data are usually published 3–4 weeks after the end of the respective month.

(*4*) 1928–(cont.) Monthly and annual data for all climatological stations are published in *Meteorological Observations for (Year)*. New Zealand Meteorol. Serv., Misc. Publ., 109.

Data from rainfall stations

(*1*) 1889–1940 Monthly and annual rainfall data published at monthly intervals in the *New Zealand Gazette* (Government Printer, Wellington).

(*2*) 1941–(cont.) Monthly and annual data published in *Meteorological Observations for (Year)*. New Zealand Meteorol. Serv. Misc., Publ., 109.

Acknowledgements

Grateful acknowledgement is made to the Director, New Zealand Meteorological Service, Wellington, for supplying much of the data used in this survey, and for allowing access to unpublished data. Special thanks are due to Messrs N. G. Robertson, J. Finkelstein, and D. C. Meldrum, of the Climatological Section of the New Zealand Meteorological Service.

All the maps were drawn by Mr. G. A. H. Kidd, Department of Geography, University of Otago, Dunedin, to whom special thanks are due.

Acknowledgement is also made to Professor R. G. Lister, Mr. W. J. Brockie, and Dr. R. P. Hargreaves, Department of Geography, University of Otago, Dunedin, for their assistance.

References

BARNETT, M. A. F., 1938. The cyclonic storms in Northern New Zealand on 2nd February and 26th March 1936. *New Zealand Meteorol. Serv., Meteorol. Office Note*, 22.

BOJKOV, R. D. and CHRISTIE, A. D., 1966. Vertical ozone distribution over New Zealand. *J. Atmospheric Sci.*, 23: 791–798.

CURRY, L., 1960. Atmospheric circulation in the southern South Pacific. *New Zealand Geograph.*, 16: 71–83.

FARKAS, E., 1964. Long-period fluctuations of upper level winds and temperatures over the South Pacific. In: J. W. HUTCHINGS (Editor), *Proc. Symp. Tropical Meteorol.* New Zealand Meteorol. Serv., Wellington, pp.180–189.

GABITES, J. F., 1953a. Mean westerly wind flow in the upper levels over the New Zealand region. *New Zealand J. Sci. Technol., B*, 34: 384–390.

GABITES, J. F., 1953b. Temperatures in the troposphere and lower stratosphere over the N. Z. region. *New Zealand J. Sci. Technol., B*, 35: 213–224.

GABITES, J. F., 1956. *A Survey of Tropical Cyclones in the South Pacific*. Tropical Cyclone Symp., Brisbane, 1956, presented paper.

HILL, H. W., 1959. Disturbances in south westerly airflow over N.Z. *New Zealand Meteorol. Serv., Tech. Note*, 130.

HOFMEYR, W. L., 1957. Upper air over the Antarctic. In: M. P. VAN ROOY (Editor), *Meteorology of the Antarctic*. Weather Bureau, Pretoria. 240 pp.

HUTCHINGS, J. W., 1950. A meridional cross-section for an oceanic region. *J. Meteorol.*, 7: 94–100.

HUTCHINGS, J. W., 1953. Tropical cyclones in the southwest Pacific. *New Zealand Geograph.*, 9: 37–57.

HUTCHINGS, J. W. and THOMPSON, W. J., 1962. The torque exerted on the atmosphere by the Southern Alps. *New Zealand J. Geol. Geophys.*, 5: 18–28.

KERR, I. S., 1944. Seasonal variation of weather types in New Zealand. *New Zealand Meteorol. Serv., Circ. Note*, 25.

KERR, I. S., 1953. Some features of upper level depressions. *New Zealand Meteorol. Serv., Tech. Note*, 106.

KERR, I. S., 1962. Characteristic weather sequences in Northland. *New Zealand Meteorol. Serv. Tech. Note*, 144.

KIDSON, E., 1932a. Climatology of New Zealand. In: W. KÖPPEN and R. GEIGER (Editors), *Handbuch der Klimatologie*. Borntraeger, Berlin, IV (S): 111–138.

KIDSON, E., 1932b. The Canterbury "Northwester". *New Zealand J. Sci. Technol.*, 14: 65–74.

KIDSON, E., 1947. Daily weather charts extending from Australia and New Zealand to the Antarctic continent. In: *Australian Antarctic Expedition, 1911–14—Sci. Rept., Ser. B.*, Vol. 7, Sydney. (Quoted from Garnier, 1958).

KIDSON, E., 1950. The elements of New Zealand's climate. In: B. J. GARNIER (Editor), *New Zealand Weather and Climate*. New Zealand Geograph. Soc. Christchurch, pp.45–83.

LAMB, A. H., 1959. The southern westerlies: A preliminary survey. *Quart. J. Roy. Meteorol. Soc.*, 85: 1–23.

PORTER, E. M., 1953. The westerly wind flow at 300 mb across Australia and New Zealand. *New Zealand Meteorol. Serv., Tech. Note*, 98.

RAYNER, J. N., 1965. *Dynamic Climatology of New Zealand*. Thesis, University of Canterbury, New Zealand, unpublished.

RAYNER, J. N., 1967. A statistical model for the explanatory description of large scale time and spatial climate. *Can. Geograph.*, 11: 67–85.

ROBERTSON, N. G., 1966. Meteorology. In: A. H. McLINTOCK (Editor), *An Encyclopaedia of New Zealand*. Government Printer, Wellington, 2: 548–552.

ROBERTSON, N. G., 1967. Climate of New Zealand. In: *New Zealand Official Year Book 1967*. Government Printer, Wellington, pp.15–20.

SEELYE, C. J., 1945. Tornadoes in New Zealand. *New Zealand J. Sci. Technol.*, 27: 166–174.

VISHER, S. S., 1922. Tropical cyclones in Australia and the South Pacific and Indian Oceans. *Monthly Weather Rev.*, 50: 288–295.

WATTS, I. E. M., 1945. Forecasting N. Z. Weather. *New Zealand Geograph.*, 1: 119–138.

WATTS, I. E. M., 1947. The relations of N. Z. weather and climate: An analysis of the westerlies. *New Zealand Geograph.*, 3: 115–129.

Chapter 9

Elements of New Zealand's Climate

W. J. MAUNDER

Meteorological observations in New Zealand

In New Zealand, the Meteorological Service organizes the collection and co-ordination of most meteorological data, and meteorological data of one or more elements are available for more than 1,400 stations, which fall into four main classes: (*1*) synoptic reporting stations; (*2*) aerodrome reporting stations; (*3*) climatological stations; and (*4*) rainfall stations.

History

As early as 1840 at Auckland and 1841 at Wellington, weather had been recorded by private observers. These and other early records were collected together, and published from 1853 onwards. Robertson commented that the inclusion of meteorological data in official statistics represented the start of a reliable record for New Zealand, but that obviously the accuracy of the information left much to be desired (ROBERTSON, 1950). In 1859 a plan was formulated for the establishment of Government Observatories. In 1862 observatories were established at Auckland, New Plymouth, Nelson, Wellington and Dunedin. Hokitika was established in 1866, and Mangonui, Napier, Christchurch, Bealey and Invercargill in 1867. Further aspects of early meteorological observations in New Zealand are given by OWEN (1946), ROBERTSON (1950), and DE LISLE (1959). The development of observations from 1870 onwards is discussed by OWEN (1946), ROBERTSON (1950) and COULTER (1964).

Reference may also be made to ROBERTSON (1959) for a description of New Zealand climates and associated maps published in the *Descriptive Atlas of New Zealand* (MCLINTOCK, 1959); to ROBERTSON (1966, 1967) for descriptions of climate in the *Encyclopedia of New Zealand*, and the *New Zealand Official Year Book*, respectively; to ROBERTSON (1957) for a contribution on climate in an earlier survey of *Science in New Zealand*; to DACRE (1950) for a bibliography of New Zealand weather and climate to 1948; to CREASI (1959) for a bibliography of climatic maps of New Zealand; and to SPARROW and HEALY (1968) for a bibliography of the meteorology and climatology of New Zealand to 1967.

The number of *climatological stations* did not reach twenty until 1906, after which there has been a generally steady increase to reach 40 in the 1920's, 60 in the 1930's, 100 in the 1940's, and 170 in 1963[1]. Of the 170 stations in operation in March, 1963, 122 were

---

[1] In May 1970 there were 260 climatological stations in operation.

below 250 m, 33 between 250–500 m, 9 between 500–750 m, and 3 between 750–1,000 m. Only three of the 170 stations were located above 1,000 m and only one (Chateau Tongariro, 1,119 m) has a reasonably long record. Black Birch Range in Marlborough (1,396 m) and Mount John in the Mackenzie Country (1,028 m) were not established until the early 1960's.

The number of stations recording *rainfall* has naturally been far greater, reaching 100 in the 1890's, 400 in the 1920's, 700 in the late 1940's and 1,300 in 1963. All provide daily rainfall data and most of them are in the charge of private voluntary observers. They give an adequate coverage of most settled areas, but in mountain and other remote areas, there are still large gaps in the network. A number of octapent *storage rain-gauges*, mostly maintained by Catchment Boards, provide some data in remote mountainous areas.

Upper air observations of winds by means of pilot balloons were started in 1929, and radar wind observations have been made since 1942. In 1967, there were five radar wind stations in New Zealand and one at Campbell Island. Upper air observations of pressure, temperature and humidity were first regularly carried out in 1942, and in 1967 there were three radiosonde stations in New Zealand and three on islands distant from the coast (Raoul, Chatham, Campbell).

TABLE I

BASIC CLIMATOLOGICAL STATIONS

Station	Latitude (S)		Longitude (E)		Height above
	(°)	(min)	(°)	(min)	M.S.L. (m)
(a) North Island					
Te Paki Te Hapua	34	30	172	49	58
Glenbervie	35	39	174	21	107
Auckland (Albert Park)*	36	51	174	46	49
Tauranga*	37	40	176	12	4
Hamilton (Ruakura)	37	46	175	20	40
Kaingaroa	38	24	176	34	544
Gisborne	38	40	177	59	4
New Plymouth (City)*	39	04	174	05	49
Chateau Tongariro*	39	12	175	32	1119
Napier*	39	29	176	55	2
Palmerston North (D.S.I.R.)	40	23	175	37	34
Masterton (Waingawa)	40	59	175	37	104
Wellington (Kelburn)*	41	17	174	46	126
(b) South Island					
Nelson (Airport)*	41	17	173	13	2
Blenheim	41	30	173	58	4
Hanmer	42	31	172	52	387
Hokitika (South)*	42	43	170	57	4
Lake Coleridge	43	22	171	32	364
Christchurch (Gardens)*	43	32	172	37	7
Lake Tekapo*	44	00	170	29	683
Milford Sound	44	41	167	55	5
Alexandra*	45	15	169	24	158
Dunedin (Musselburgh)*	45	55	170	31	2
Invercargill (Airport)*	46	25	168	19	−0.3

* Full climatological data are given in special tables for these stations in the Appendix.

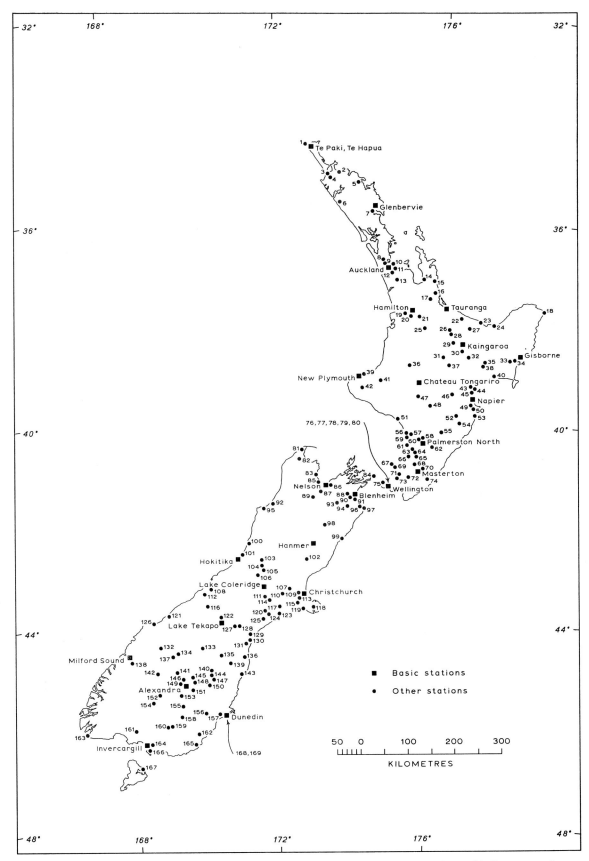

Fig.1. Location of stations mentioned in the tables and text. All 24 "basic" stations (see Table I) are named on the map; all other stations are given in Table II. Full climatological data are given in the Appendix.

231

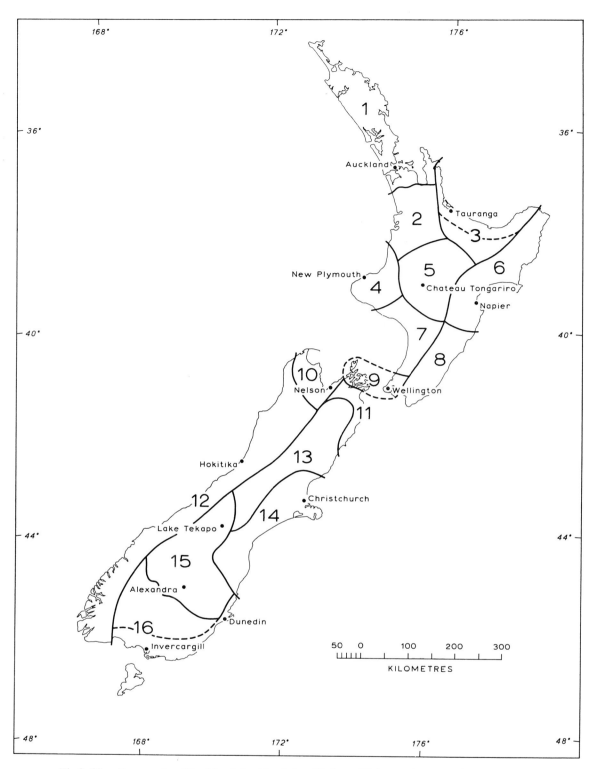

Fig.2. Climatic areas. *1* = Northland–Auckland. *2* = Waikato–Waitomo. *3* = Bay of Plenty: (a) coastal; (b) inland. *4* = Taranaki. *5* = Central districts of the North Island. *6* = Gisborne, northern and central Hawkes Bay. *7* = Wanganui–Manawatu. *8* = Southern Hawkes Bay and Wairarapa. *9* = Wellington–Cook Strait. *10* = Nelson. *11* = Coastal Marlborough. *12* = West coast of the South Island. *13* = Inland Marlborough and Canterbury High Country. *14* = Canterbury Plains and North Otago. *15* = Mackenzie Country–Central Otago. *16* = Southland–East Otago: (a) coastal; (b) inland. Full climatological data are given for the named stations in the Appendix.

Climatological data

A basic list of 24 stations has been used, selected so as to give a good coverage (Table I). Unless otherwise stated, all the climatic data used refer to the present site of the station. Table II lists all stations mentioned in either the tables or the text, their location on the map (Fig.1) and the climatic area in which they are found (Fig.2).

Except in the case of rainfall averages which refer to the period 1921–1950, all averages and extremes are based on the total period of record, which varies from less than 10 years to more than 100 years. This should be borne in mind when comparing places which do not have the same period of record. In those cases where there has been a site change at a particular station, the average generally refers to each particular site, but all sites are normally used together for extremes. Observations are usually made at 09h00 New Zealand standard time.

TABLE II

STATION LOCATIONS AND CLIMATIC AREAS

Station	Map no.[1]	Climatic area[2]	Station	Map no.[1]	Climatic area[2]
Adair	131	14	Chateau Tongariro	*	5**
Akaroa (see Onawe)	118	14	Christchurch (Airport)	109	14
Albert Park (See Auckland)	*	1	Christchurch (Gardens)	*	14
Alexandra	*	15	Christchurch (Wigram)	113	14
Appleby	85	10	Clyde	147	15
Arapuni	25	2	Cromwell	142	15
Arthurs Pass	104	13**			
Ashburton	124	14	Dannevirke	55	8
Auckland (Albert Park)	*	1	Darfield	110	14
Auckland (Mechanics Bay)	10	1	Dawson Falls	42	4**
Awarua	166	16	Downertown	92	12**
			Dunedin (Beta Street)	169	16
Bainham	82	10**	Dunedin (Gardens)	168	16
Balclutha	162	16	Dunedin (Musselburgh)	*	16
Balmoral	102	13	Duntroon	139	14
Bealey	105	13			
Black Birch Range	94	13**	Earncleugh	149	15
Blenheim	*	11	East Cape	18	6
Blenheim (Omaka)	90	11	East Gore	159	16
Blenheim (Woodbourne)	88	11	Esk Forest	43	6
Brothers	84	9	Eyrewell	107	14
Cambridge	21	2	Fairlie	127	14
Cape Campbell	97	11	Featherston	72	8
Cape Reinga	1	1	Flock House	56	7

TABLE II *(continued)*

Station	Map no.[1]	Climatic area[2]	Station	Map no.[1]	Climatic area[2]
Fox Glacier	112	12**	Lincoln	115	14
Foxton	61	7			
Franz Josef	108	12	Makara	75	9
			Mangahao (Hydro)	64	7**
Garston	152	16	Mangahao (Upper)	65	7**
Gisborne	*	6	Mangere	12	1
Glenbervie	*	1	Mangonui	2	1
Golden Downs	89	10	Manorburn Dam	151	15**
Gore	160	16	Manutuke	34	6
Gore (East)	159	16	Masterton	70	8
Greymouth	100	12	Masterton (Waingawa)	*	8
Gwavas (Forest)	52	6	Mechanics Bay	10	1
			Methven ("Rudstone")	111	14
Haast	121	12	Mid Dome	154	16
Halfmoon Bay	167	16	Milford Sound	*	12**
Hamilton (Ruakura)	*	2	Milson (see Palmerston North)	58	7
Hamilton (Rukuhia)	20	2	Minginui	32	5
Hanmer (Forest)	*	13**	Moa Creek	148	15
Harper River	106	13	Moa Flat	155	16
Hastings	49	6	Molesworth	98	13**
Havelock North	50	6	Motueka	83	10
Hawea Flat	134	15	Mt. Aspiring	132	15**
Highbank	114	14	Mt. Cook (see The Hermitage)	116	15**
Hinds	125	14	Mt. John	122	15**
Hokitika	101	12	Musselburgh (see Dunedin)	*	16
Hokitika South	*	12			
Holdsworth	68	8**	Napier	*	6
Homer Tunnel	138	12**	Naseby Forest	140	15**
			Nelson (Airfield)	*	10
Invercargill (Airport)	*	16	Nelson (Cawthron Institute)	86	10
Invercargill (City)	164	16	New Plymouth	*	4
			New Plymouth Airfield	39	4
Jacksons Bay	126	12	Ngaumu (Forest)	74	8
Kaihoka	81	10	Oamaru	143	14
Kaikoura West	99	11	Ohakea	57	7
Kaingaroa	*	3	Omaka (see Blenheim)	90	11
Kaitaia	4	1	Omarama	133	15**
Kapiti Island	67	7	Onawe	118	14
Karioi	47	5**	Onepoto	38	6
Kawerau	27	3	Ophir	145	15
Kelburn (see Wellington)	*	9	Opotiki	24	3
Kerikeri	5	1	Otara	11	1
Kuripapanga	46	6	Otautau	161	16
			Otira	103	12**
Lake Coleridge	*	13**	Owaka	165	16
Lake Ellesmere	119	14			
Lake Grassmere	96	11	Paerata	13	1
Lake Mahinerangi	156	15	Pahiatua	62	8
Lake Tekapo	*	15**	Palmerston North		
Lake Waikaremoana	35	6	(Boys' High School)	60	7
Lake Waitaki	135	14	Palmerston North (D.S.I.R.)	*	7
Lambrook	128	13	Palmerston North (Milson)	58	7
Levin	66	7	Paraparaumu	69	7

TABLE II *(continued)*

Station	Map no.[1]	Climatic area[2]	Station	Map no.[1]	Climatic area[2]
Patearoa	150	15	Upper Hutt	71	9
Pendarves	123	14			
Petane	45	6	Waerenga-o-Kuri	33	6
Puysegur Point	163	12	Waihi	16	3
			Waihopai	93	11
Queenstown	143	15	Waimarama	53	6
			Waimate	136	14
Rakaia	117	14	Waimihia	37	5**
Ranfurly	144	15	Waingawa (see Masterton)	*	8
Riverhead	8	1	Waiotapu	29	3
Rotoehu (Forest)	22	3	Waipapakauri	3	1
Rotorua	26	3	Waipiata	147	15
Rotorua (Whakarewarewa)	28	3	Waipoua (Forest)	6	1
Roxburgh Hydro	153	15	Waipukurau	54	6
Ruakura (see Hamilton)	*	2	Wairapukao	30	5
Rukuhia	20	2	Wairoa	40	6
			Wakefield	87	10
Shannon	63	7	Wallaceville	73	9
			Wanaka	137	15
Taieri	157	16	Wanganui	51	7
Taihape ("Hiwi")	48	5	Wellington (Hawkins Hill)	78	9
Tairua (Forest)	15	3	Wellington (Kelburn)	*	9
Tangimoana	59	7	Wellington (Moa Point)	80	9
Tangoio	44	6	Wellington (Rongotai)	79	9
Tapanui	158	16	Wellington (Thorndon)	76	9
Tara Hills (see Omarama)	133	15**	Wellington (Tinakori Hill)	77	9
Taumarunui	36	5	Westport	95	12
Taupo	31	5	Whakarewarewa (see Rotorua)	28	3
Tauranga	*	3	Whakatane	23	3
Te Aroha	17	2	Whangarei	7	1
Te Paki Te Hapua	*	1	Whatawhata	19	2
Te Wera (Forest)	41	4	Whenuapai	9	1
Thames	14	1	Wigram (see Christchurch)	113	14
The Hermitage	116	15**	Winchmore	120	14
Timaru (Airport)	129	14	Wither Hills	91	11
Timaru (City)	130	14	Woodbourne (see Blenheim)	88	11

[1] Location on map, Fig.1.
[2] Climatic area as discussed in Chapter 10. *Key*: 1 = Northland–Auckland; 2 = Waikato–Waitomo; 3 = Bay of Plenty; 4 = Taranaki; 5 = Central districts of the North Island; 6 = Gisborne, northern and central Hawkes Bay; 7 = Wanganui–Manawatu; 8 = Southern Hawkes Bay and Wairarapa; 9 = Wellington–Cook Strait; 10 = Nelson; 11 = Coastal Marlborough; 12 = West coast of the South Island; 13 = Inland Marlborough and Canterbury High Country; 14 = Canterbury Plains and North Otago; 15 = Mackenzie Country–Central Otago; 16 = Southland–East Otago.
* Basic station, named on map Fig.1 (see Table I).
** Upland climate.

Pressure and winds

The *monthly average pressure* from Kew type barometers, reduced to mean sea-level and standard gravity (but not corrected for diurnal variations) taken at 09h00 only for nine selected stations ranges from 1018.5 mbar at Tauranga in September, to 1007.6 mbar at Invercargill in November. The average for the year varies from 1016.0 mbar at Auckland to 1011.6 mbar at Taieri (Dunedin). The month of highest average pressure is September at Tauranga, New Plymouth, Nelson, Taieri and Invercargill; March at Hokitika and Christchurch; and April at Auckland and Wellington. Lowest average pressures, on the other hand, occur in January at Auckland, but in either November or December at the other eight stations.

There are, of course, appreciable day-to-day and week-to-week variations, resulting from the passage of anticyclones, troughs and depressions. The average pressures discussed above are, therefore, of only secondary importance (see KIDSON, 1931b).

Dines pressure tube anemometers are used for continuous *wind* records, most of the climatological wind data being obtained from an analysis of the anemographs. Owing to the irregular surface it may be difficult to secure a satisfactory exposure of the anemometer, and the records sometimes do not give a true representation of conditions in the immediate neighborhood.

Prevailing airflow

Data on various aspects of wind are available for 25–30 stations, mainly located in relatively open places, such as at airfields, most of them on or near the coast. Few records extend over more than 20 years.

The wind above 2,000 m over New Zealand is generally westerly, but owing to the physical nature of the country wind conditions near the surface are extremely complicated—both in direction and speed—and in many areas representative observations are difficult to obtain. For example, wind observations at Kelburn (Wellington) at 09h00, 12h00 and 15h00 for the 10-year period, 1939–1948, show that 46.6% of the wind directions are north or northwest, while 28.8% are south or southeast. This marked predominance of either northerly or southerly winds is a consequence of the situation of Wellington adjacent to Cook Strait. A similar pattern occurs at many other New Zealand stations.

Generally, however, the prevailing westerly airflow during its passage over the mountainous areas from west to east, turns northeastwards west of the ranges, and swings back towards the southeast, east of the ranges. There is, therefore, an increase in a southwesterly flow west of the ranges, and a compensating increase in northwesterlies in eastern areas, especially inland. On the eastern coasts of the South Island, however, the wind flow, in such conditions, is generally northeasterly, mainly because of the funnel effect of Cook Strait with its strong northerly flow. If the westerly flow is sufficiently strong, föhn conditions prevail in Canterbury; then the western coast experiences heavy rain and mild temperatures, while Canterbury gets high temperatures but little if any rain. KIDSON (1950) quotes the case of January and February, 1931, when a protracted period of westerly conditions brought 590 mm of rain to Hokitika and 1,655 mm to Otira, on the western side of the Southern Alps, but only 76 mm to Christchurch.

Cook Strait acts as a natural air funnel, the flow through it in westerly conditions being nearly always from the north or northwest, even though the unimpeded wind direction may be from the southwest. For the same reason, a strong southeasterly flows through Cook Strait in southerly to easterly conditions. A similar funnel effect occurs in Foveaux Strait, where the general westerly airflow is increased.

Day-to-day weather fluctuations give marked variations to the general wind flow described above, strong winds from almost any quarter being experienced in most districts at some time during the year. With high-index situations (i.e., large north–south pressure differences) the general flow is between northwest and southwest, but in low index situations (i.e., small or negative north–south pressure differences) the flow is generally from the east and southeast, or south. The broad pattern is disturbed near ground level by the presence of daytime sea breezes, which occur especially in the Bay of Plenty, Hawkes Bay, Manawatu and Tasman Bay areas, and on the eastern coast of the South Island. Sea breezes are most marked during the warmer months, and depending on the locality may extend inland for 10–30 km. These sea-breezes generally have a speed of 5–8 m/sec, but speeds of over 12 m/sec are not uncommon. Mountain and valley winds are also well marked locally, especially in the South Island.

Mean wind speed

The mean wind speed is available for just over 20 stations. The Wellington area has the highest mean wind speed, the former Moa Point station recording an average of 8 m/sec over a period of 12 years. New Plymouth averages 5.1 m/sec; Ohakea, 4.8 m/sec; and Mechanics Bay, 4.6 m/sec; all have a westerly exposure. By contrast, the inland station of Rotorua is the least windy, with an average speed of 2.0 m/sec.

On a *monthly* basis, the average speed varies from 9.0 m/sec at Moa Point in October, to 1.7 m/sec at Rotorua in July. Most stations have their highest mean wind speeds in either October, November or December, these times corresponding to the general southward movement of the pressure systems resulting normally in a high north–south pressure gradient. (Table III of Chapter 8 shows that anticyclones move more rapidly then than at any other time.) The lowest speeds occur between March and August, autumn usually having the lowest wind speeds in the north, these lighter winds being partly accounted for by the most southerly position of the anticyclones in February and March. In the south, however, winter is generally the period of lowest wind speeds, a time in which sea breezes are at their minimum, and "cold" anticyclones with light winds often form in the cold air to the rear of a cold front. However, most southern areas are subject to rapid northwest–southwest changes of airflow in winter, and strong cold southerly winds often occur for brief periods.

There is not a large difference between the mean wind speed in the windiest and least windy month, the ratio varying from 2.3 to 1 at Nelson and 1.8 to 1 at Invercargill, to 1.2 to 1 at Kaitaia and Moa Point. As to the *daily* variations, speed in the early afternoon (the time of the maximum wind for all stations) is generally twice that recorded during the early morning.

Detailed information showing wind roses of 09h00 surface wind directions at 64 stations for January, April, July and October is given in McINTOSH (1958); upper level wind data are also available (N. Z. METEOROLOGICAL SERVICE, 1962).

The mountainous and irregular terrain of many areas of New Zealand usually results in gustiness rather than high average wind speeds. Consequently some parts of New Zealand have a large number of days on which high wind gusts occur; in Wellington, for example, wind gusts at 18 m/sec or more occur on over 150 days each year.

Wind gusts in the Cook Strait area are the highest in New Zealand, Wellington City sites recording a peak of 50 m/sec from the northwest, and 55 m/sec from the south. At 491 m, however, observations at Hawkins Hill (near Wellington) show wind gusts in excess of 65 m/sec.

The occurrence of extreme surface winds in New Zealand is discussed by DE LISLE (1965) who analysed the maximum wind gusts for 33 stations. The geographical distribution of maximum gusts, with an average return period of 20 years, shows a peak of 41–45 m/sec for the Cook Strait and Foveaux Strait area together with the north of Northland. By contrast, the Bay of Plenty–Rotorua area has a similar gust probability of less than 32 m/sec.

Rainfall

Rainfall is usually measured in standard 127 mm (5 inches) gauges. Several stations are also equipped with continuous rainfall recorders.

In 1963, rainfall was recorded at about 1,300 stations; a large number of these have records of less than 20 years, but a few have records of over 100 years, including Auckland, Christchurch, Wellington and Dunedin.

The duration of rainfall was available for 19 stations, over periods of up to 28 years, and the average intensity of rainfall was based on these records. The diurnal variation of rainfall has been calculated for 16 stations, with records of from 5 to 27 years.

Average yearly rainfall

The average yearly rainfall (1921–1950) ranges from 7,094 mm (279.3 inches) at Homer Tunnel, Fiordland to 335 mm (13.2 inches) at Alexandra, Central Otago.

The map (Fig.3A) shows how remarkably complete is the *control of rainfall by relief*. Whenever a mountain range is exposed to winds from a westerly or northwesterly quarter, the annual rainfall is generally in excess of 2,500 mm, and in some areas it probably exceeds 8,000 mm. By contrast, in the lee of most mountain ranges the rainfall is comparatively low, with less than 500 mm being recorded in some areas of Central and North Otago. The main exception to the general west–east rainfall contrast occurs in the Gisborne hill country and the Coromandel Peninsula in the North Island, where the higher rainfalls for easterly situations are mainly due to the effect of orography on easterly to southerly airflows.

More detailed information on the regional patterns of mean annual rainfall is available (MAUNDER, 1966a), this publication including a two-colour map of rainfall distribution using small circles for portraying the point rainfalls at several hundred stations. An earlier publication (SEELYE, 1945), on maps of average annual rainfall in New Zealand is also available.

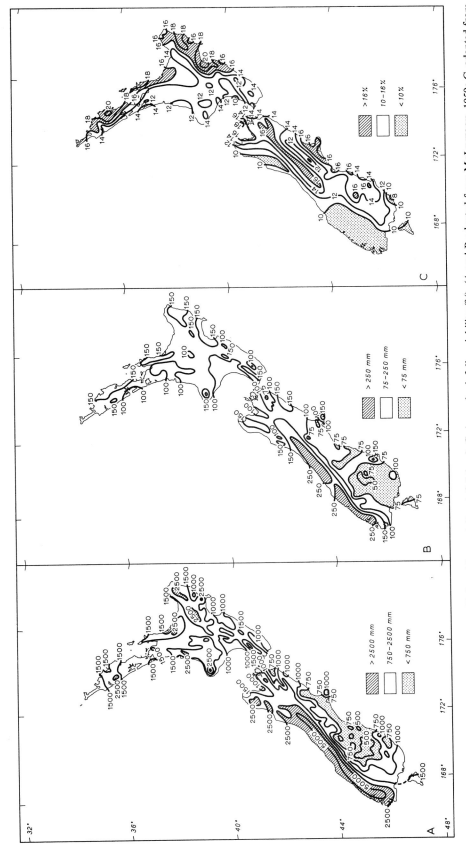

Fig.3. Rainfall. A. Mean annual rainfall (mm). B. Highest 10-year daily rainfall (mm). C. Annual rainfall variability (%). (A and B adapted from MCLINTOCK, 1959; C, adapted from SEELYE, 1940.)

Areas which are exposed to the west and southwest are subject to much changeable showery weather, and *at least 0.13 mm* (0.005 inches) of rain is recorded on about half the days of the year. At a few stations, rain occurs on more than 200 days, the 251 days at Puysegur Point in the southwest of the South Island and the 222 days at Halfmoon Bay in Stewart Island, being among the highest. By contrast, some stations east of the main ranges in the North Island average fewer than 120 rain days, and a few stations in interior locations in the South Island have measurable rain on fewer than 100 days, such stations including Lambrook (near Fairlie) with 94 rain days, Lake Waitaki in North Otago (95) and Ophir in Central Otago with an average of only 82 rain days. At most stations there is an increase in the number of rain days in the winter and spring months, although in the south of the South Island there is little seasonal variation.

Highest yearly rainfalls

The highest rainfall recorded in a calendar year at any New Zealand station is 9,670 mm (380.7 inches) (36% above normal), which occurred at Homer Tunnel in 1940. Alexandra, on the other hand, has never reached an annual rainfall of 500 mm.

At Napier, the highest year's rainfall (1,466 mm) is 85% above the average of 792 mm, while at Rotorua, the peak rainfall of 2,581 mm in 1962 is 78% above the average of 1,450 mm. These large departures from the average are a reflection of the relatively high variability of the easterly conditions which on occasion bring heavy rain to eastern North Island areas. By contrast, stations with a westerly exposure have much lower percentage departures from average; Chateau Tongariro recorded a peak rainfall of 3,518 mm which is only 22% above the average, while in the Invercargill area the highest rainfalls are about 30% above average. Three stations (out of 85 with long records) have had over 7,620 mm (300 inches) in a year—Dawson Falls on Mt. Egmont, and Milford Sound and Homer Tunnel in Fiordland. At four stations (all in Central Otago) an annual rainfall of 762 mm (30 inches) has never been reached; at Alexandra the highest annual rainfall on record is 495 mm. Milford Sound, on the other hand—less than 160 km away— has not recorded less than 4,900 mm (193 inches) in any of the 34 years to 1962. Four of the 85 stations with long records have not recorded less than 2,540 mm (Otira, Arthurs Pass, Fox Glacier, Milford Sound, all with a westerly exposure) and a further nine stations have not recorded less than 1,270 mm—the stations being Kerikeri, Waihi, Chateau Tongariro and Dawson Falls, in the North Island; and Westport, Greymouth, Hokitika, Puysegur Point and The Hermitage in the South Island.

Lowest yearly rainfalls

Lowest rainfalls of less than 380 mm (15 inches) have been recorded at 13 of the 85 stations (Cape Campbell, Lake Grassmere, Christchurch, Lincoln, Ashburton, Fairlie, Timaru, Lake Tekapo, Duntroon, Oamaru, and three stations in Central Otago). All are east of the main Alpine ranges. At six stations less than 330 mm has been recorded in a year and at Alexandra only 210 mm, 63% of normal, the lowest annual rainfall for any station in New Zealand. This record dry calendar year occurred in 1964, but an even drier 12-month period occurred at Alexandra from November 1, 1963–October, 31, 1964 when only 167 mm (6.59 inches) was recorded.

Variability of yearly rainfalls

The variability of annual rainfall (up to 1935) has been discussed by SEELYE (1940), who showed that the percentage variability for most of the country lies between 10–20% (Fig.3C), the lowest being 8.7% at Waikawa on the southern coast of Southland, the highest 21% at Petane in Hawkes Bay. Generally where the main rainfalls are derived from westerlies, the variability is low.

Highest monthly rainfalls

The highest monthly average (1921–1950) rainfall is 711 mm (28.0 inches) at Homer Tunnel in January and October, contrasting with Alexandra's highest mean monthly rainfall of 46 mm (1.8 inches) also in January.

Winter is the time of maximum rainfall for most North Island stations, June being the wettest month at 26 of the 52 representative North Island stations, and July at 14; the only exception to a winter maximum is Chateau Tongariro with its maximum in October, due to the strong westerly flow in this month. In the South Island, by contrast, the month of maximum rainfall varies considerably—with maxima occurring in each of the twelve months. Stations in the south, especially inland, have their maxima during the *summer* months, corresponding to maximum daytime heating, whereas stations in the west and north have their maxima during the *spring*—mainly in the month of October, corresponding to the peak westerly wind flow. Stations in the northeast (Lake Grassmere, Kaikoura, Hanmer, Christchurch, Akaroa) have their maxima during *late autumn and early winter*, a feature which may be associated with the minimum of westerly airflow over the South Island at that time (see KIDSON, 1931b).

Lowest monthly rainfalls

The amount of rain during the driest month varies from 427 mm (16.8 inches) at Homer Tunnel in June, to 15 mm (0.6 inches) at Alexandra in August. In the North Island and the northeast of the South Island, March or December is the month of lowest average rainfall; in western South Island areas, June or July; and in the southeast of the South Island, July, August or September are the dry months.

Monthly rainfall differences

At most stations the average rainfall of the wettest month is nearly twice the average rainfall of the driest month. At Alexandra in Central Otago, however, the ratio is 3.0 to 1, and at Gisborne, 2.5 to 1, while at Nelson and Hokitika it is 1.4 to 1. At five out of 85 representative stations rainfall during a calendar month has exceeded 1,270 mm (50 inches); the stations include Fox Glacier (1,288 mm), The Hermitage (1,290 mm), Otira (1,504 mm) and Milford Sound (1,755 mm). At the Homer Tunnel in Fiordland a rainfall of 1,847 mm (representing 26% of the average annual rainfall) was recorded in the 29 days of February, 1940, giving an average of 64 mm/day. Maps of extreme monthly rainfall are available (SEELYE, 1946a) for all 12 months. Several stations sheltered from westerly rains have not recorded 203 mm (8 inches) or more in any one month; they include Blenheim and Molesworth in Marlborough, and Hawea Flat, Clyde and Alexandra in Central Otago. In addition, at 21 of the 85 stations there has been at least one month with no rain. These stations include Auckland, Waihi, East Cape, Napier, Taupo, Masterton, New Plymouth and Wellington, all in the North Island; and Nelson, Cape Campbell, Arthurs Pass, Lake Coleridge, Lake Tekapo, Waipiata, and Oamaru, in the South Island. These stations have average annual rainfalls ranging from 447 mm at Waipiata to nearly 4,000 mm at Arthurs Pass; hence, the occurrence of a completely rainless month is possible at places having widely varying average rainfalls.

Variability of monthly rainfalls

The variability of monthly rainfall has been analysed by SEELYE (1946b). From monthly records for the period 1911–1940 at 91 stations (47 in the North Island and 44 in the South Island), the general average variability for all places is 44%, from 77% at Waimarama (Hawkes Bay) in January to 18% at Puysegur Point in May. SEELYE (1946b) found that for New Zealand as a whole, February was the month of most variable rainfall (although stations in eastern-central areas of the South Island have their greatest variability in July), whereas September has the lowest variability. The highest variability occurs from December-March in the northeastern areas of both islands together with Hawkes Bay and Nelson, from April–June on the eastern coast south of Gisborne, and from July–October on the eastern coast of the South Island. Low variability, on the other hand, is a very persistent feature in the area south of Westport and south of Nugget Point in South Otago.

SEELYE (1946) also analysed the "runs" of months, all having rainfall departures either above or below normal. For the 30 year period, 1911–1940, he found that for an average locality, a sequence of negative departures beyond six months occurred about four times; whereas, a rainfall above normal for more than six consecutive months was a rare event, occurring less than once. The persistence of daily rainfall has been investigated by FINKELSTEIN (1967).

Highest daily rainfalls

The month of occurrence of the highest daily rainfall varies, although autumn (March, April, May) is the time of the maximum at many stations.

The frequency of heavy daily rainfalls at selected stations in New Zealand has been analysed by SEELYE (1950a), who found that once a year a daily rainfall of 51 mm (2 inches) is likely in most of the North Island low country, while in the east of the South Island, by contrast, amounts are generally under 51 mm for the majority of places; he also calculated that a daily fall of over 254 mm (10 inches) could be expected once in 10 years at Waihi, in the North Island, and at stations in the southwest of the South Island, whereas falls of less than 75 mm were calculated (on the basis of one fall in 10 years), for most stations in Manawatu, Central Otago and eastern Southland (Fig.3B). The highest daily fall recorded in New Zealand actually occurred at the Public Works Camp, Milford Sound, on April 17, 1939, when 559 mm (22 inches) was recorded. Further details of high intensity rainfalls are available in a Meteorological Service publication (ROBERTSON, 1963).

The pattern of the highest daily rainfalls shows that stations with high average annual rainfalls also have high daily rainfalls, most places adjacent to western facing mountain ranges having experienced daily rainfalls in excess of 400 mm. By contrast, at most areas east of the ranges in the South Island, a daily rainfall of more than 100 mm is unusual. These areas include Central Otago, which unlike many low-rainfall areas of the world has very few vigorous thunderstorms with intense rains (BROWNE, 1959).

Duration of rainfall

The duration of rainfall is available for 18 stations, none of which is in inland areas of the South Island; it is taken as the total time during which rain falls at a rate of at least 0.10 mm/h (0.004 inches/h) and varies from 405 h (4.6% of the time) at Ohakea to 995 h (11% of the time) at Hokitika. At Hokitika, in October, rain occurs 13.6% of the time (3.3 h/day), compared with 9.8% (2.4 h/day) in January and March. By contrast at Ohakea during December (the "wettest" month at that station), the duration of rain is only 5.9%. The most frequent month with the highest rainfall duration is July, followed by May, June, and August.

In January, Auckland has a rainfall duration of 19 h (representing rain 2.6% of total time, or 0.6 h/day), the lowest record of 18 stations. Hokitika, by contrast, records rain in January and March for 73 h (9.8% of the time), or 2.4 h/day. Except for Hokitika, most stations have rain for a total of about one hour per day during the "least rainy" month. March is the month of lowest rainfall duration at 10 of the 18 stations, and only at Dunedin and Invercargill, where the lowest duration occurs in late winter and early spring, does the month of lowest duration occur outside the "summer" period—November–March.

Diurnal rainfall regimes

For a small country in the middle latitudes, New Zealand shows a considerable variation in the diurnal rainfall regimes from one part of the country to another. The north generally has a maximum during the afternoons (convective activity); the west and south of the South Island and Cook Strait have a rainfall maximum during the night (usually in the early morning); at eastern stations the time of the maximum is variable. Because of the physical nature of New Zealand, the pattern is far from simple when considered in any detail (MAUNDER, 1956, 1957). The minimum rainfall generally occurs 12–18 h before the maximum.

Rainfall intensities

Stations in the north and west usually have a higher intensity[1] than stations in the east and south; the highest for the year being at Auckland (Mechanics Bay) 3.0 mm/h, Hokitika, 2.8 mm/h, and Tauranga, 2.6 mm/h. The lowest are at Taieri (near Dunedin), 1.5 mm/h, Blenheim, 1.5 mm/h, and Christchurch, 1.3 mm/h. The pattern is generally a reflection of orographic and convective influences. Hokitika has a relatively high

---

[1] Calculated by dividing the total rainfall by the duration.

intensity due to its location on the windward side of the Southern Alps, and Auckland and Tauranga have similar high intensities due in the main to thermal influences. The stations with the lowest intensities—Blenheim, Christchurch and Taieri—are near the eastern coast of the South Island, subject to little orographic rain, and apart from cold frontal passages, to little convective rain. In these areas light rain and drizzle (frequently from easterly–southerly conditions) make up a large proportion of the time rain is falling.

The month of highest average intensity varies from September to March, 12 of the 19 stations having their highest average intensity in one of the two warmest months—January or February—when the thermal activity is at its highest. Auckland has the highest average intensity for any month of 4.7 mm/h in February, followed by Ruakura with 4.5 mm/h also in February. Christchurch has no month with a higher average intensity than the 1.7 mm/h occurring in November.

The month with the lowest average intensity occurs between May and August at 15 of the 19 stations, varying from 2.3 mm/h in July at Hokitika, and 2.3 mm/h at Tauranga in August, to 1.0 mm/h at Taieri in July, and 0.9 mm/h at Christchurch, also in July. The lowest intensities thus occur in the main in the colder months when thermal activity is at its lowest, and when light rain and drizzle are frequent.

Rainfall trends

The secular trend of rainfall from 1863–1947 was analysed by SEELYE (1950b). He summarised into an annual index the rainfall throughout New Zealand, the average rainfall in the period 1911–1940 being taken as 100. The fluctuations in the index from year to year were resolved into two periods, a longer one, which was correlated with the 11 year sunspot cycle, and a shorter more irregular period averaging about three years. The indices for the period 1863–1947 showed (SEELYE, 1950b) a declining tendency at an average rate of 4% per 100 years in the North Island, and 2% per 100 years in the South Island. However, a readjustment of the method of evaluating the indices prior to 1890, increased these declining trends to 7% and 3% per 100 years.

However, DE LISLE (1956) in analysing secular trends of west coast (South Island) rainfall, showed that there was no evidence of long-period (about 60 years) secular trends in seasonal rainfall. *Shorter period changes* in spring (a general decrease from about 1912) were found to be significant, and some variations in winter and summer were also found to be significant, but none in autumn. De Lisle suggested that the rainfall variations in winter, spring and summer depended on variations in the low level westerly winds, and that the spring variations (a general decrease from about 1912), implied a general poleward (southward) displacement of the spring position of the subtropical anticyclone tracks over the last 40 years. Further aspects of the fluctuations of the seasonal pressure patterns in the vicinity of New Zealand are discussed by DE LISLE (1957), who concluded that the spring and winter rainfall variations on the western coast of the South Island were associated with an irregular southward movement of the mean subtropical anticyclone and of the mean westerly wind belt from about 1918 to the mid-1930's.

A filter analysis of long term rainfall variations has been made by DE LISLE (1961) who states that fluctuations of periods of over 12 years in three northern areas were found to correlate with one another, but no relations were found between any of the other areas.

An analysis of variations in monthly and seasonal rainfall at Dunedin from 1913 to 1961 has also been made (MAUNDER, 1962b).

Droughts

Some general information on the occurrence of absolute droughts ("a period of at least 15 consecutive days in which there is no measurable rain"), and partial droughts ("a period of at least 29 consecutive days during which the mean daily rainfall does not exceed 0.25 mm per day") in New Zealand is given by BONDY (1950), who examined 46 stations with records varying from 29 to 83 years. Major droughts include 59 consecutive days at Motueka without rain in the summer 1907–1908; 64 consecutive days without rain at East Cape in the summer 1927–1928; and 53 consecutive days without rain at Clyde in the late summer and autumn of 1898. A partial drought at Clyde in 1937 had only 43 mm of rain in 176 consecutive days, and at "Marshlands" Blenheim, only 34 mm was recorded in 150 consecutive days. In addition, 1964 brought record dry conditions to parts of North and Central Otago, Alexandra recording only 30 mm of rain in the 5 months June–October 1964. "Dry years" in New Zealand are discussed by KIDSON (1931c). BONDY (1950) divided New Zealand into three areas of *drought seasonality*. In the north, central and southwest of the North Island, and in the northeast of the South Island, February is the most drought-prone month (which as shown in Chapter 8, Table III is the month in which the anticyclones reach their most southerly position); in the east of the North Island, December or January has the greatest number of drought days; whereas in central and southern Canterbury and Otago, winter is the most drought-prone period.

Between 10 and 30% of the total number of absolute droughts lasted longer than three weeks, high frequencies being recorded at Napier (30%) and Clyde (31%), and small frequencies in areas well exposed to westerly and southwesterly conditions, such as the south and west of the South Island, Invercargill having no absolute drought longer than three weeks in 51 years of record. At Mount Vernon station in southern Hawkes Bay, 13% of the absolute droughts lasted more than four weeks, and 6% persisted beyond five weeks, a six week drought also having been recorded in the Nelson, North Otago and Central Otago areas of the South Island, all of which are sheltered from westerly and southwesterly flows.

The average number of absolute droughts per 10 years varies from 30.0 at Clyde in Central Otago and 20.0 at Napier in Hawkes Bay to the lee of the main mountain ranges, to 3.4 at Dunedin and only 2.7 at Invercargill, these latter two stations being subject to disturbed westerly or southwesterly conditions much of the time.

Partial droughts (BONDY, 1950) are comparatively frequent in portions of the Hawkes Bay, Wairarapa, Nelson, Marlborough, South Canterbury, North Otago, and Central Otago areas (all east of the main ranges). Clyde in Central Otago recorded 99 partial droughts in 53 years of record, 22% of these droughts being over nine weeks and 7%, over three months.

The occurrence of agricultural drought has also been investigated, RICKARD (1960a) giving details of agricultural droughts at Ashburton (central Canterbury) during the previous 44 seasons.

Water vapour and cloudiness

The average annual *water vapour pressure* at 09h00 for 13 stations varies from over 13 mbar in the north to less than 8 mbar at Lake Tekapo in the inland Mackenzie Country Basin. There is a general north–south decrease, as well as a decrease with increasing altitude. For coastal locations, the annual variation is from 13.5 mbar at Auckland, to 9.7 mbar at Dunedin. The late summer month of February has the highest vapour pressure at all 13 stations (except Invercargill where it occurs in January), while the mid-winter month of July is the month of lowest vapour pressure at all 13 stations. The difference between the February and July vapour pressures varies from 7.4 mbar at Hokitika, to 5.0 mbar at Chateau Tongariro, the three inland stations (Chateau Tongariro, Lake Tekapo, Alexandra) having the smallest range, except for a 5.2 mbar difference at Dunedin in coastal Otago.

The highest monthly vapour pressure varies from 17 mbar at Auckland to 10 mbar at Lake Tekapo, these two stations also having the highest and lowest vapour pressures in July (10.6 mbar and 4.9 mbar).

The average *cloudiness* at 09h00 for thirteen representative stations varies from 5.9 oktas (74%) at Invercargill, to 4.1 oktas (51%) at Napier. The exposed western stations of Wellington, Hokitika, and Invercargill record an average cloud cover of 70% or more, Invercargill with 74% being cloudier than Hokitika, although the average annual rainfall at Hokitika is more than double the average annual rainfall at Invercargill.

Solar radiation and sunshine

Solar radiation

Solar radiation is measured at four stations by means of pyrheliometers (DE LISLE, 1966). The solar radiation measured is the total incoming radiation (sun and sky) received on a horizontal surface. During 1954–1965, the mean varied from 365 Ly/day at Ohakea to 304 Ly/day at Invercargill. In January, Ohakea and Wellington both have a higher average than Whenuapai, which in turn is higher than Christchurch and Invercargill, the Ohakea (585 Ly/day)—Invercargill (509 Ly/day) ratio being 1.15 to 1. In mid-winter, however, Invercargill has an average daily incoming radiation of 88 Ly, which is only 51% of that received at Whenuapai.

The ratio between the winter and summer incoming radiation decreases from north to south. At Whenuapai, the average July incoming radiation is about one-third of the January average, this proportion decreasing to one-fourth at Ohakea, under one-fourth at Wellington, and one-sixth at Invercargill. The highest mean daily total (in the period 1954–1959) occurred at Ohakea in 1957 with a January average of 720 Ly/day in 323 h of sunshine, the lowest at Invercargill in June, 1955 (74 Ly/day).

The mean daily insolation in New Zealand has been mapped by DE LISLE (1966) from an analysis of measured radiation, and sunshine data; the maps show that in spring and summer, the latitudinal gradient of monthly mean radiation is small, whereas in autumn and winter differences of up to 100 Ly/day in mean monthly totals occur.

Sunshine duration

The duration of bright sunshine is recorded at over 60 stations by Campbell-Stokes recorders. Owing to topographical features, many stations are not able to record all the possible sunshine; Waipoua, Wallaceville, The Hermitage, Waimate and Queenstown are most affected by neighbouring hills.

The average (1935–1960) annual duration (Fig. 4A) varies from 2,433 h at Blenheim to 1,522 h at The Hermitage (Mt. Cook). The duration at Blenheim represents an average of 6.7 h/day, or 57% of the possible duration of sunshine. Nelson averages 6.6 h/day (57%), whereas Invercargill records 4.5 h (38%) and The Hermitage only 4.2 h/day (36%). At The Hermitage, however, some sunshine is intercepted by nearby mountain ranges. January is usually the most sunny month (10–12% of the yearly total) and June the least sunny (4–6% of the yearly total).

The relatively marked difference in the average sunshine duration between areas sheltered from the southwest such as Nelson, Blenheim and Tauranga, and areas which are exposed to these influences such as coastal Southland and Otago, is a major feature of the climate of New Zealand; another feature is the relatively small difference between the sunshine recorded on the western and eastern coasts of the South Island. Indeed, some areas of the western coast of the South Island (Westport, 1,961 h; Hokitika, 1,855 h) have more sunshine than coastal Otago and Southland (Dunedin, 1,734 h; Invercargill, 1,661 h), despite the fact that the rainfall on the western coast is two or three times that of coastal Otago and Southland. The average annual sunshine duration at Westport (1,961 h) is also only a little less than that recorded at such localities as Wellington (2,012 h), Christchurch (1,990 h), and Hamilton (1,982 h). Even Alexandra (2,081 h), which has the lowest annual average rainfall of any New Zealand station, is not much sunnier than Westport.

Temperature

In 1963, screen temperatures (in Stevenson screens, 1.22 m above the ground) were being recorded at 170 climatological stations, 161 of which had records of five or more years up to 1960. Several stations have kept records over 50 years, and temperatures have been recorded in Wellington and Auckland for over 100 years.

There are many areas for which representative temperatures are difficult if not impossible to obtain, because of the irregular nature of the terrain. The temperature data for some stations may, therefore, reflect site and location anomalies, and may not be representative of the surrounding area. Such anomalies—which are important locally—come particularly to notice when comparing the temperature records of a locality which has had recordings taken at different sites, e.g., the Cawthron Institute and Airport sites at Nelson, and the Botanical Gardens, Beta Street and Musselburgh sites at Dunedin.

Extreme temperatures

The *highest screen temperature* recorded in New Zealand (up to December 1966) is the 38.4°C (101.2°F) at Ashburton on the eastern side of central South Island on January 19, 1956, during northwesterly föhn conditions. On January 22, in the same year,

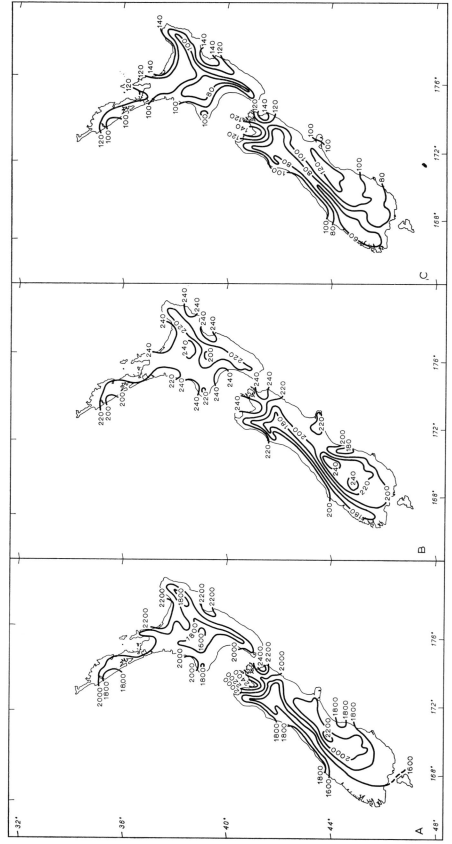

Fig.4. Sunshine (h). A. Average annual duration. B. Average January duration. C. Average June duration. (Compiled from New Zealand Meteorological Service data.)

248

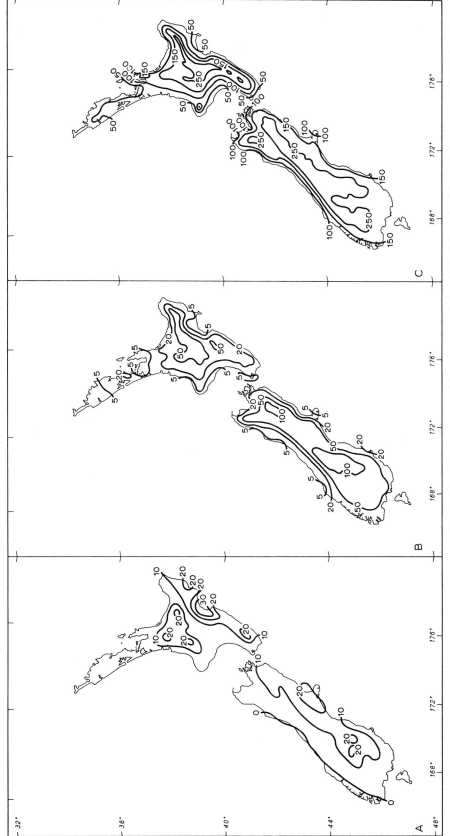

Fig. 5. Temperature aspects. A. Days per year with temperature above 27°C. B. Days per year with temperature below 0°C. C. Length of (screen) frost season (days). (A, adapted from COULTER, 1963; B, compiled from New Zealand Meteorological Service data; C, adapted from McLINTOCK, 1959.)

249

Gisborne (on the eastern side of the North Island) recorded a temperature of 37.7°C (99.8°F) under similar conditions, this temperature being the highest recorded in the North Island. By contrast, Jacksons Bay in South Westland (period 1938–1949) did not record a temperature higher than 25.5°C, and at Makara (on the western side of the North Island, adjacent to Cook Strait) the highest temperature in the period 1955–1966 was only 25°C. In general Canterbury and Central Otago areas have the highest extremes. The *lowest screen temperature* recorded in New Zealand (up to December, 1966) was −19.7°C (−3.5°F), at Ophir in Central Otago on July 2, 1943. By contrast at Cape Reinga in the period 1950–1960, the temperature did not fall below 2°C. At two other stations, Kaitaia and Mechanics Bay (Auckland), the temperature has not fallen below 0.5°C, while at Auckland City's various sites in the 99 year period 1868–1966 the lowest minimum recorded is −0.1°C (31.9°F). The lowest minimum recorded in the North Island is −13.5°C at the Chateau Tongariro (at 1,119 m), while the highest extreme minimum in the South Island has been −1.5°C at Greymouth and −2°C at Westport, on the western coast, and −2°C Beta Street (Dunedin).

The *extreme range* of temperature varies from 25°C (45°F) at Cape Reinga and Makara, to 55°C (99°F) at Ophir, and 53°C at Manorburn Dam. All but one (Queenstown) of the ten Central Otago stations have recorded an extreme range of 44.5°C or more.

Annual extreme temperatures

The *mean annual maximum* temperature varies from 34.5°C (94°F) at Pendarves and 33.5°C at Christchurch Airport and Roxburgh Hydro to 24.5°C at Hokitika and Hokitika South, 24°C at Makara and 23.5°C (74°F) at Chateau Tongariro and Jacksons Bay. The main feature of the mean annual maximum temperature variations is the relatively low maxima over all the North Island (except for eastern districts) and on the western coast of the South Island, all areas west of the main mountain ranges and subject to southwesterly conditions. These contrast with high maxima of about 32°C in practically all eastern areas to the lee of the main mountain ranges, which are subject to föhn conditions. The number of days each year with a temperature of over 27°C (80°F) is shown in Fig.5A which shows the importance of west–east rather than north–south differences. The *mean annual minimum* temperature varies from 5°C at Cape Reinga, to −14.5°C (6°F) at Manorburn Dam. At 11 of the 161 stations with records of 5 or more years, the average annual minimum temperature is at or above the freezing point (0°C); six of these 11 stations are in Northland and Auckland (most northern area of New Zealand and well exposed to oceanic influences) and four in the southwestern area of the North Island (subject to persistent winds). Cape Reinga and Ophir have the smallest and largest *mean annual range* of temperature—20.5°C at Cape Reinga (4.5°C less than the extreme range) and 42°C at Ophir (13°C less than the extreme range).

The distribution of the mean annual range based on data from 80 stations with records of over 10 years (ROBERTSON, 1952) shows a very marked difference between the western and eastern coasts (south of East Cape). This is partly accounted for by the greater cloudiness of the western coast, another important factor being the hot, dry föhn wind which occurs at times east of the main ranges, and which produces temperatures 11°–14°C higher than on the western coast.

Mean daily temperatures

The *mean daily maximum* temperature for the year at 161 stations has a variation of 9°C, ranging from 20°C at Kerikeri to 11°C at Chateau Tongariro; both stations are in the North Island, the latter at almost 1,200 m. In the South Island, the range is from 18.5°C at Riwaka (near Motueka), Blenheim and Woodbourne, in the north of the South Island to 11.5°C at Manorburn Dam in Central Otago.

The *mean daily minimum* temperature at 161 stations varies from 13°C to 1°C (a difference of 12°C compared with a difference of 9°C in the average daily maximum). The highest are 13°C at Cape Reinga and 12°C at Mechanics Bay compared with 1.5°C at Naseby Forest and 1°C at Manorburn Dam. No station south of New Plymouth records an average daily minimum of more than 9.5°C. The 17 stations that record an average daily minimum of less than 4.5°C include Chateau Tongariro and Karioi in the North Island, all three stations in the Mackenzie Country, over half the stations in Central Otago and four stations in South Otago–inland Southland.

Of the 161 stations in New Zealand, 88 have a *mean daily range* of 9–10.5°C (16–19°F), 37 of 8.5°C or less and 51 of 11°C or more. The smallest daily range, 5°C, occurs at Cape Reinga and Makara; Wellington records 6°C. The smallest daily range in the South Island is 6.5°C at Greymouth. A daily range of 12°C or more occurs at 15 stations, from Arapuni in the Waikato to Garston in inland Southland. The largest average daily range, 13.5°C, occurs at Kaingaroa, Wairapukao, Fairlie and Earnscleugh.

The *mean annual temperature* reduced to sea level is shown in Fig.6A. Mean temperatures are discussed by KIDSON (1931d).

January temperatures

The *mean daily maximum* in January for 161 stations varies from 26°C at Kawerau to 16.5°C at Chateau Tongariro (Fig.7A)—both in the North Island. The highest in the South Island, however, is only 1.5°C less than that for Kawerau and the lowest only 1°C more than that at the Chateau Tongariro.

The pattern of *mean daily minima* in January (Fig.7B) generally follows the pattern of mean daily maxima—with higher minimum temperatures in the north and lower ones in the south and in inland areas. At Cape Reinga and Mechanics Bay (Auckland), the average daily minimum is 16°C, the highest in New Zealand. In the south 12 stations record an average daily minimum of 8.5°C or less, with Naseby Forest recording the lowest in New Zealand, 6°C.

The *mean daily range* of temperature in January (Fig.7C) is 1.5°–3°C more than the mean daily range for the year. Cape Reinga with a range of only 5°C has the smallest diurnal range, compared with 14.5°C at Waiotapu, Wairapukao, Minginui (in the central area of the North Island), and Fairlie, Naseby Forest and Earnscleugh (in inland areas of the South Island). The highest range of 15°C occurs at Tara Hills (Omarama) in the Mackenzie Country. This range may be compared with the highest daily range over the year of 13.5°C.

The mean *January temperature reduced to sea-level* is shown in Fig.6B (see also KIDSON, 1931d).

252

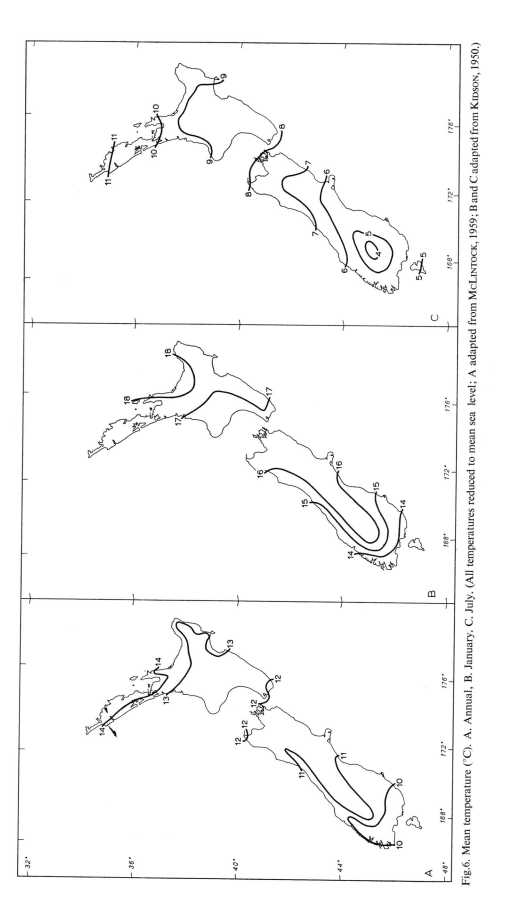

Fig. 6. Mean temperature (°C). A. Annual. B. January. C. July. (All temperatures reduced to mean sea level; A adapted from MCLINTOCK, 1959; B and C adapted from KIDSON, 1950.)

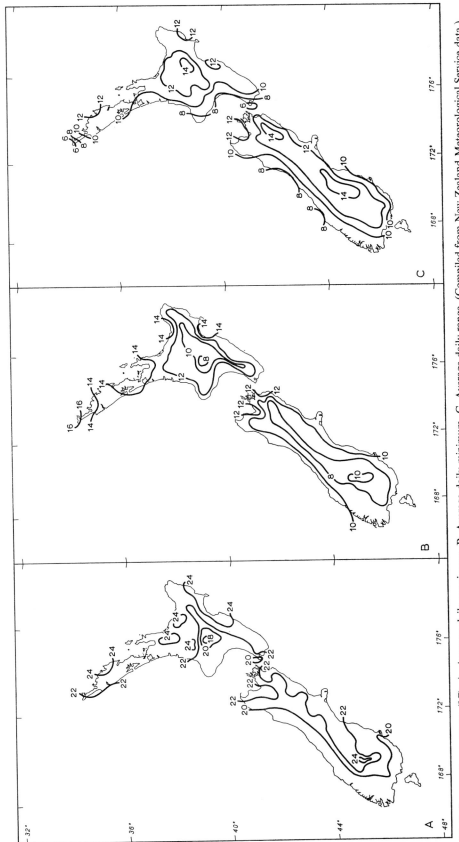

Fig. 7. January temperature (°C). A. Average daily maximum. B. Average daily minimum. C. Average daily range. (Compiled from New Zealand Meteorological Service data.)

253

July temperatures

Mean daily maxima in July (Fig.8A) vary from 15.5°C at three stations in Northland (Te Paki Te Hapua, Waipapakauri, Kerikeri) to 3.5°C at Manorburn Dam in Central Otago. In the South Island, Blenheim has the highest mean daily maximum (13°C), exceeded, however, at 37 stations in the northern half of the North Island.

Fourteen New Zealand stations have a July *mean daily minimum* of 6°C or more (Fig. 8B) and 39 stations, 4.5°C (40°F) or more (the 39 stations include all 19 stations in Northland and Auckland). The highest mean daily minimum, 9.5°C, is recorded in the far north at Cape Reinga, the lowest (5.5°C) at Manorburn Dam.

July *mean daily range* (Fig.8C) varies from 4°C at Makara and 5°C at Wellington to 11°C or more at 14 stations, including 12°C at Wairapukao, Wakefield, Fairlie and Timaru Airfield. At 26 stations the July mean daily range is 7°C or less. The mean July temperature reduced to sea level is shown in Fig.6C (see also KIDSON, 1931d).

Temperatures and rainfall

The temperatures associated with rainfall in New Zealand during the period 1931–1940 have been examined by BONDY and SEELYE (1947). As southerlies are the most effective rain-bearing winds in districts east of the main mountain ranges, temperatures in eastern districts are rather low during rain, with winter temperatures in the east being 1°C colder than in the west in wet weather. In summer, parts of the eastern coast have daily maximum temperatures in rain up to 3°C cooler than their normal level, the difference from normal being about half this amount in winter. Wet days at Christchurch and Dunedin in summer also have average minimum temperatures 1°C cooler than usual, although in the Waikato in summer, rain "increases" the minimum temperature by 2°C. In winter, the minima for wet days in the Bay of Plenty, Taupo, Nelson, Marlborough and Central Otago are at least 3°C above the normal level, this being a result of the reduction in radiation cooling by the associated cloud cover, but the "mildness" of wet winter nights is much less in Wellington, and in coastal Otago and Southland.

Temperatures associated with rainfall show the greatest contrast across the Southern Alps, the daily maxima for wet days at Christchurch for the decade 1931–1940 being 1.7°C colder than at Hokitika, and the daily minima 2.5°C colder. The temperatures on wet days at Hokitika between March–September are comparatively warm, whereas at Christchurch, except possibly in August, wet days are comparatively cold (BONDY and SEELYE, 1947).

Ground frosts

Grass minimum temperatures are recorded by means of a thermometer exposed horizontally 25 mm (1 inch) above a level grass surface, a "ground frost" occurring when the grass minimum temperature is −0.9°C (30.3°F) or lower.

The *extreme grass minimum* temperatures for 113 stations for various periods up to 1960, show a considerable variation, from −4.5°C at Albert Park (Auckland), Wairoa and New Plymouth to −23.6°C at Lake Tekapo, the most severe ground frost recorded. At 27 of the 113 stations the grass minimum temperature in January has not been less

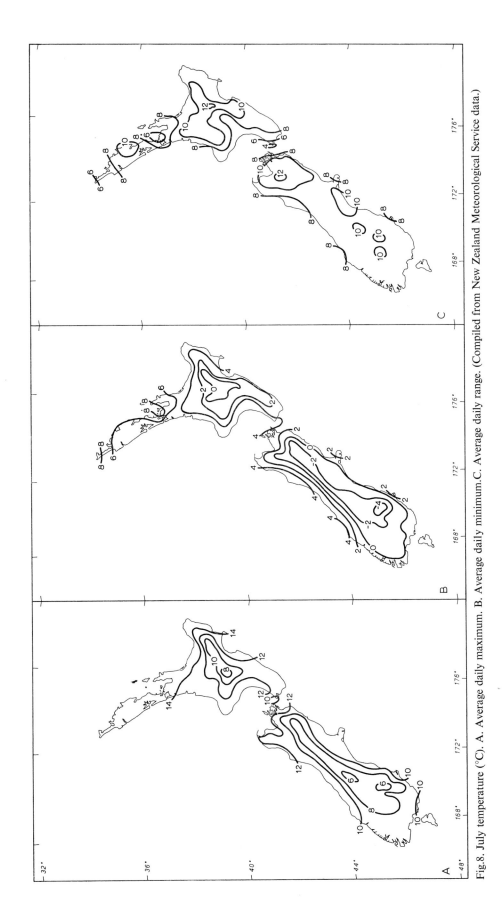

Fig.8. July temperature (°C). A. Average daily maximum. B. Average daily minimum. C. Average daily range. (Compiled from New Zealand Meteorological Service data.)

than 0°C. Of these, 20 are in the North Island (mostly coastal) and seven in the South Island—two in Nelson, two on the western coast and three in Canterbury. The pattern of extreme minimum grass temperatures in winter is similar to that for the year; the yearly extreme usually occurs in July, June or August.

There is a considerable variation in the annual *frequency of ground frosts*, from an average of 3 days a year at Kaitaia and Albert Park (Auckland) to 204 at Lake Tekapo and 218 at Molesworth (the average at the new site at Lake Tekapo in the period 1950–1960 is, however, only 154 days). At 30 (of 131) stations, ground frosts occur on less than 10% of days, these stations including only one (Greymouth) in the South Island, but 18 (out of 25) in the Northland, Auckland and Waikato areas.

There is also considerable local variation: Nelson Airfield records an average of 99 days compared with 61 days at Cawthron Institute in Nelson; while at Dunedin, the (present) Musselburgh site records ground frosts on 94 days, compared with just over half this number (48) at the former Botanical Gardens site. In both examples the stations are within 8 km of each other.

Screen frosts

Records of the average number of days with screen frost are available for 137 stations— 78 in the North Island and 59 in the South Island; Christchurch has the longest record, over 60 years. The frequency varies from 0 at Kaitaia, Mechanics Bay (Auckland) and Albert Park (Auckland)—coastal stations in the north of the North Island—and in the exposed area around Cook Strait, to 130 at Naseby Forest and 138 at Manorburn Dam, both in the Central Otago basin country (Fig.5B).

At 37 stations (31 in the North Island), screen frosts occur on less than 10 days a year. However, screen frosts occur on the average on at least one day in five at 22 stations, and on at least one day in four at 10 stations. The length of the "frost-liable" season is shown in Fig.5C.

Evaporation and the water balance

Three types of evaporimeter are currently used, but all but two stations have either a New Zealand-type sunken-pan evaporimeter or a raised-pan type.

FINKELSTEIN (1961) investigated the various measurements from over 30 stations, located mainly in the lower rainfall areas, and interpreted these measurements to obtain the "open water" evaporation for 39 stations. The estimated average annual *open water evaporation* of the stations varies from 500 mm at Lake Mahinerangi in Otago to 1,100 mm at Lake Grassmere in Marlborough.

Evaporation

Maximum evaporation is usually recorded in January, although it occurs in December at a few stations; lowest values are reported in the mid-winter months of June–July. January evaporation totals are mainly at least five times the July totals, and about 75% of the annual total occurs during the six warm months October–March, whereas less than 15% occurs in the four months May–August (FINKELSTEIN, 1961).

The highest value of 1,100 mm at Lake Grassmere (over twice the rainfall) is associated with the high wind speeds adjacent to Cook Strait, combined with high sunshine hours. In the South Island values of 890–965 mm occur in parts of the high country and upper plains of Canterbury and northern Otago, and in the Wairau Valley. In Central Otago (Alexandra) evaporation is over twice the rainfall, whereas in western areas of the South Island and in the North Island, the evaporation is generally lower than the rainfall. The annual evaporation at Hokitika is only 19% of the rainfall, and at The Hermitage (Mount Cook), also with a high average annual rainfall, only 17%. In January, the open water evaporation usually exceeds the rainfall and varies from 94 mm at Waerenga-okuri and Onepoto, to 170 mm at Omarama and 173 mm at Lake Grassmere. In June, by contrast, the evaporation is less than the rainfall at all stations.

Water need

The comparison of water need (allowing for soil moisture supply) and rainfall has been estimated for a number of stations in New Zealand by GARNIER (1951), using THORN-THWAITE's (1948) method.[1] He recognised 4 moisture deficiency areas (Fig.9A).

Most of the North Island—except for small areas near Gisborne, Napier, Palmerston North and Masterton—and more than half of the South Island (mainly in the west) have no significant moisture deficiency.

Areas in which an appreciable amount of soil moisture is used up during the warmer months, but in which the rainfall deficit is usually not sufficient to cause an actual moisture deficiency have a moisture problem which, according to GARNIER (1951) is not usually serious, but could cause difficulty in unfavourable years. Fairlie is a typical station in this category (Fig.9B).

In another group are stations which have a winter surplus and a summer deficiency of moisture; these can normally expect to have a period in which soil-moisture reserves are entirely used up. They include a small area around Napier, a larger area about and southeast of Blenheim, the coastal Canterbury Plains, the Mackenzie Country and much of North Otago and bordering Central Otago.

At Hastings (Fig.9B), for example, there is a short period of moisture surplus in July–September, followed by a time of soil-moisture usage. GARNIER (1951) indicates that the moisture reserve is normally used up by the end of December and then a deficiency occurs which lasts until late March. At Christchurch a surplus occurs from July–September and a deficiency from February–April.

There are also parts of New Zealand in which a period of moisture surplus does not normally occur. The driest part is in Central Otago, with a second area where conditions of moisture deficiency are less severe in the vicinity of Waimate and Timaru. The actual period of deficiency at Timaru is February–April; at Waimate, March–April and at Alexandra, it is October–April (GARNIER, 1951). During the other months of the year there are periods of soil-moisture usage and soil-moisture storage.

The water balance regimes of New Zealand have also been studied by CRITCHFIELD (1966) who assessed in particular the degree of analogy between the climatic water balances of New Zealand and the central Pacific coast of North America.

[1] Reference should also be made to the important studies of GABITES (1956), RICKARD (1960a, b, 1961) and COULTER (1966).

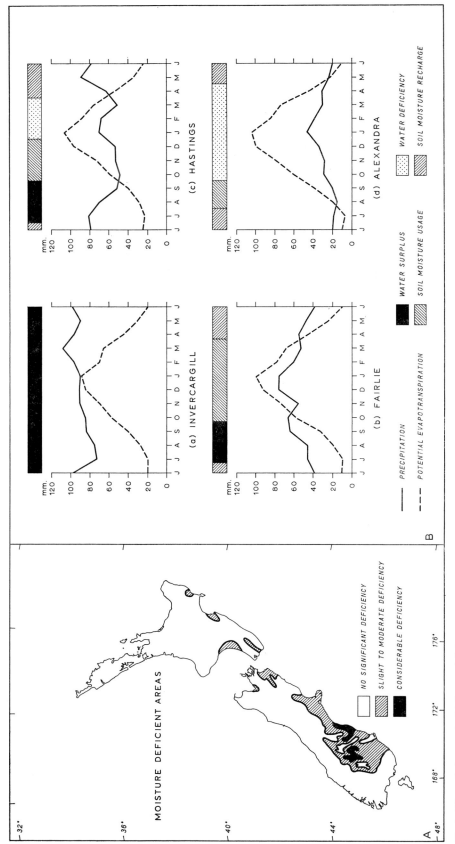

Fig. 9. Water balance. A. Moisture deficient areas (adapted from GARNIER, 1951). B. Water balance diagrams (compiled from data in GARNIER, 1951, 1958).

258

Special phenomena

Thunderstorms

Thunderstorms are usually not very numerous. In general the frequency increases from east to west and with altitude. The greatest frequency occurs on the western coast of the South Island, with 15–25 thunder days each year and in parts of Northland with about 15–20 days each year. Average values at some stations include 16.5 days at Hokitika, 11.8 at New Plymouth and 9.3 at Invercargill, while Christchurch averages 3.5 days, Dunedin 3.3 and Alexandra, 3.0. In the north and west of the North Island the frequency is highest in winter, whereas in the east of New Zealand, the frequency is usually highest during the summer. The majority of thunderstorms occurs in the afternoon and evening in association with cold fronts or northwesterly orographic conditions, though some, especially in inland areas, are associated with convective activity (KIDSON and THOMSON, 1931).

Hail

The frequency of hail generally increases from north to south, with a pronounced increase in frequency in areas exposed to southwesterly conditions. It normally occurs in association with cold fronts or instability showers in disturbed southwesterly conditions, especially in winter and spring. Severe local hail storms occur on occasion in many areas and although the stones very seldom exceed 2 cm in diameter, KIDSON (1932) states that some as large as 7 cm in diameter have been reported.

Snow

In the North Island there is a small permanent snowfield above about 2,500 m on Mount Ruapehu (2,797 m), and in the South Island the permanent snowline is at about 2,000 m. During the winter, however, the snow line usually descends in the North Island to about 1,000 m and in the South Island to about 800 m. Snow is occasionally experienced at lower altitudes and may lie on the ground for several days in many eastern areas of the South Island, except those immediately adjacent to the coast.
Snowfalls are most frequent in July and August and rare in January and February, but may occur at any time, especially in inland areas of the South Island, although snow does not lie for long except in the colder months.
The annual number of days with snow is 17.2 at Chateau Tongariro (9.6 in June, July, August), 11.8 at Lake Tekapo, 9.7 at Dunedin and 3.2 at Invercargill. At most coastal stations in the North Island and in the north and west of the South Island, snow very rarely occurs (KIDSON, 1932).

Fog

Fog is a very local phenomenon and occurs principally at night and in the early morning. Most fogs experienced in New Zealand are radiation fogs, especially in eastern areas of the South Island, but sea fogs also occur, on the eastern coast of the South Island, with an easterly to southerly airflow. Occasionally very humid air in advance of a wave depression or semi-tropical cyclone will create fairly widespread fog and low-cloud conditions especially in areas open to the north and east.

The human climates

The relationship between degree-days and domestic heating requirements in New Zealand has been studied by BASTINGS and SIMMONS (1950). Using 15.6°C (60°F) as the base, they found that the number of degree-days per year near sea-level increase steadily from north to south at a rate of roughly 200 degree-days per degree of latitude, values for eastern coast stations exceeding those at a similar latitude on the western coast by some 200 degree-days. The annual degree-days suggested a division of New Zealand into five zones. Zone 1: (less than 1,800 degree-days) the area north of Hamilton together with the western coastal strip north of Egmont (Taranaki). Zone 2: (1,800–2,400 degree-days) the coastal area of the North Island, excluding Zone 1, with northern and western coastal strips of the South Island to 41½°S; also inland low-lying areas in the Auckland provincial district. Zone 3: (2,400–3,000 degree-days) coastal portions of Westland, Marlborough and north Canterbury to 43°S; Banks Peninsula; upland North Island below 200 m. Zone 4: (3,000–3,600 degree-days) coastal south Canterbury and Otago to 46°S; upland North Island between 200–400 m. Zone 5: (3,600 degree-days or more) Central Otago; Southland; high country in the rest of the South Island, upland North Island above 400 m.

A *human* classification of the climate of 22 stations in New Zealand has been suggested by MAUNDER (1962a). In this classification, 13 aspects of the climate (five major aspects), were considered and each aspect was "weighted" according to its probable importance in day-to-day living.

Alexandra has the most favourable "rain" rating and Hokitika the least favourable; Nelson and Blenheim the most favourable "sunshine" rating, Invercargill the least favourable; Kaitaia and Auckland the most favourable "temperature" rating, Alexandra and Lake Tekapo the least favourable; while stations in the north of the North Island have the least favourable "humidity" rating and Wellington the least favourable "wind" rating.

The complete "human climatic index" suggests that Blenheim, Nelson and Napier have the most favourable "human climates", while Invercargill, Hokitika and Hamilton have the least favourable. Of the four main centres, Auckland is suggested as most and Wellington as least favourable, with Christchurch and Dunedin falling between.

An earlier approach to human climates in New Zealand (MACKY, 1937) assessed the invigorating effect of the weather. The data showed that the air is generally more "cooling" at Wellington than in most other parts of New Zealand (13 stations were considered), Wellington being followed by The Hermitage, Palmerston North and Lincoln.

Applied aspects

Applied aspects of New Zealand's climate are not discussed in this volume, but reference may be made to the following selected papers:
Agriculture: COULTER (1964), CURRY (1955, 1962), MAUNDER (1966c). *Livestock*: CURRY (1959, 1963),

MITCHELL et al. (1959). *Dairy production:* MAUNDER (1966). *Wool production:* COLLIN (1962), MAUNDER (1967a). *Meat production:* MAUNDER (1967b). *Plant ecology:* COULTER (1966), WARDLE and MARK (1956), ZOTOV (1938). *Climatic change:* CUMBERLAND (1962), DE LISLE (1956, 1957, 1959, 1961), GAGE (1965), GRANT (1963), HARRIS (1949), HOLLOWAY (1964), JOHNSON (1959), MAUNDER (1962b), RAINS (1967), SEELYE (1950b). *Air pollution:* GABITES (1965), SPARROW (1968). *Forestry:* FAHEY (1964), HOLLOWAY (1954). *Agricultural incomes:* MAUNDER (1965, 1966, 1968). *Soil moisture:* RICKARD (1960b, 1961), GABITES (1956). *Power transmission:* ROBERTSON (1952). *Soil climates:* HURST (1951). *Domestic heating:* BASTINGS and SIMMONS (1950). *Pastures:* BROUGHAM (1959), MITCHELL (1959). *Floods:* SOIL CONSERVATION AND RIVERS CONTROL COUNCIL (1957).

References

BASTINGS, L. and SIMMONS, P. E., 1950. Climatic zones and domestic heating in New Zealand. *New Zealand J. Sci. Technol. B*, 32: 44–53.

BONDY, F., 1950. Droughts in New Zealand. *New Zealand J. Sci. Technol., B*, 32: 1–10.

BONDY, F. and SEELYE, C. J., 1947. Temperatures associated with rainfall in New Zealand. *New Zealand J. Sci. Technol., B*, 28: 253–258.

BROUGHAM, R. W., 1959. The effects of season and weather on the growth rate of a ryegrass and clover pasture. *New Zealand J. Agr. Res.*, 2: 283–296.

BROWNE, M. L., 1959. Typical rainfall patterns over Central Otago. *New Zealand J. Geol. Geophys.*, 2: 88–94.

COLLIN, F. H., 1962. The drought in Hawke's Bay and some conclusions drawn from it. *Sheepfarming Annual*, 25: 139–149.

COULTER, J. D., 1964. The climate of New Zealand in relation to agriculture. *Proc. New Zealand Inst. Agr. Sci.*, 9: 41–62.

COULTER, J. D., 1966. Dry spells in New Zealand as a factor in plant ecology. *Proc. New Zealand Ecol. Soc.*, 13: 4–8.

CREASI, V. J., 1959. *Bibliography of Climatic Maps for New Zealand*. U.S. Weather Bureau, Washington, D.C., 25 pp.

CRITCHFIELD, H. J., 1966. Water balance analogues in the marine climates of New Zealand and North America. *New Zealand Geograph.*, 22: 111–124.

CUMBERLAND, K. B., 1962. Climatic change or cultural interference. In: M. McCASKILL (Editor), *Land and Livelihood: Geographical Essays in Honour of George Jobberns*. New Zealand Geograph. Soc., Christchurch, pp.88–142.

CURRY, L., 1955. Some seasonal rhythms in New Zealand agriculture. *Proc. New Zealand Geograph. Conf., 1st, Christchurch*, pp.51–53.

CURRY, L., 1959. *Climate and Livestock in New Zealand—A Functional Geography*. Thesis, University of New Zealand, unpublished.

CURRY, L., 1962. The climatic resources of intensive grassland farming: the Waikato, New Zealand. *Geograph. Rev.*, 52: 174–194.

CURRY, L., 1963. Regional variation in the seasonal programming of livestock farms in New Zealand. *Econ. Geograph.*, 39: 95–118.

DACRE, J. C., 1950. Climatology and meteorology of the New Zealand area: A bibliography. In: B. J. GARNIER (Editor), *New Zealand Weather and Climate—New Zealand Geograph. Soc., Misc. Ser.*, 1: 141–147.

DE LISLE, J. F., 1956. Secular variations of west coast rainfall in New Zealand and their relation to circulation changes. *New Zealand J. Sci. Technol., B*, 37: 700–715.

DE LISLE, J. F., 1957. Fluctuations of seasonal pressure patterns in the vicinity of New Zealand. *New Zealand J. Sci. Technol., B*, 38: 400–415.

DE LISLE, J. F., 1959. Is our climate changing? *New Zealand Family Doctor*, August, pp.27–30.

DE LISLE, J. F., 1961. A filter analysis of New Zealand's annual rainfall. *New Zealand J. Sci.*, 4: 296–308.

DE LISLE, J. F., 1965. Extreme surface winds over New Zealand. *New Zealand J. Sci.*, 8: 422–430.

DE LISLE, J. F., 1966. Mean daily insolation in New Zealand, *New Zealand J. Sci.*, 9: 992–1005.

FAHEY, B. D., 1964. Throughfall and interception of rainfall in a stand of radiata pine. *J. Hydrol. New Zealand*, 3: 17–26.

FINKELSTEIN, J., 1961. Estimation of open water evaporation in New Zealand. *New Zealand J. Sci.*, 4: 506–522.

FINKELSTEIN, J., 1967. Persistence of daily rainfall at some New Zealand stations. *J. Hydrol. New Zealand*, 6: 33–45.

GABITES, J. F., 1956. The estimation of natural evaporation and water need. In: *Proc. Conf. Soil Moisture—New Zealand Dept. Sci. Ind. Res., Inform. Ser. Bull.*, 12: 122–131.

GABITES, J. F., 1965. Air pollution and the siting of industry. *Proc. New Zealand Geograph. Conf., 4th, Christchurch*, pp.161–165.

GAGE, M., 1965. Some characteristics of Pleistocene cold climates in New Zealand. *Trans. Roy. Soc. New Zealand (Geol.)*, 3: 11–21.

GARNIER, B. J. (Editor), 1950. *New Zealand Weather and Climate—New Zealand Geograph. Soc., Misc. Ser.*, 1: 154 pp.

GARNIER, B. J., 1951. The application of the concept of potential evapotranspiration to moisture problems in New Zealand. *New Zealand Geograph.*, 7: 43–61.

GARNIER, B. J., 1958. *The Climate of New Zealand.* Arnold, London, 191 pp.

GRANT, P. J., 1963. Forests and recent climatic history of the Huiarau Range, Urewera region, North Island. *New Zealand J. Botan.*, 2: 143–172.

HARRIS, W. F., 1949. Post-glacial chronology and climate history. *New Zealand J. Sci. Technol.*, B, 30: 240–242.

HOLLOWAY, J. T., 1954. Forests and climates in the South Island of New Zealand. *Trans. Roy. Soc. New Zealand*, 82: 329–410.

HOLLOWAY, J. T., 1964. The forest of the South Island: the status of the climatic change hypothesis. *New Zealand Geograph.*, 20: 1–9.

HURST, F. D., 1951. Climates prevailing in the yellow-grey earth and yellow-brown earth zones in New Zealand. *Soil Sci.*, 72: 1–19.

JOHNSON, J. A., 1959. *Recent Climatic Changes in the South Island—A Geographic Analysis.* Thesis, University of New Zealand, unpublished.

KIDSON, E., 1931a. The annual variation of rainfall in New Zealand. *New Zealand J. Sci. Technol.*, 12: 268–271.

KIDSON, E., 1931b. The annual variation of pressure in New Zealand. *New Zealand J. Sci. Technol.*, 13: 39–44.

KIDSON, E., 1931c. Dry years in New Zealand. *New Zealand J. Sci. Technol.*, 13: 79–84.

KIDSON, E., 1931d. Mean temperatures in New Zealand. *New Zealand J. Sci. Technol.*, 13: 140–153.

KIDSON, E., 1932. The frequency of frost, snow and hail in New Zealand. *New Zealand J. Sci. Technol.*, 14: 42–54.

KIDSON, E. and THOMSON, A., 1931. The occurrence of thunderstorms in New Zealand. *New Zealand J. Sci. Technol.*, 12: 193–206.

MACKY, W. A., 1937. Some comparisons of the invigorating effect of the climate in different parts of N. Z. *New Zealand J. Sci. Technol.*, 19: 164–172.

MAUNDER, W. J., 1956. *A study of the Diurnal Variation of Rainfall in New Zealand.* Thesis, University of New Zealand, unpublished.

MAUNDER, W. J., 1957. Diurnal variation of rainfall in New Zealand. *New Zealand Geograph.*, 13: 151–160.

MAUNDER, W. J., 1962a. A human classification of climate. *Weather*, 17: 3–12.

MAUNDER, W. J., 1962b. Monthly and seasonal rainfalls at Dunedin 1913–1961. *New Zealand Geograph.*, 18: 184–202.

MAUNDER, W. J., 1965. *The Effect of Climatic Variations on some Aspects of Agricultural Production in New Zealand, and an Assessment of their Significance on the National Agricultural Income.* Thesis, University of Otago, Dunedin, unpublished.

MAUNDER, W. J., 1966a. New Zealand in maps, 5. Average annual rainfall patterns. *New Zealand Geograph.*, 22: 177–181.

MAUNDER, W. J., 1966b. Climatic variations and agricultural production in New Zealand. *New Zealand Geograph.*, 22: 55–69.

MAUNDER, W. J., 1966c. Climatic variations and dairy production in New Zealand. *New Zealand Sci. Rev.*, 24: 69–73.

MAUNDER, W. J., 1967a. Climatic variations and wool production—A New Zealand Review. *New Zealand Sci. Rev.*, 25 (4): 35–39.

MAUNDER, W. J., 1967b. Climatic variations and meat production—A New Zealand review. *New Zealand Sci. Rev.*, 25 (5): 9–12.

MAUNDER, W. J., 1968. The effect of significant climatic factors on agricultural production and incomes —A New Zealand example. *Monthly Weather Rev.*, 96: 39–46.

McINTOSH, C. B., 1958. Maps of surface winds in New Zealand. *New Zealand Geograph.*, 14: 75–81.

McLINTOCK, A. H. (Editor), 1959. *A Descriptive Atlas of New Zealand.* Government Printer, Wellington, 109 pp., 48 maps.

MITCHELL, K. J., 1959. Results of climate studies and their implications for seasonal productivity of pastures. *Proc. Conf. New Zealand Grassland Assoc.*, 21: 108–114.

MITCHELL, K. J., WALSHE, T. O. and ROBERTSON, N. G., 1959. Weather conditions associated with outbreaks of facial eczema. *New Zealand J. Agr. Res.*, 2: 548–604.

NEW ZEALAND METEOROLOGICAL SERVICE, 1962. Summaries of upper wind observations at selected stations. *New Zealand Meteorol. Serv. Misc. Publ.*, 114 (1), (2), (4), (5), (6).

OWEN, J. B., 1946. *A Brief History of the Development of Meteorology in New Zealand*. Thesis, University of New Zealand, unpublished.

RAINS, R. B., 1967. The late Pleistocene glacial sequence of the High Peak Valley, Canterbury. *New Zealand J. Geol. Geophys.*, 10: 1145–1158.

RICKARD, D. S., 1960a. The occurrence of agricultural drought at Ashburton, New Zealand. *New Zealand J. Agr. Res.*, 3: 431–441.

RICKARD, D. S., 1960b. The estimation of seasonal soil moisture deficits and irrigation requirements for Ashburton, New Zealand, 1. Soil moisture deficits. *New Zealand J. Agr. Res.*, 3: 820–828.

RICKARD, D. S., 1961. The estimation of seasonal soil moisture deficits and irrigation requirements for Ashburton, New Zealand, II. Irrigation requirements. *New Zealand J. Agr. Res.*, 4: 667–675.

ROBERTSON, N. G., 1950. The organization and development of weather observations in New Zealand. In: B. J. GARNIER (Editor), *New Zealand Weather and Climate—New Zealand Geograph. Soc. Misc. Ser.*, 1: 7–25.

ROBERTSON, N. G., 1952. Notes on weather factors affecting electric power transmission in New Zealand. *Trans. Elec. Supply Auth. Engs. Inst. New Zealand*, 22: 97–102.

ROBERTSON, N. G., 1957. Climate. In: F. R. CALLAGAN (Editor), *Science in New Zealand*. Government Printer, Wellington, pp.20–25.

ROBERTSON, N. G., 1959. The climate of New Zealand. In: A. H. MCLINTOCK (Editor), *A Descriptive Atlas of New Zealand*. Government Printer, Wellington, pp.19–22.

ROBERTSON, N. G., 1963. The frequency of high intensity rainfalls in New Zealand. *New Zealand Meteorol. Serv., Misc. Publ.*, 118: 84 pp.

ROBERTSON, N. G., 1966. Climate. In: A. H. MCLINTOCK (Editor), *An Encyclopaedia of New Zealand*. Government Printer, Wellington, 1: 359–363.

ROBERTSON, N. G., 1967. Climate of New Zealand. In: *New Zealand Official Yearbook 1967*. Government Printer, Wellington, pp.15–20.

SEELYE, C. J., 1940. Variability of annual rainfall in New Zealand. *New Zealand J. Sci. Technol.*, B, 22: 18–21.

SEELYE, C. J., 1945. *Maps of Average Rainfall in New Zealand*. New Zealand Meteorol. Service, Wellington, 16 maps.

SEELYE, C. J., 1946a. *Maps of Extreme Monthly Rainfall in New Zealand*. New Zealand Meteorol. Service, Wellington, 16 maps.

SEELYE, C. J., 1946b. Variations of monthly rainfall in New Zealand. *New Zealand J. Sci. Technol.*, B, 27: 397–405.

SEELYE, C. J., 1950a. The frequency of heavy rainfalls in New Zealand. *Trans. Roy. Soc. New Zealand*, 77: 66–70.

SEELYE, C. J., 1950b. Fluctuations and secular trend of New Zealand rainfall. *New Zealand J. Sci. Technol.*, B, 31: 11–24.

SOIL CONSERVATION AND RIVERS CONTROL COUNCIL, 1957. *Floods in New Zealand, 1920–1953*. Government Printer, Wellington.

SPARROW, C. J., 1968. A survey of the New Zealand air pollution literature and a bibliography. *Public Health*, 8 (2): 25–33.

SPARROW, C. J. and HEALY, T. R., 1968. *Meteorology and Climatology of New Zealand—A Bibliography*. Oxford Univ. Press for Univ. of Auckland, Auckland, 64 pp.

THORNTHWAITE, C. W., 1948. An approach towards a rational classification of climate. *Geograph. Rev.*, 38: 55–94.

WARDLE, P. and MARK, A. F., 1956. Vegetation and climate in the Dunedin district. *Trans. Roy. Soc. New Zealand*, 84: 33–44.

ZOTOV, V. D., 1938. Some correlations between vegetation and climate in New Zealand. *New Zealand J. Sci. Technol.*, 19: 474–487.

Chapter 10

Climatic Areas of New Zealand

W. J. MAUNDER

During the past 25 years several attempts to classify the climate of New Zealand have been published. These include those of KIDSON (1932a, 1937), SWENEY (1946) and GARNIER (1950a) using the classification of THORNTHWAITE (1931), GARNIER (1951) using the classification of THORNTHWAITE (1948); and ROBERTSON (1957); GARNIER (1958) and MAUNDER (1962a). GARNIER (1950a) and others have also commented on the application of Köppen's classification to New Zealand.

In addition, a much earlier (1857) classification of New Zealand coastal climates has been compared by HARGREAVES and MAUNDER (1964) with some present day classifications, and aspects of the seasonal climates of New Zealand are described by GARNIER (1950b).

Climatic classifications

Of the objective classifications the Köppen system classifies 46 out of 50 stations throughout New Zealand as *Cfb*, the only exceptions being Alexandra and Ophir (*BSk'*), Waipiata (*Cfs*) and Chateau Tongariro (*Cfc*). By contrast, at least 17 subtypes can be recognized using the THORNTHWAITE (1931) system (GARNIER, 1950a).

Two recent classifications—ROBERTSON (1957) and GARNIER (1958) are basically descriptive. Robertson divides New Zealand into fifteen climatic districts, using mean rainfall and mean temperature and their seasonal variation as the basic elements, but modified to take account of additional weather elements such as wind, sunshine and humidity, wherever they were considered to exert a major influence on the local climate. Garnier, on the other hand, divides New Zealand into nine climatic regions, based mainly on the combination of westerly, easterly, subtropical, antarctic, inland and altitudinal influences.

An indication of some of the differences between the classifications is given in Table I. This table shows that according to the classifications of Köppen, THORNTHWAITE (1931, 1948) and GARNIER (1958), Wanganui and Nelson have the same "climate", whereas differences occur using the classifications of ROBERTSON (1957), MAUNDER (1962a), and the "climatic areas" which are discussed in this section. On the other hand, Wellington and Nelson have the same "climate" according to the Köppen and THORNTHWAITE (1931) system, but are different when using the other classifications. A further approach is offered by GARNIER (1958) who classifies Gisborne, Napier and Blenheim, *with* Wellington. In the South Island, with its more distinctive local climates, the differences from one classification to another are small.

TABLE I

CLIMATIC CLASSIFICATIONS OF SELECTED STATIONS[1]

Station	A	B	C	D	E	F	G
Kaitaia	Cfb	BB'ra	$AB_1'ra'$	A	Northern N.Z.	77	1
Auckland	Cfb	BB'ra	$B_3B_2'ra'$	A	Northern N.Z.	78	1
Tauranga	Cfb	BB'ra	$B_4B_1'ra'$	B	Northern N.Z.	71	3
Gisborne	Cfb	BB'ra	$B_2B_2'ra'$	C	Eastern N.I.	74	6
Napier	Cfb	CB'rb	$C_2B_2'ra'$	C_0	Eastern N.I.	69	6
Wanganui	Cfb	BB'ra	$B_1B_1'ra'$	D	Middle N.Z.	79	7
Wellington	Cfb	BB'ra	$B_4B_1'ra'$	D	Eastern N.I.	90	9
Nelson	Cfb	BB'ra	$B_1B_1'ra'$	B	Middle N.Z.	65[2]	10
Blenheim	Cfb	CB'da	$C_1B_1'sa'$	C_0	Eastern N.I.	66	11
Hokitika	Cfb	AC'ra	$AB_1'ra'$	E	Western S.I.	97	12
Christchurch	Cfb	CC'rb	$C_2B_1'ra'$	F	Eastern S.I.	83	14
Lake Tekapo	Cfb	CC'rb	$C_2B_1'ra'$	F_2	Inland S.I.	74	15
Queenstown	Cfb	BC'rb	$B_1B_1'ra'$	F_2	Inland S.I.	85	15
Alexandra	BSk'	DC'db	$DB_1'da'$	F_0	Inland S.I.	72	15
Dunedin	Cfb	BC'ra	$B_2B_1'ra'$	G	Southern N.Z.	85	16
Invercargill	Cfb	BC'ra	$B_4B_1'ra'$	G_2	Southern N.Z.	101	16

[1] A = Köppens's classification, quoted from GARNIER, 1950a.
 B = THORNTHWAITE's (1931), quoted from GARNIER, 1950a.
 C = THORNTHWAITE's (1948) classification, quoted from GARNIER, 1958.
 D = ROBERTSON's (1957) classification.
 E = GARNIER's (1958) classification.
 F = MAUNDER's (1962) classification.
 G = "climatic area" as discussed in this chapter.
[2] Adjusted from MAUNDER, 1962a.

Climatic areas

The climatic areas into which New Zealand is here divided have been arrived at after consideration of the climatic aspects discussed in Chapter 9. It should be noted that except where the boundary follows a marked orographical feature, the change in climate from one area to the next is gradual.

Sixteen areas have been differentiated (Chapter 9, Fig.2) together with one undefined area of "upland climates" which has few climatological records but embraces parts of many of the sixteen basic areas. The number of climatic areas could be readily made more or less than the seventeen finally adopted. For example, the subdivision of the Bay of Plenty area (area 3) and of the Mackenzie Country–Central Otago area (area 15), or the combining of Gisborne, Hawkes Bay and Wairarapa (areas 6 and 8) are, perhaps, justifiable additions or contractions.

Aspects of the regional climates are discussed in detail in several publications and reference may be made to the following:

Northland/Auckland: COULTER (1966b), DE LISLE (1964, 1965), GLUCKMAN (1963), KERR (1962), WOOD (1918); *Waikato/Waitomo/Bay of Plenty:* COULTER (1966b), DE LISLE (1963, 1967), DEVEREAUX (1910); *Gisborne/Hawkes Bay:* COULTER (1962), GRANT (1966), KIDSON (1930, 1939); *Manawatu:* COULTER (1966a), SIMMERS (1962); *Wairarapa:* KIDSON (1933); *Wellington/Cook Strait:* GABITES (1960), KIDSON and CRUST (1932), MEL-

DRUM (1930a, b), SEELYE (1944); *Nelson:* DE LISLE (1962); *West Coast of the South Island:* DE LISLE (1956), TOWN AND COUNTRY PLANNING BRANCH (1959); *Canterbury:* CURRY (1958), EDIE et al. (1945), FARKAS (1958), HILL (1961), KIDSON (1932b), NEALE (1961), RAYNER and SOONS (1965), RICKARD (1960); *Otago:* BROWNE (1959), GARNIER (1948), MAUNDER (1962b, 1965, 1966); *Upland areas:* COULTER (1967), GRANT (1963), HOLLOWAY (1954), MARK (1965), MORRIS (1965); and the "Problem Climates": TREWARTHA (1961).

References

BROWNE, M. L., 1959. Typical rainfall patterns over Central Otago. *New Zealand J. Geol. Geophys.*, 2: 88–94.

COULTER, J. D., 1962. Easterly and westerly rainfalls in Hawke's Bay. *J. Hydrol. New Zealand*, 1: 3–4.

COULTER, J. D., 1966a. Climate of the Horowhenua lowlands. *Wellington Botan. Soc. Bull.*, 33: 41–52.

COULTER, J. D., 1966b. Flood and drought in Northern New Zealand. *New Zealand Geograph.*, 22: 22–34.

COULTER, J. D., 1967. Mountain climate. *Proc. New Zealand Ecol. Soc.*, 14: 40–57.

CURRY, L., 1958. Canterbury's grassland climate. *Proc. New Zealand Geograph. Conf., 2nd, Christchurch*, pp.58–63.

DE LISLE, J. F., 1956. Secular variations of west coast rainfall in New Zealand and their relation to circulation changes. *New Zealand J. Sci. Technol., B*, 37: 700–715.

DE LISLE, J. F., 1962. The climate of Nelson. *New Zealand Meteorol. Serv., Tech. Note*, 147.

DE LISLE, J. F., 1963. The climate and weather of the Bay of Plenty region. In: *Nat. Res. Surv. II—Bay of Plenty*. Govt. Printer, Wellington, pp.45–56.

DE LISLE, J. F., 1964. The climate and weather of Northland. In: *Nat. Res. Surv. III—Northland*. Govt. Printer, Wellington, pp.38–49.

DE LISLE, J. F., 1965. The climate of Auckland. In: L. O. KERMODE (Editor), *Science in Auckland—Handbook New Zealand Sci. Congr., 11th, Auckland*, pp. 31–35.

DE LISLE, J. F., 1967. The climate of the Waikato Basin. *Earth Sci. J.*, 1: 2–16.

DEVEREUX, H. B., 1910. The remarkable rainfall and meteorology of Waihi. *Trans. New Zealand Inst.*, 42: 408–411.

EDIE, E. G., SEELYE, C. J. and RAESIDE, J. D., 1945. Notes on the Canterbury floods of February, 1945. *New Zealand J. Sci. Technol., B*, 27: 406–420.

FARKAS. E., 1958. Mountain waves over Banks Peninsula. *New Zealand J. Geol. Geophys.*, 1: 677–683.

GABITES, J. F., 1960. Climate of Wellington. In: F. R. CALLAGHAN (Editor), *Science in Wellington*. Reed, Wellington, pp.33–35.

GARNIER, B. J., 1948. The climate of Otago. In: B. J. GARNIER (Editor), *The Face of Otago*. Otago Centennial History Publication, Dunedin, pp.18–25.

GARNIER, B. J., 1950a. The climates of New Zealand: according to Thornthwaite's (1931) classification. In: B. J. GARNIER (Editor), *New Zealand Weather and Climate—New Zealand Geograph. Soc., Misc. Ser.*, 1, pp. 84–104.

GARNIER, B. J., 1950b. The seasonal climates of New Zealand. In: B. J. GARNIER (Editor), *New Zealand Weather and Climate—New Zealand Geograph. Soc., Misc. Ser.*, 1, pp.105–140.

GARNIER, B. J., 1951. Thornthwaite's new (1948) system of climatic classification and its application to New Zealand. *Trans. Roy. Soc. New Zealand*, 79: 87–103.

GARNIER, B. J., 1958. *The Climate of New Zealand*. Edward Arnold, London, 191 pp.

GLUCKMAN, A. J., 1963. *Climates of Auckland*. Thesis, University of Auckland, Auckland (unpublished).

GRANT, P. J., 1963. Forests and recent climatic history of the Huiarau Range, Urewera region, North Island. *New Zealand J. Botan.*, 2: 143–172.

GRANT, P. J., 1966. Variations of rainfall frequency in relation to erosion in eastern Hawke's Bay. *J. Hydrol. (New Zealand)*, 5: 73–86.

HARGREAVES, R. P. and MAUNDER, W. J., 1964. An early classification of New Zealand coastal climates. *Prof. Geograph.*, 16: 6–10.

HILL, H. W., 1961. Northwesterly rains in Canterbury. *New Zealand Meteorol. Serv., Tech. Note*, 136.

HOLLOWAY, J. T., 1954. Forests and climates in the South Island of New Zealand. *Trans. Roy. Soc. New Zealand*, 82: 329–410.

KERR, I. S., 1962. Characteristic weather sequences in Northland. *New Zealand Meteorol. Serv., Tech. Note*, 144.

KIDSON, E., 1930. The flood rains of 11th March, 1924, in Hawke's Bay. *New Zealand J. Sci. Technol.*, 12: 53–60.

KIDSON, E., 1932a. Climatology of New Zealand. In: W. KÖPPEN and R. GEIGER (Editors), *Handbuch der Klimatologie*. Borntraeger, Berlin, Band IV, Teil S, pp.111–138.

KIDSON, E., 1932b. The Canterbury "Northwester". *New Zealand J. Sci. Technol.*, 14: 65–74.

KIDSON, E., 1933. The Wairarapa floods of August, 1932. *New Zealand J. Sci. Technol.*, 14: 220–226.

KIDSON, E., 1937. The Climate of New Zealand. *Quart. J. Roy. Meteorol. Soc.*, 62: 83–92.

KIDSON, C., 1939. The Climate of the Heretaunga Plains. In: *Land Utilisation of the Heretaunga Plains—New Zealand Dept. Sci. Ind. Res., Res. Bull.*, 70: 12–17.

KIDSON, E. and CRUST, A. G. C., 1932. The diurnal variation of temperature at Wellington. *New Zealand J. Sci. Technol.*, 13: 278–283.

MARK, A. F., 1965. Vegetation and mountain climate. In: R. G. LISTER and R. P. HARGREAVES (Editors), *Central Otago—New Zealand Geograph. Soc., Misc. Ser.*, 5, pp.69–91.

MAUNDER, W. J., 1962a. A human classification of climate. *Weather*, 17: 3–12.

MAUNDER, W. J., 1962b. Monthly and seasonal rainfalls at Dunedin 1913–1961. *New Zealand Geograph.*, 18: 184–202.

MAUNDER, W. J., 1965. Climatic character. In: R. G. LISTER and R. P. HARGREAVES (Editors), *Central Otago—New Zealand Geograph. Soc., Misc. Ser.*, 5, pp.46–68.

MAUNDER, W. J., 1966. Waitaki River catchment—climate. *Soil Conserv. Newsletter*, 1: 2–9.

MELDRUM, D. C, 1930a. Hourly sunshine at Wellington, January to June, 1929. *New Zealand J. Sci. Technol.*, 11: 382–384.

MELDRUM, D. C., 1930b. Hourly sunshine at Wellington, latter half of 1929. *New Zealand J. Sci. Technol.*, 12: 142–144.

MORRIS, J. Y., 1956. Climate investigations in the Craigieburn Range, New Zealand. *New Zealand J. Sci.*, 8: 556–582.

NEALE, A. A., 1961. Situations associated with significant rainfalls at Christchurch. *New Zealand Meteorol. Serv., Tech. Note*, 138.

RAYNER, J. N. and SOONS, J. M., 1965. The storm in Canterbury of 12–17 July, 1963. *New Zealand Geograph.*, 21: 12–25.

RICKARD, D. S., 1960. The occurrence of agricultural drought at Ashburton, New Zealand. *New Zealand J. Agric. Res.*, 3: 431–441.

ROBERTSON, N. G., 1957. Climatic districts of New Zealand. *Proc. New Zealand Ecol. Soc.*, 4: 6–22.

SEELYE, C. J., 1944. Wellington City rainfall. *New Zealand J. Sci. Technol.*, B, 26: 36–46.

SIMMERS, I., 1962. Analysis of some heavy rainfalls in the Rangitikei district. *J. Hydrol. New Zealand*, 1: 23–29.

SWENEY, H. M., 1946. *The Climates of New Zealand According to the Thornthwaite (1931) System*. Thesis, University of New Zealand (unpublished).

THORNTHWAITE, C. W., 1931. The climates of North America according to a new classification. *Geograph. Rev.*, 21: 663–655.

THORNTHWAITE, C. W., 1948. An approach towards a rational classification of climate. *Geograph. Rev.*, 38: 55–94.

TOWN AND COUNTRY PLANNING BRANCH, 1959. Climate. In: *West Coast Region—Nat. Res. Surv., I*. Govt. Printer, Wellington, pp.31–38.

TREWARTHA, G. T., 1961. *The Earth's Problem Climates*. Univ. Wisc. Press, Madison, Wisc., pp.82–86.

WOOD, J., 1918. Phenomenal rainfall and floods of the North Auckland district. *New Zealand J. Sci. Technol.*, 1: 293–296.

Climatic Tables for Australia and New Zealand

Geographical arrangement of the tables

I–III, XII–XXVIII	Western Australia
IV–V, XXIX–XL	Northern Territory and South Australia
VI–X, XLI–LXVIII	Queensland, New South Wales, Victoria
XL, LXIX–LXXI	Tasmania
LXXII–LXXXIV	New Zealand

Within each section the tables are arranged from north to south.

TABLE I

FREQUENCY DISTRIBUTION OF MONTHLY RAINFALL, ROEBOURNE, W.A., AUSTRALIA[1]
Latitude 20°46′S, longitude 117°09′E, elevation 12 m

Rainfall (mm)	Jan.	Feb.	Mar.	Apr.	May	June	July	Aug.	Sept.	Oct.	Nov.	Dec.	Total
508–609	–	–	–	1	–	–	–	–	–	–	–	–	1
407–508	–	–	1	–	–	–	–	–	–	–	–	–	1
305–406	3	1	3	1	–	–	–	–	–	–	–	–	8
203–304	4	4	10	–	1	–	–	–	–	–	–	–	19
151–202	7	6	4	3	1	3	–	–	–	–	–	–	24
102–151	6	13	7	6	7	1	1	–	–	–	–	1	42
51–101	11	18	11	4	15	6	6	3	–	–	–	6	80
26–50	13	8	7	7	7	27	8	3	3	1	1	6	91
13–25	6	10	7	3	15	17	14	7	1	1	4	4	89
1–12	29	27	23	32	23	23	31	37	13	11	11	38	298
0	21	13	27	43	31	23	40	50	83	87	84	45	547
Total	100	100	100	100	100	100	100	100	100	100	100	100	1200

[1] The dotted lines show where the interval changes; $N = 71$.

TABLE II

FREQUENCY DISTRIBUTION OF MONTHLY RAINFALL, DEAKIN, W.A., AUSTRALIA[1]
Latitude 30°48′S, longitude 129°00′E, elevation 152 m

Rainfall (mm)	Jan.	Feb.	Mar.	Apr.	May	June	July	Aug.	Sept.	Oct.	Nov.	Dec.	Total
151–202	–	3	–	–	–	–	–	–	–	–	–	–	3
102–151	–	–	–	–	–	–	–	–	–	–	–	–	0
51–101	7	–	13	–	7	13	–	3	–	13	3	3	62
26–50	7	17	7	10	17	–	13	7	7	13	13	20	131
13–25	7	17	23	30	20	27	23	30	7	20	23	13	240
1–12	49	33	27	30	43	43	57	50	59	34	48	37	510
0	30	30	30	30	13	17	7	10	27	20	13	27	254
Total	100	100	100	100	100	100	100	100	100	100	100	100	1200

[1] The dotted lines show where the interval changes; $N = 33$.

TABLE III

FREQUENCY DISTRIBUTION OF MONTHLY RAINFALL, PERTH, W.A., AUSTRALIA[1]

Rainfall (mm)	Jan.	Feb.	Mar.	Apr.	May	June	July	Aug.	Sept.	Oct.	Nov.	Dec.	Total
407–508	–	–	–	–	–	1	–	–	–	–	–	–	1
305–406	–	–	–	–	1	2	2	2	–	–	–	–	7
203–303	–	–	–	–	10	32	24	13	–	–	–	–	79
151–202	–	1	–	–	23	31	32	23	5	1	–	–	116
102–151	–	–	2	6	32	24	29	34	23	9	–	–	159
51–101	1	2	4	34	23	10	13	23	53	39	6	2	210
26–50	5	7	22	29	9	–	–	4	15	37	28	16	172
13–25	16	13	23	12	2	–	–	–	2	12	27	23	130
1–12	72	67	44	18	–	–	–	1	2	2	38	55	299
0	6	10	5	1	–	–	–	–	–	1	4	27
Total	100	100	100	100	100	100	100	100	100	100	100	100	1200

[1] The dotted lines show where the interval changes; $N = 82$. For position and other data see Table XXV.

TABLE IV

FREQUENCY DISTRIBUTION OF MONTHLY RAINFALL, DARWIN, N.T., AUSTRALIA[1]

Rainfall (mm)	Jan.	Feb.	Mar.	Apr.	May	June	July	Aug.	Sept.	Oct.	Nov.	Dec.	Total
610–711	6	4	–	–	–	–	–	–	–	–	–	–	10
508–609	19	11	4	1	–	–	–	–	–	–	–	4	39
407–508	21	15	7	–	–	–	–	–	–	–	–	10	53
305–406	25	22	27	2	–	–	–	–	–	–	2	12	90
203–304	15	29	29	12	1	–	–	–	–	1	10	36	133
151–202	11	14	12	7	–	–	–	–	–	1	19	16	80
102–151	1	4	10	10	1	–	–	–	–	9	29	19	83
51–101	2	–	10	28	7	–	1	1	6	32	30	2	119
26–50	–	–	–	14	7	1	–	1	15	26	7	–	71
13–25	–	1	1	15	6	10	–	–	11	14	2	1	61
1–12	–	–	–	10	32	23	20	20	42	10	1	–	158
0	–	–	–	1	46	66	79	78	26	7	–	–	303
Total	100	100	100	100	100	100	100	100	100	100	100	100	1200

[1] The dotted lines show where the interval changes; $N = 82$. For position and other data see Table XXIX.

TABLE V

FREQUENCY DISTRIBUTION OF MONTHLY RAINFALL, ADELAIDE, S.A., AUSTRALIA[1]

Rainfall (mm)	Jan.	Feb.	Mar.	Apr.	May	June	July	Aug.	Sept.	Oct.	Nov.	Dec.	Total
151–202		1	–	–	3	3	–	1	–	–	–	–	8
102–151		–	3	8	18	18	16	7	3	2	2	–	77
51–101	9	13	9	25	41	48	53	54	42	33	17	12	356
26–50	18	9	20	35	28	27	23	33	37	42	28	29	329
13–25	22	18	28	20	7	3	7	3	16	16	30	25	195
1–12	46	54	37	11	3	1	1	2	2	7	23	33	220
0	5	5	3	1	–	–	–	–	–	–	–	1	15
Total	100	100	100	100	100	100	100	100	100	100	100	100	1200

[1] The dotted lines show where the interval changes; $N = 119$. For position and other data see Table XXXIX.

TABLE VI

FREQUENCY DISTRIBUTION OF MONTHLY RAINFALL, TULLY, QLD., AUSTRALIA[1]
Latitude 17°54′S, 145°54′E, elevation 15 m

Rainfall (mm)	Jan.	Feb.	Mar.	Apr.	May	June	July	Aug.	Sept.	Oct.	Nov.	Dec.	Total
1778–	–	3	3	–	–	–	–	–	–	–	–	–	6
1270–1778	10	13	23	6	–	–	–	–	–	–	–	–	52
890–1269	26	27	13	10	–	–	–	–	–	–	–	3	79
712–889	23	16	13	13	–	–	–	–	–	–	–	–	65
610–711	3	10	10	3	10	–	–	–	–	3	3	3	45
508–609	10	10	6	6	6	6	3	–	–	–	–	6	53
407–508	3	6	10	16	27	13	–	3	–	3	6	–	87
305–406	3	6	19	3	13	6	6	3	3	3	6	16	87
203–304	13	3	3	24	16	23	16	29	23	6	13	6	175
151–202	6	3	–	10	3	19	29	10	13	6	17	17	133
102–151	–	3	–	–	19	13	26	13	13	20	13	16	136
51–101	–	–	–	6	–	10	10	13	27	16	27	17	126
25–50	–	–	–	3	6	10	10	13	6	3	6	10	67
13–25	3	–	–	–	–	–	–	3	6	17	6	6	41
1–12	–	–	–	–	–	–	–	10	3	20	3	–	36
0	–	–	–	–	–	–	–	3	6	3	–	–	12
Total	100	100	100	100	100	100	100	100	100	100	100	100	1200

[1] The dotted lines show where the interval changes; $N = 31$.

TABLE VII

FREQUENCY DISTRIBUTION OF MONTHLY RAINFALL, BRISBANE, QLD., AUSTRALIA[1]

Rainfall (mm)	Jan.	Feb.	Mar.	Apr.	May	June	July	Aug.	Sept.	Oct.	Nov.	Dec.	Total
610–711	1	1	–	–	–	–	–	–	–	–	–	–	2
508–609	2	2	2	–	–	–	–	–	–	–	–	–	6
407–508	2	6	2	–	–	–	–	–	–	–	–	1	11
305–406	4	7	6	4	1	1	–	2	–	–	1	4	30
203–304	15	10	12	7	5	3	2	2	–	5	7	13	81
151–202	17	13	16	9	5	7	7	1	–	3	9	8	95
102–151	25	18	18	14	12	15	9	6	7	7	19	27	177
51–101	21	26	24	29	27	22	20	15	35	40	38	27	324
26–50	11	13	11	16	20	19	21	26	25	27	14	12	215
13–25	1	4	8	9	14	13	19	21	23	9	9	5	135
1–12	1	–	1	12	15	20	20	24	10	9	2	3	117
0	–	–	–	–	1	–	2	3	–	–	1	–	7
Total	100	100	100	100	100	100	100	100	100	100	100	100	1200

[1] The dotted lines show where the interval changes; $N = 106$. For position and other data see Table LIII.

TABLE VIII

FREQUENCY DISTRIBUTION OF MONTHLY RAINFALL, SYDNEY, N.S.W., AUSTRALIA[1]

Rainfall (mm)	Jan.	Feb.	Mar.	Apr.	May	June	July	Aug.	Sept.	Oct.	Nov.	Dec.	Total
610–711	–	–	–	1	–	1	–	–	–	–	–	–	2
508–609	–	1	1	1	3	–	–	–	–	–	–	–	6
407–508	–	2	3	2	–	2	–	–	–	–	–	–	9
305–406	2	2	3	5	9	6	3	1	1	–	–	1	33
203–304	7	6	11	8	8	12	17	6	3	5	6	5	94
151–202	11	12	9	13	8	9	9	7	6	5	2	5	96
102–151	16	16	19	18	13	17	12	11	12	11	17	10	172
51–101	31	26	27	25	30	18	30	31	33	32	36	35	354
26–50	23	20	22	15	17	21	11	20	25	30	16	23	243
13–25	7	7	4	8	7	10	6	11	12	13	11	17	113
1–12	3	8	1	4	5	4	12	13	8	4	12	4	78
0	–	–	–	–	–	–	–	–	–	–	–	–	–
Total	100	100	100	100	100	100	100	100	100	100	100	100	1200

[1] The dotted lines show where the interval changes; $N = 99$. For position and other data see Table LXI.

TABLE IX

FREQUENCY DISTRIBUTION OF MONTHLY RAINFALL, MOUNT BUFFALO, VIC., AUSTRALIA[1]
Latitude 36°48′S, longitude 146°48′E, elevation 1,332 m

Rainfall (mm)	Jan.	Feb.	Mar.	Apr.	May	June	July	Aug.	Sept.	Oct.	Nov.	Dec.	Total
712–889							3						3
610–711	–	–	–	–	–	–	–	–	–	–	–	–	–
508–609	–	–	–	–	3	8	3	5	–	3	–	–	22
407–508	–	–	–	–	5	8	3	–	3	8	–	–	27
305–406	–	8	3	11	11	14	11	25	11	8	5	3	110
203–304	3	5	11	11	19	20	38	25	28	14	19	14	207
151–202	8	–	11	16	8	11	14	11	19	22	11	5	136
102–151	17	14	17	8	26	20	17	14	11	17	17	22	200
51–101	33	29	19	37	17	11	11	14	22	14	37	31	275
26–50	17	31	20	8	5	8	–	3	3	11	3	17	126
13–25	5	5	11	3	–	–	–	–	3	–	8	5	40
1–12	17	5	8	3	3	–	–	3	–	3	–	3	45
0	–	3	–	3	3	–	–	–	–	–	–	–	9
Total	100	100	100	100	100	100	100	100	100	100	100	100	1200

[1] The dotted lines show where the interval changes; $N = 35$.

TABLE X

FREQUENCY DISTRIBUTION OF MONTHLY RAINFALL, MELBOURNE, VIC., AUSTRALIA[1]

Rainfall (mm)	Jan.	Feb.	Mar.	Apr.	May	June	July	Aug.	Sept.	Oct.	Nov.	Dec.	Total
203–304	–	–	–	–	–	–	–	–	–	–	1	–	1
151–202	1	3	3	1	–	–	1	–	1	3	1	3	17
102–151	10	10	11	14	4	2	2	2	5	11	13	9	93
51–101	25	21	25	34	46	41	39	38	48	55	41	39	452
26–50	33	25	32	31	33	50	48	44	40	22	26	29	413
13–25	16	22	15	18	14	7	10	15	6	8	15	16	162
1–12	15	19	14	1	3	–	–	1	–	1	3	4	61
0	–	–	–	1	–	–	–	–	–	–	–	–	1
Total	100	100	100	100	100	100	100	100	100	100	100	100	1200

[1] The dotted lines show where the interval changes; $N = 102$.
Normal data for Melbourne will be found in Table LXVIII.

TABLE XI

FREQUENCY DISTRIBUTION OF MONTHLY RAINFALL, HOBART, TASMANIA, AUSTRALIA[1]

Rainfall (mm)	Jan.	Feb.	Mar.	Apr.	May	June	July	Aug.	Sept.	Oct.	Nov.	Dec.	Total
203–304	–	–	1	1	–	1	–	–	–	–	–	–	3
151–202	–	–	1	1	1	5	1	1	1	3	4	1	19
102–151	10	5	5	8	4	7	4	4	7	11	4	11	80
51–101	36	26	25	36	32	36	41	29	28	45	43	32	409
26–50	24	35	42	32	37	37	37	43	47	33	35	34	435
13–25	22	18	22	14	21	11	16	21	14	7	9	14	189
1–12	8	16	4	8	5	3	1	3	3	1	5	8	65
0	–	–	–	–	–	–	–	–	–	–	–	–	–
Total	100	100	100	100	100	100	100	100	100	100	100	100	1200

[1] The dotted lines show where the interval changes; $N = 75$. Normal data for Hobart will be found in Table LXX.

TABLE XII

CLIMATIC TABLE FOR KUNMUNYA MISSION (FORMERLY PORT GEORGE IV MISSION) W.A., AUSTRALIA
Latitude 15°25′S, longitude 125°43′E, elevation 594 m

Month	Mean daily temp. (°C)	Mean daily temp. range (°C)	Temperatures extremes (°C)		Mean precip. (mm)	Max. 24 h precip. (mm)	Relative humidity (%)			Evap. (mm)*
			highest	lowest			daily index	mean 09h00	mean 15h30	
Jan.	28.6	7.9	39	16	384	193	76	72	61	165
Feb.	28.5	7.9	39	17	282	178	77	71	69	150
Mar.	28.4	9.1	38	16	254	197	76	70	63	175
Apr.	27.1	13.6	39	8	54	72	87	56	55	175
May	24.4	15.9	38	5	24	69	62	49	33	190
June	22.2	16.5	38	2	19	59	60	46	34	165
July	20.9	18.2	36	3	8	19	60	45	34	170
Aug.	22.8	18.1	37	5	1	1	62	48	24	180
Sept.	25.6	16.6	39	9	2	5	61	49	–	215
Oct.	28.1	13.1	41	14	10	7	61	52	38	250
Nov.	29.6	11.1	42	15	47	70	64	57	46	235
Dec.	29.6	8.9	41	17	194	113	69	64	63	200
Annual	26.3	13.1	42	2	1280	197	67	57	–	2270
Years	30	30	23	23	27	11	33	23	1	–

Month	Number of days				Mean cloudi-ness (oktas)	Mean sun-shine (h)*	Wind		Probability of precipitation (%)		
	precipitation		thunder	fog			most freq. dir. 09h00	mean speed 09h00 (kt)	<12.5 mm	<25 mm	<50 mm
	>0.25 mm	>2.5 mm									
Jan.	18	12	17	0	5.2	190	SW	8	0	0	0
Feb.	16	12	19	0	4.9	175	SW	6	0	0	3
Mar.	15	12	12	<1	4.0	210	SW	4	0	0	0
Apr.	4	2	5	<1	1.9	260	SE	4	29	47	62
May	2	1	<1	0	1.6	280	SE	5	71	73	82
June	2	1	0	0	1.5	270	SE	8	73	83	88
July	1	1	0	1	1.0	290	E	8	85	88	94
Aug.	0	0	<1	0	0.7	290	SW	6	97	97	100
Sept.	1	1	1	1	0.9	280	SE, SW	6	91	94	100
Oct.	1	<1	5	0	1.6	310	SW	5	82	91	94
Nov.	4	1	15	0	2.3	290	SW	4	29	41	65
Dec.	11	9	21	<1	4.2	235	SW	4	3	16	18
Annual	75	52	95	3	2.5	3080	SW	6	–	–	–
Years	34	10	7	5	23	–	5	5	34	34	34

* Estimated by extrapolation from maps.

TABLE XIII

CLIMATIC TABLE FOR WYNDHAM, W.A., AUSTRALIA
Latitude 15°27'S, longitude 128°07'E, elevation 7 m.

Month	Mean station pressure (mbar)		Mean daily temp. (°C)	Mean daily temp. range (°C)	Temp. extremes		Precipitation			Max. precip. 24 h. (mm)	Relative humidity (%)			Evap. (mm)*
	09h00	15h00			highest (°C)	lowest (°C)	mean (mm)	highest (mm)	lowest (mm)		daily index	mean 09h00	mean 15h00	
Jan.	1006.3	1002.5	31.2	8.7	45.3	18.7	202	718	13	307	66	67	54	190
Feb.	1006.6	1002.8	30.8	8.8	43.9	16.7	163	364	14	150	67	69	54	170
Mar.	1007.9	1004.0	30.8	8.8	42.2	18.3	122	446	3	317	63	65	49	185
Apr.	1010.8	1006.9	29.9	9.7	41.7	17.2	34	238	0	440	46	46	38	180
May	1012.9	1008.9	27.3	9.8	39.4	11.1	10	58	0	69	41	42	37	182
June	1014.6	1010.7	24.9	9.9	37.8	10.0	10	120	0	113	40	42	37	165
July	1015.2	1011.0	24.2	10.4	35.7	8.9	5	133	0	86	38	40	35	180
Aug.	1014.5	1010.2	26.1	10.6	38.9	8.3	<1	21	0	11	40	43	39	190
Sept.	1013.3	1008.6	29.0	10.4	41.1	15.6	2	35	0	35	44	44	43	220
Oct.	1011.0	1006.3	31.3	9.6	43.9	18.3	9	82	0	57	52	51	47	250
Nov.	1008.8	1004.2	32.2	9.5	45.3	14.4	42	132	1	85	55	56	50	246
Dec.	1007.1	1003.1	31.9	9.1	44.4	18.3	104	291	7	110	60	55	52	200
Annual	1010.7	1006.6	29.1	9.6	45.3	8.3	703	1353	365	440	51	52	45	2352
Years	29	29	30	30	70	70	30	55	55	80	30	42	42	—

TABLE XIII (*continued*)

Month	Number of days						Mean cloudiness, all types		Most freq. wind dir.		Mean speed (knots)		Probability of precipitation (%)		
	precipitation		max. temp.		thunder	fog	09h00	15h00	09h00	15h00	09h00	15h00	<12.5 mm	<25 mm	<50 mm
	>0.25 mm	>2.5 mm	>32.2°C	>37.8°C											
Jan.	13	7	29	17	15	0	3.4	3.8	SW	N	6	8	1	3	9
Feb.	10	6	26	12	12	0	3.4	3.8	SW	N	6	7	0	1	11
Mar.	9	7	29	16	8	0	2.6	3.4	SW	N	5	7	5	8	17
Apr.	3	1	26	7	3	0	1.6	2.2	SE, SW	N	6	6	49	73	88
May	1	<1	26	1	0	0	1.1	1.4	SE	N	7	6	82	88	98
June	1	<1	12	0	0	0	1.0	1.1	SE, S	S, SE	8	6	92	96	97
July	0	<1	13	0	0	0	0.6	0.6	SE, SW	SE, N	7	6	95	96	96
Aug.	0	<1	24	<1	1	0	0.5	0.5	SE	N	6	6	99	100	100
Sept.	1	0	29	4	4	0	0.6	1.0	SW,N,SE	N	6	7	90	97	100
Oct.	1	<1	31	17	6	0	1.2	1.6	NW, N	N	5	10	68	85	97
Nov.	5	3	29	22	12	0	1.9	2.6	SW	N	5	9	13	39	60
Dec.	9	5	29	18	13	0	2.8	3.5	SW	N	4	8	5	8	27
Annual	53	31	304	115	74	0	2.0	2.2	SW	N	6	7	—	—	—
Years	30	10	42	42	5	10	30	30	5	5	5	5	75	75	75

* Estimated from maps, standard Australian tank.

TABLE XIV

CLIMATIC TABLE FOR BROOME, W.A., AUSTRALIA
Latitude 17°57′S, longitude 122°15′E, elevation 11 m

Month	Mean sta. press. (mbar)				Mean daily temp. (°C)	Mean daily temp. range (°C)	Temp. extr. (°C)		Precipitation (mm)			Max. precip. 24 h
	09h00	15h00	highest[1]	lowest[1]			highest	lowest	mean	highest	lowest	
Jan.	1006.5	1003.6	1015.7	985.5	29.6	6.7	44.2	17.8	146	825	3	356
Feb.	1006.7	1003.7	1016.6	993.9	29.7	7.1	42.7	15.0	134	599	2	302
Mar.	1008.1	1004.8	1016.6	988.6	29.7	8.6	41.7	12.8	102	600	0	269
Apr.	1011.0	1007.5	1019.0	997.4	28.1	12.1	41.7	12.2	29	232	0	181
May	1013.2	1009.6	1022.2	1004.9	24.7	12.9	38.3	7.3	33	131	0	119
June	1015.1	1011.2	1023.8	1006.3	21.7	12.8	36.2	5.6	21	246	0	143
July	1015.6	1011.8	1024.7	1007.0	20.8	13.8	35.0	3.3	7	59	0	55
Aug.	1015.0	1011.2	1023.9	1007.2	22.5	13.9	38.1	4.8	2	95	0	37
Sept.	1013.6	1009.9	1021.1	1005.9	25.0	13.2	39.7	8.9	1	22	0	21
Oct.	1011.5	1008.1	1018.5	1003.8	27.4	10.2	42.8	11.6	1	10	0	12
Nov.	1009.1	1005.7	1017.6	970.0	29.3	8.9	44.0	14.7	7	279	0	140
Dec.	1007.4	1004.1	1014.8	992.0	30.1	7.7	44.8	17.2	34	368	0	143
Annual	1011.1	1007.6	1024.7	970.0	25.6	10.6	44.8	3.3	517	1084	142	356
Years	30	30	29	29	30	30	74	74	30	52	52	74

Month	Relative humidity (%)							Mean evap. (mm)*	Visibility frequency (%) 09h00			
	daily index	09h00			15h00				< 5 km	< 10 km	< 20 km	< 30 km
		highest monthly	mean	lowest monthly	highest monthly	mean	lowest monthly					
Jan.	75	83	72	61	80	67	51	158	7	7	23	93
Feb.	76	86	73	57	79	67	53	142	4	7	36	96
Mar.	73	79	69	55	74	59	43	152	4	4	23	96
Apr.	56	68	54	36	56	45	29	206	0	0	8	89
May	54	68	53	37	57	44	30	173	0	0	2	94
June	53	69	54	39	63	46	29	160	0	0	6	96
July	52	75	52	38	65	43	29	160	0	4	7	100
Aug.	54	73	54	45	61	42	30	170	0	0	7	100
Sept.	56	67	52	40	60	44	28	196	4	4	22	100
Oct.	64	73	57	46	71	52	43	188	0	0	7	100
Nov.	66	69	60	48	66	57	46	201	0	0	7	100
Dec.	70	79	66	54	73	62	53	196	0	0	0	93
Annual	63	70	60	53	60	52	45	2094	2	2	12	96
Years	30	30	44	30	30	44	30	estim.*	unsp.	unsp.	unsp.	unsp.

TABLE XIV *(continued)*

Month	Number of days					thunder	fog	clear	Mean cloudiness		
	precipitation		max. temp.		min. temp.				all types		low
	>0.25 mm	>2.5 mm	>37.8 °C	>32.2 °C	<21.1 °C				09h00	15h00	
Jan.	11	6	3	30	0	10	0	6	4.1	3.7	1.8
Feb.	9	5	1	26	<1	10	0	7	4.2	3.6	1.8
Mar.	7	6	3	25	2	6	0	14	3.2	3.0	1.4
Apr.	3	1	2	26	7	2	0	22	1.8	2.1	0.7
May	2	1	0	20	14	1	1	22	1.9	2.0	1.0
June	2	1	0	6	23	<1	3	26	1.4	1.4	0.2
July	2	<1	0	6	30	0	4	26	1.3	1.3	0.2
Aug.	0	<1	0	10	30	0	2	26	1.0	1.0	0.3
Sept.	1	0	1	16	20	<1	2	26	1.1	0.9	0.6
Oct.	0	0	5	18	7	1	0	23	1.6	1.1	0.8
Nov.	1	<1	3	26	0	4	0	20	1.7	1.6	0.5
Dec.	5	2	3	29	1	8	0	12	3.3	2.9	1.4
Annual	43	22	21	247	134	41	12	227	2.2	2.1	1.0
Years	30	10	4	4	4	30	5	10	29	29	2

Month	Low cloud frequency (%) 09h00			Wind				Probability of precipitation (%)		
				most freq. dir.		mean speed (kt.)				
	<2,400 m	<900 m	<300 m	09h00	15h00	09h00	15h00	<12.5 mm	<25 mm	<50 mm
Jan.	100	90	10	SW	W	5	8	8	14	19
Feb.	93	86	0	SW	SW	5	8	0	8	25
Mar.	50	46	8	SE	SW	5	7	25	28	33
Apr.	23	15	0	E,SE	SE,SW	6	6	47	58	64
May	11	9	0	SE	S	6	5	69	94	100
June	19	2	2	SE	S	7	6	61	64	83
July	20	13	7	SE	S	6	6	83	92	97
Aug.	11	7	3	SE	SW	6	7	95	97	97
Sept.	11	7	7	SE	SW	7	8	92	100	100
Oct.	32	25	3	S,SW	SW	7	9	100	100	100
Nov.	37	37	11	SW,W	W	6	9	78	87	95
Dec.	60	53	3	SW	W	6	9	20	39	42
Annual	39	32	4	SE	SW	6	7	—	—	—
Years	unsp.	unsp.	unsp.	5	5	5	5	36	36	36

[1] Mean for 09h00 and 15h00.
* Estimated, Fitzpatrick method.

TABLE XV

CLIMATIC TABLE FOR HALLS CREEK, W.A., AUSTRALIA
Latitude 18°13'S, longitude 127°46'E, elevation 374 m.

Month	Mean daily temp. (°C)	Mean daily temp. range (°C)	Extreme temp. (°C)		Precipitation (mm)			Max. precip. 24 h. (mm)	Relative humidity (%)			Mean evap.* (mm)
			highest	lowest	mean	highest	lowest		daily index	mean 09h00	15h00	
Jan.	30.3	12.3	44.3	15.6	123	578	5	211	54	51	36	234
Feb.	29.8	12.7	43.8	12.2	109	373	3	130	54	52	36	211
Mar.	28.6	13.6	41.9	11.0	51	369	0	174	50	46	33	211
Apr.	25.3	16.3	39.9	7.2	15	164	0	147	40	35	27	218
May	21.4	16.5	37.2	2.4	12	167	0	61	41	37	29	173
June	18.6	16.7	35.0	+0.2	8	87	0	36	44	40	32	145
July	17.7	18.1	34.0	−1.1	9	80	0	48	42	37	30	152
Aug.	20.6	18.8	37.8	+0.4	3	56	0	52	39	33	26	188
Sept.	24.3	18.8	40.2	3.0	3	53	0	31	35	29	24	244
Oct.	28.8	16.0	42.8	8.9	9	104	0	36	38	32	26	285
Nov.	30.7	14.6	43.7	11.7	22	200	0	50	38	35	28	310
Dec.	30.8	13.3	44.2	12.1	59	230	3	120	46	43	32	285
Annual	25.6	15.7	44.3	−1.1	423	1,068	214	211	45	39	30	2,656
Years	30	30	41	41	30	41	41	41	30	41	41	estim.*

TABLE XV (continued)

Month	Number of days								Mean cloudiness (oktas)		Mean sun-shine (h)	Wet periods			Global radiation (cal./cm² day)**	Probability of precipitation (%)		
	precipitation		max. temp.			min. temp.		thunder	09h00	15h00		number	>1 day	>5 days		<12.5 mm	<25 mm	<50 mm
	>0.25 mm	>2.5 mm	>37.8 °C	>32.2 °C	>26.7 °C	<2.2 °C	<10 °C											
Jan.	11	8	17	28	31	0	0	9	4.3	5.3	7.8	4.9	2.7	0.4	416	1	6	22
Feb.	10	5	8	26	27	0	0	6	3.8	4.7	7.8	4.4	2.8	0.2	500	6	7	20
Mar.	6	5	5	27	31	0	0	3	3.1	4.4	8.1	3.3	1.7	0.1	456	20	34	54
Apr.	2	1	1	23	30	0	1	1	1.9	2.7	9.0	1.1	0.3	<0.1	504	65	79	91
May	2	0	0	9	27	0	5	<3	1.6	1.8	9.1	0.7	0.2	<0.1	444	74	84	94
June	1	0	0	9	16	0.4	15	0	1.5	1.6	9.0	0.6	0.4	0	433	83	91	96
July	1	1	0	1	20	0.4	19	<1	1.4	1.5	9.6	0.3	0.2	<0.1	442	86	91	94
Aug.	1	0	0	8	28	0.1	13	<1	1.0	1.3	9.8	0.1	0.1	0	522	96	97	99
Sept.	0	0	1	22	30	0	4	<1	1.0	1.5	10.3	0.4	0.1	0	616	91	94	98
Oct.	2	1	11	29	31	0	<1	3	1.6	3.0	9.5	1.7	0.7	0	596	71	84	93
Nov.	5	1	19	29	30	0	0	4	2.3	4.6	9.4	3.4	1.1	<0.1	566	30	54	81
Dec.	8	4	16	29	31	0	0	8	3.0	4.4	8.3	4.8	1.2	0.2	528	12	20	46
Annual	49	26	78	232	332	0.9	57	37	3.2	3.1	9.0	25.7	11.5	1.0	502	—	—	—
Years	30	unsp.	10	10	10	10	10	14	1	1	1	unsp.	unsp.	unsp.	estim.** 68	68	68	68

* Estimated, Fitzpatrick method.
** Estimated, Black method.

TABLE XVI

CLIMATIC TABLE FOR MARBLE BAR, W.A., AUSTRALIA
Latitude 21°11'S, longitude 119°42'E, elevation 181 m.

Month	Mean station pressure (mbar)		Mean daily temp. (°C)	Mean daily temp. range (°C)	Extreme temp. (°C)		Precipitation (mm)*			Max. precip. 24 h. (mm)	Relative humidity (%)			Evap.* (mm)	Number of days				
	09h00	15h00			highest	lowest	mean	highest	lowest		daily index	mean 09h00	mean 15h00		precip. >0.25 mm	max. temp. >32.2 °C	>37.8 °C	min. temp. <2.2 °C	thunder
Jan.	987.2	983.7	33.7	15.2	49.2	18.9	82	310	0	146	42	43	27	271	7	30	28	0	10
Feb.	987.2	983.7	33.3	14.9	48.3	13.9	70	187	0	119	44	44	27	227	6	26	22	0	9
Mar.	989.3	985.9	32.1	14.5	46.7	15.3	58	389	0	305	35	39	25	238	4	29	19	0	4
Apr.	993.0	989.5	28.5	15.3	45.0	11.1	21	241	0	136	38	34	26	199	2	26	9	0	1
May	995.4	991.7	23.7	14.8	39.4	5.6	28	103	0	74	44	42	30	156	2	10	<1	0	0
June	997.4	993.6	19.9	14.6	35.6	1.1	19	159	0	105	48	42	29	117	2	<1	0	0	0
July	997.6	994.2	19.2	15.7	35.0	2.2	11	134	0	63	42	41	29	129	1	1	0	<1	0
Aug.	996.8	993.1	21.6	16.7	37.2	3.9	5	34	0	32	38	34	24	152	1	7	0	1	0
Sept.	995.0	991.1	25.4	18.9	42.7	5.6	2	24	0	24	34	29	21	202	0	23	2	0	0
Oct.	992.2	988.3	29.1	17.5	45.6	10.0	5	116	0	84	31	28	19	262	0	26	13	0	0
Nov.	989.6	985.8	32.5	17.1	47.2	14.4	8	61	0	60	31	27	20	284	1	30	24	0	3
Dec.	987.5	983.8	33.8	16.3	48.3	17.2	31	243	0	150	36	32	21	296	3	31	29	0	1
Annual	992.3	988.7	27.7	15.9	49.2	1.1	340	742	71	305	39	36	24	2535	29	240	145	1	28
Years	26	25	30	30	65	65	30	48	48	70	37	27	27	15**	30	65	65	65	4

TABLE XVI *(continued)*

Month	Mean cloudiness		Wind prev. dir.		Probability of precipitation (%)		
	09h00	15h00	09h00	15h00	<12.5 mm	<25 mm	<50 mm
Jan.	2.0	3.0	E	E	19	32	43
Feb.	2.1	2.9	SE	NE	13	13	35
Mar.	1.8	2.6	SE	SE	38	45	64
Apr.	1.4	1.9	SE	E, SE	64	78	90
May	1.8	2.0	ESE	E	64	71	81
June	1.4	1.4	SE	E	55	64	80
July	1.1	1.2	SE	E	64	77	87
Aug.	0.7	0.8	SE	E	77	93	100
Sept.	0.6	0.8	SE	NW	100	100	100
Oct.	0.6	1.1	SE	NW	90	97	97
Nov.	0.9	1.8	SE	EW	68	87	97
Dec.	1.1	2.4	S	W	29	58	81
Annual	1.3	1.8	—	—	—	—	—
Years	29	28	5	5	31	31	31

* The frequency distribution of the monthly rainfall at Roebourne, at the same latitude on the coast, is given in Table I.
** Australian copper tank evaporimeter, sunken.

TABLE XVII

CLIMATIC TABLE FOR ONSLOW, W.A., AUSTRALIA
Latitude 21°43′S, longitude 114°57′E, elevation 4.3 m

Month	Mean sta. press. (mbar)		Mean daily temp. (°C)	Mean daily temp. range (°C)	Temp. extremes (°C)		Precipitation (mm)			Max. precip. (24 h.)	Relative humidity (%)				Number of days			
	09h00	15h00			highest	lowest	mean	highest	lowest		daily index	mean 09h00	mean 15h00	lowest 15h00	precipitation		max. temp.	
															>0.25 mm	>2.5 mm	>32.2 °C	>37.8 °C
Jan.	1006.2	1003.9	29.6	12.3	47.7	15.8	21	261	0	237	57	52	52	38	2	2	26	9
Feb.	1006.2	1003.9	29.7	12.1	48.3	16.6	46	323	0	356	59	54	54	43	3	2	24	7
Mar.	1008.2	1005.8	29.1	12.2	46.4	14.7	63	415	0	314	58	54	53	39	3	4	28	9
Apr.	1012.0	1009.8	26.4	13.8	43.8	10.0	18	279	0	157	56	52	50	33	1	2	17	2
May	1014.1	1011.4	22.4	13.3	38.3	5.6	50	259	0	238	58	56	52	37	3	2	3	0
June	1016.2	1013.5	19.0	13.1	32.2	2.9	40	157	0	111	60	60	54	43	3	2	<1	0
July	1017.2	1014.4	18.0	14.3	32.3	3.1	20	221	0	90	58	51	57	39	3	1	0	0
Aug.	1016.4	1013.7	19.3	14.7	35.3	4.4	9	107	0	64	55	52	47	33	2	2	1	0
Sept.	1015.1	1012.1	21.6	15.7	38.3	5.5	<1	107	0	7	51	46	45	32	0	1	5	<1
Oct.	1012.8	1010.0	23.8	15.5	44.6	7.4	<1	15	0	7	50	44	44	32	0	1	13	2
Nov.	1009.9	1007.3	26.6	15.1	46.1	10.0	3	30	0	30	52	44	48	30	1	0	19	6
Dec.	1007.7	1005.2	28.3	13.8	47.5	12.5	3	61	0	50	55	47	50	35	1	1	25	10
Annual	1011.8	1009.2	24.5	13.8	48.3	3.1	274	717	14	356	56	52	50	—	22	18	161	44
Years	30	30	30	30	50	50	30	57	57	79	42	42	42	46	30	10	50	50

TABLE XVII *(continued)*

Month	Number of days				Visibility 09h00 frequency (%)			Mean cloudiness			Low cloud frequency (%) 09h00			Wind				Probability of precipitation (%)		
								all types		low				most freq. dir.		mean speed (kt.)				
	thunder	clear	cloudy	fog	<10 km	<20 km	<50 km	09h00	15h00		<2400 m	<900 m	<300 m	09h00	15h00	09h00	15h00	<12.5 mm	<25 mm	<50 mm
Jan.	7	17	3	0	7	19	96	1.8	1.8	0.6	41	22	4	S	SW	8	10	65	75	85
Feb.	4	17	2	0	24	44	92	1.9	2.1	1.0	12	12	4	S	W	8	11	62	70	87
Mar.	4	17	3	0	8	40	85	1.8	1.9	0.7	4	0	0	S	W	7	10	57	70	80
Apr.	1	18	2	1	4	8	69	1.6	1.8	0.2	15	12	8	S,E	N	8	9	75	82	88
May	1	19	3	0	4	8	73	1.8	2.0	0.6	19	15	0	E,SE,S	N	8	8	55	63	73
June	1	21	3	0	0	0	93	1.7	1.7	0.6	19	19	0	E,S	N	9	8	25	40	58
July	0	23	2	1	2	14	85	1.4	1.4	0.6	19	15	6	S	N	8	8	53	72	92
Aug.	0	23	2	1	10	30	87	1.0	1.0	0.9	19	13	2	S	N,W	8	9	85	90	92
Sept.	<1	26	<1	0	0	12	96	0.7	0.7	0.1	4	0	0	S	W	10	11	100	100	100
Oct.	<1	26	<1	0	0	12	100	0.7	0.5	0.1	32	4	0	S	W	8	11	97	100	100
Nov.	1	24	<1	0	8	15	100	0.8	0.7	0.2	23	0	0	S	W	8	11	100	100	100
Dec.	2	23	<1	0	22	50	92	1.0	1.1	0.3	15	11	0	S	W	8	11	95	95	97
Annual	21	254	22	1	7	21	89	1.4	1.4	0.5	19	10	2	S	W	8	10	—	—	—
Years	3	10	10	5	unsp.	unsp.	unsp.	30	30	3	unsp.	unsp.	unsp.	5	5	5	5	40	40	40

285

TABLE XVIII

CLIMATIC TABLE FOR MUNDIWINDI, W.A., AUSTRALIA
Latitude 23°52'S, longitude 120°52'E, elevation 408 m

Month	Mean sta. level press.[1] (mbar)		Temperature (°C)				Precipitation (mm)				Relative humidity (%)			
	09h00	15h00	mean	daily range	extreme max.	extreme min.	mean	highest	lowest	max. in 24 h	daily index	09h00	15h00	
Jan.	1009	1004	30.6	14.9	44.4	13.9	31	321	0	70	29	29	17	
Feb.	....	....	29.8	14.4	44.4	12.8	29	325	0	71	31	31	19	
Mar.	....	....	27.5	13.9	42.3	9.4	33	267	0	175	36	35	21	
Apr.	1016	1012	23.1	14.7	40.6	3.9	28	138	0	57	35	34	19	
May	....	....	18.1	14.7	36.4	-1.7	24	121	0	56	43	42	25	
June	....	....	13.8	15.0	29.8	-4.4	22	113	0	40	51	49	28	
July	1021	1017	13.2	15.9	30.6	-5.3	11	70	0	43	47	45	24	
Aug.	....	....	15.4	16.4	37.3	-3.3	7	53	0	39	40	38	19	
Sept.	....	....	19.6	17.8	37.2	-1.7	1	61	0	34	29	26	13	
Oct.	1015	1010	23.3	17.1	40.7	3.3	2	93	0	53	24	22	12	
Nov.	....	....	27.5	16.4	43.3	7.8	7	53	0	58	23	20	11	
Dec.	....	....	29.4	15.9	44.4	11.7	20	160	0	114	27	24	15	
Annual	....	....	22.6	15.6	44.4	-5.3	215	816	26	175	33	33	19	
Years	—	—	31	31	31	31	30	31	31	31	29	29	29	

TABLE XVIII (*continued*)

Month	Evap.[2] (mm)	Number of days					Sun-shine[2] (h)	Probab. of precip. (%)		
		precip. >0.25 mm	max. temp.		min. temp.			<12.5 mm	<25 mm	<50 mm
			>32.2 °C	>37.8 °C	<2.2 °C	<0 °C				
Jan.	335	6	29	20	0	0	292	24	47	77
Feb.	279	5	25	16	0	0	255	25	41	70
Mar.	253	5	25	10	0	0	260	39	53	66
Apr.	226	3	12	<1	0	0	262	57	71	84
May	129	3	1	0	0.4	0.2	250	61	66	84
June	76	3	0	0	3.0	1.6	237	49	72	88
July	109	2	0	0	3.5	2.9	272	80	90	98
Aug.	126	2	<1	0	2.2	0.8	285	78	88	98
Sept.	175	1	6	0	0.6	0	298	96	98	98
Oct.	253	1	16	1	0	0	320	80	92	96
Nov.	292	2	25	10	0	0	322	71	82	96
Dec.	330	3	29	20	0	0	318	37	76	90
Annual	2583	36	168	77	9.7	5.5	3371	—	—	—
Years	—	30	31	31	10	10	—	51	51	51

[1] ... = not available.

[2] Estimated by interpolation from maps.

TABLE XIX

CLIMATIC TABLE FOR GILES, W.A., AUSTRALIA
Latitude 25°02'S, longitude 128°17'E, elevation 580 m

Month	Mean sta. press. (mbar)	Mean daily ampl. (mbar)	Mean daily temp. (°C)	Mean daily temp. range (°C)	Temperatures extremes (°C) highest*	lowest*	Mean precip. (mm)	Max. precip. 24 h* (mm)	Rel. mean humidity* (%) 09h00	15h00	Mean evapor. (mm) **	Precip.* >0.25 mm
Jan.	942.2	3.5	30.7	13.4	45.0	10.0*	19	31	25	22	406	4
Feb.	942.9	3.4	39.7	13.4	46.1	12.2	21	79	31	25	371	3
Mar.	944.9	3.3	27.9	13.4	43.3	11.1	8	102	32	27	376	4
Apr.	948.0	3.0	23.3	12.9	39.4	5.0	12	76	32	29	282	3
May	949.5	2.2	17.4	12.0	33.3	−0.6	19	18	44	29	185	4
June	950.7	2.2	14.9	11.9	31.7	−2.8	10	41	52	32	141	3
July	951.4	2.3	13.3	12.4	31.7	−4.4	15	5	46	29	145	2
Aug.	949.9	3.2	15.5	13.5	33.9	−1.7	18	5	36	24	178	2
Sept.	948.2	3.3	20.2	14.1	38.3	+1.1	3	5	27	15	344	2
Oct.	945.9	3.4	24.4	14.7	40.0	3.9	7	10	31	17	323	3
Nov.	944.8	3.2	27.6	14.4	43.9	7.2	11	48	27	18	366	4
Dec.	943.0	3.7	28.7	13.9	43.9	9.4	31	46	28	17	421	3
Annual	946.8	3.1	22.8	13.3	46.1	−4.4	174	102	34	24	3429	37
Years	9	9	9	9	13	13	9	2	5	5	5	11

Month	Probab. precip. (%) <12.5 mm	<25 mm	<50 mm	Mean sun-shine (h)***	Wind preval. dir. 14h	20h	02h	07h	mean speed (kt.) 14h	20h	02h	07h
Jan.	50	70	100	310	E	SE	ESE	E	2	3	3	6
Feb.	60	90	90	265	ESE	SSE	ESE	E	2	3	4	6
Mar.	60	90	100	280	ESE	S	ESE	E	2	3	4	6
Apr.	60	70	100	275	SE	SSE	ESE	E	2	3	5	6
May	30	60	90	245	E	SE	E	E	2	1	3	4
June	60	80	90	240	NE	SE	ESE	E	3	1	3	4
July	60	70	100	260	N	E	E	E	1	1	2	4
Aug.	60	70	90	270	N	S	ESE	ESE	0	1	3	4
Sept.	91	100	100	290	WNW	S	ESE	E	1	2	3	4
Oct.	82	91	100	300	WNW	S	ESE	E	3	2	3	5
Nov.	54	73	91	315	NW	S	ESE	E	0	3	4	7
Dec.	36	63	91	320	NNE	SSE	ESE	E	1	3	4	5
Annual				3370	NE	SSE	ESE	E	1	2	3	5
Years	10	10	10	4	4	4	4	4	4	4	4	4

  * At Warburton Range Mission, 26°05'S, 126°36'E, altitude 365 m.
  ** Standard Australian galvanized iron tank, sunken.
  *** Estimated by interpolation from maps.

TABLE XX

CLIMATIC TABLE FOR YALGOO, W.A., AUSTRALIA
Latitude 28°21'S, longitude 116°41'E, elevation 318 m

Month	Mean sta. press. level (mbar)* 09h00	Mean sta. press. level (mbar)* 15h00	Mean daily temp. (°C)	Mean daily temp. range (°C)	Temperature extreme (°C) highest	Temperature extreme (°C) lowest	Precipitation (mm) mean	Precipitation (mm) highest	Precipitation (mm) lowest	Precipitation (mm) max. 24 h	Relative humidity (%) daily index	Relative humidity (%) 09h00	Relative humidity (%) 15h00
Jan.	1010	1006	28.8	16.1	46.1	10.0	14	101	0	61	38	40	25
Feb.	...	...	28.6	15.8	46.8	10.6	20	178	0	84	41	44	28
Mar.	...	...	26.1	14.3	43.9	8.9	24	120	0	99	45	51	33
Apr.	1017	1014	21.6	14.3	40.6	4.4	21	70	0	89	52	56	37
May	...	...	16.6	13.1	34.4	0.6	28	141	0	64	61	66	44
June	...	...	13.4	11.8	28.9	−1.1	40	105	5	51	68	76	54
July	1021	1017	12.4	11.8	27.2	−1.7	38	82	9	30	70	77	54
Aug.	...	...	13.5	13.1	33.9	−0.6	26	112	4	61	66	71	47
Sept.	...	...	16.6	15.3	35.6	+1.1	9	64	1	38	55	58	38
Oct.	1015	1012	19.6	16.0	40.0	−1.1	7	37	2	36	47	49	31
Nov.	...	...	23.9	16.4	42.8	+6.1	9	55	0	33	41	42	26
Dec.	...	...	27.1	16.8	45.0	+9.4	10	34	0	51	37	39	24
Annual	...	...	20.7	14.6	46.7	−1.7	246	497	94	99	49	56	37
Years	—	—	30	30	58	58	30	29	29	58	43	43	42

Month	Evapor. (mm) (**)	Evapor. (mm) (*)	Number of days with precip. >0.25 mm	Number of days with min. temp. <2.2°C	Number of days with min. temp. <0°C	Probability of precipitation (%) <12.5 mm	Probability of precipitation (%) <25 mm	Probability of precipitation (%) <50 mm	Sunshine monthly* (h)
Jan.	510	395	3	0	0	69	83	86	350
Feb.	442	350	3	0	0	52	72	93	300
Mar.	417	280	3	0	0	55	66	86	280
Apr.	296	210	3	0	0	65	83	96	255
May	189	130	5	0.1	0	28	58	79	230
June	125	80	8	1.5	0.3	17	28	52	210
July	131	70	9	2.3	0	10	41	76	230
Aug.	167	95	7	1.1	0	21	55	83	250
Sept.	235	132	3	0.1	0	59	86	93	270
Oct.	327	205	2	0	0	79	86	100	305
Nov.	412	280	2	0	0	86	93	96	330
Dec.	493	358	2	0	0	76	93	96	360
Annual	3744	2585	50	5.1	0.3	–	–	–	3370
Years	19*	–	30	10	10	31	31	31	–

* Estimated by interpolation from maps; ... = not available.
** Old records from portable evaporating dish, 20.3 cm in diameter, situated at Cue, 27°27'S, 117°52'E, 454 m; reading some 40% higher than standard Australian tank.

TABLE XXI

CLIMATIC TABLE FOR WILUNA, W.A., AUSTRALIA
Latitude 26°37'S, longitude 120°15'E, elevation 518 m

Month	Mean station pressure (mbar)		Temperature (°C)				Precipitation (mm)				Relat. humidity (%)			Mean evapor. (mm)	
	09h00	15h00	mean daily	daily range	extreme highest	lowest	mean	highest*	lowest*	max. 24 h	daily index	mean 09h00	mean 15h00	tank[1]	dish[2]
Jan.	951.3	948.3	30.1	15.2	46.7	9.4	38	214	0	82	36	33	21	341	494
Feb.	952.1	948.9	29.3	15.0	45.6	12.2	34	134	0	97	37	35	23	285	427
Mar.	954.1	951.0	26.9	14.6	43.3	7.1	45	154	0	77	41	39	26	252	389
Apr.	957.1	954.1	21.7	14.4	40.1	3.9	18	138	0	74	45	44	27	181	256
May	958.2	955.3	16.7	13.9	35.0	-0.3	24	131	0	59	53	53	35	118	178
June	958.9	956.4	13.1	13.4	30.0	-3.3	24	156	0	68	57	61	39	82	121
July	959.7	956.7	12.1	13.9	30.6	-3.2	12	51	0	30	56	63	38	90	129
Aug.	959.2	956.0	14.2	15.2	34.4	-2.3	9	77	0	42	50	50	29	116	162
Sept.	957.8	954.7	17.9	16.6	37.6	-0.6	3	36	0	32	43	37	23	170	236
Oct.	955.4	952.3	21.6	16.5	41.1	+1.1	4	26	0	78	39	33	20	244	333
Nov.	953.6	950.2	25.5	16.3	44.4	3.6	6	94	0	32	35	30	19	299	412
Dec.	951.5	948.7	28.7	16.1	45.0	7.8	17	104	0	72	34	29	19	344	493
Annual	955.7	952.7	21.5	15.1	46.7	-3.3	234	516	49	97	42	39	25	2522	3633
Years	30	30	30	30	66	66	30	>50	>50	66	36	30	30	7	15

TABLE XXI (*continued*)

Month	Number of days with — max. temp. >37.8 °C	>32.2 °C	min. temp. <7.2 °C	<2.2 °C	<0 °C	precip. >0.25 mm	>2.5 mm	thunder	Number of wet periods >0 days	>2 days	>5 days	Number of periods with rainfall >12.5 mm	>25 mm	>50 mm	Mean cloudiness 09h00	15h00	Prev. wind dir. 09h00	15h00	Probability of precipitation (%) <12.5 mm	<25 mm	<50 mm
Jan.	20.1	28.9	0	0	0	4	2	2	1.00	0.64	0.21	0.67	0.40	0.12	1.8	2.5	NE	SE	48	63	82
Feb.	11.5	22.7	0	0	0	3	1	4	1.38	0.48	0.12	0.48	0.36	0.17	1.7	2.3	NE	NE,SE	48	63	78
Mar.	5.6	20.7	0	0	0	5	3	3	1.78	0.64	0.24	0.57	0.48	0.21	1.8	2.4	NE	SE	59	63	70
Apr.	0.2	8.3	0.3	0	0	3	2	1	1.26	0.43	0.07	0.48	0.24	0.12	2.0	2.2	NE	E	52	63	82
May	0	0.5	8.9	0.8	0	4	1	<1	2.05	0.67	0.17	0.57	0.33	0.12	2.2	2.2	NE	NE	41	59	78
June	0	0	17.2	4.0	1	4	1	<1	1.76	0.64	0.24	0.40	0.24	0.12	2.1	2.1	NW	NW	37	55	89
July	0	0	20.8	7.6	2	4	1	0	1.90	0.48	0.10	0.19	0.05	0	1.8	2.0	NE	W	63	89	96
Aug.	0	0	16.9	2.3	1	3	2	0	1.64	0.38	0.05	0.24	0.07	0	1.5	1.4	NE	NW	78	89	96
Sept.	0	2.7	7.2	0.9	0	1	0	0	0.71	0.19	0	0.10	0.02	0	1.0	1.1	NE,SE	SW	93	96	96
Oct.	0.8	10.6	1.4	0	0	1	0	1	0.83	0.07	0	0.10	0.02	0.02	1.0	1.5	NE	SW	89	100	100
Nov.	5.3	19.4	0	0	0	2	1	3	1.07	0.36	0.12	0.19	0.10	0	1.2	1.8	NE	SW	82	92	96
Dec.	14.0	26.3	0	0	0	3	2	1	1.50	0.48	0.10	0.31	0.17	0.07	1.5	2.2	NE	E	48	81	88
Annual	57.5	140.1	72.7	15.6	4	37	16	17.78	16.88	5.46	1.42	4.30	2.48	0.95	1.6	2.0	—	—	—	—	—
Years	38	38	38	10	38	30	unsp.	38	38	38	38	38	38	38	30	30	5	5	27	27	27

* At Meekatharra, 26°35'S, 118°30'E, 511 m.
1 Standard Australian copper tank, sunken.
2 Portable evaporating dish, 20.3 cm in diameter.

TABLE XXII

CLIMATIC TABLE FOR GERALDTON, W.A., AUSTRALIA
Latitude 28°45'S, longitude 114°36'E, elevation 4 m

Month	Mean daily sta. press. (mbar)		Mean daily temp. (°C)	Mean daily temp. range (°C)	Temp. extremes		Precipitation (mm)									Relative humidity (%)		Evaporation (mm)	
	09h00	15h00			highest (°C)	lowest (°C)	Mean	highest	lowest	max. 24 h	standard dev. σ	lower quartile	median	upper quartile	daily index	09h00	15h00	(*)	(**)
Jan.	1008.7	1007.3	24.1	10.7	47.7	8.9	7	96	0	79	15	0	1	7	61	52	49	187	299
Feb.	1009.1	1007.2	24.3	10.4	46.4	10.6	10	117	0	82	20	0	1	7	61	54	50	159	251
Mar.	1011.0	1009.0	23.5	10.3	44.3	8.3	16	169	0	94	30	<1	4	14	61	56	51	160	221
Apr.	1014.1	1012.3	21.5	10.9	39.4	5.4	30	93	0	69	24	3	17	36	60	59	51	130	167
May	1015.3	1013.3	18.6	9.6	34.8	2.1	66	328	6	78	51	35	62	89	64	66	54	90	102
June	1016.3	1014.5	16.6	8.8	28.8	0.8	113	328	34	109	55	69	113	161	67	75	57	67	60
July	1017.2	1015.3	15.4	8.9	27.2	0.8	96	205	18	51	53	76	93	114	68	75	58	68	59
Aug.	1017.1	1015.3	15.8	9.3	31.6	1.7	64	242	8	93	46	38	54	86	68	72	57	86	71
Sept.	1016.7	1014.7	16.8	9.7	35.8	1.8	26	105	1	43	23	18	26	46	67	64	54	92	100
Oct.	1014.6	1012.7	18.1	10.1	40.3	3.3	18	70	0	71	12	7	13	25	65	56	53	117	154
Nov.	1012.2	1010.6	20.8	10.3	42.7	5.6	7	40	0	36	8	1	5	8	65	50	51	141	219
Dec.	1010.0	1008.6	22.5	10.3	45.0	7.7	5	32	0	51	7	0	1	5	63	52	53	169	284
Annual	1013.5	1011.7	19.8	9.8	47.7	0.8	457	855	262	109	—	—	—	—	65	60	53	1465	1997
Years	30	30	30	30	70	70	30	67	67	80	—	74	74	74	42	29	29		13

TABLE XXII (*continued*)

Month	Number of days									Mean cloudiness (okta)				Wind				Probability of precipitation (%)		
	precipitation		max. temp.		min. temp. (<2.2 °C)	thunder	cloudy	clear	fog	all types		low[1]		most freq. dir.		mean speed (knots)		<12.5 mm	<25 mm	<50 mm
	>0.25 mm	>2.5 mm	>32.2 °C	>37.8 °C						09h00	15h00	09h00	15h00	09h00	15h00	09h00	15h00			
Jan.	2	1	8.2	3.4	0	1	3	13	0	2.3	1.9	1.9	1.0	S	S	10	16	88	96	98
Feb.	2	<1	9.8	3.5	0	<1	2	15	2	2.2	1.9	1.6	1.0	S	S	8	15	85	94	99
Mar.	3	2	9.8	2.1	0	0	3	15	1	2.3	2.1	1.6	1.4	SE,E,S	S	8	14	72	84	94
Apr.	5	2	3.9	0.3	0	1	4	14	3	2.5	2.6	1.7	1.8	E	S	9	12	46	59	87
May	10	6	1.1	0	0.1	<1	5	13	1	2.8	3.0	1.8	2.1	E	S	9	10	8	14	39
June	13	9	0	0	0.1	1	6	11	3	3.0	3.1	2.6	2.9	E	S,NW,W	9	8	0	1	11
July	15	10	0	0	0.6	0	4	12	0	2.9	3.0	2.4	2.3	E	S	8	7	1	2	7
Aug.	12	8	0	0	0.1	0	4	13	0	2.6	2.6	2.6	3.0	E	W,S	7	9	3	5	44
Sept.	8	4	0.1	0	0.1	<1	4	15	0	2.4	2.1	3.6	2.7	E	SSW	8	11	15	48	81
Oct.	7	2	1.3	0.1	0	0	4	13	2	2.5	2.2	2.7	2.2	S,SE	S,SW	8	13	47	79	96
Nov.	3	1	4.9	0.9	0	<1	3	13	0	2.3	1.8	1.9	2.2	S	S	10	13	86	93	100
Dec.	2	1	5.0	1.8	0	1	3	13	1	2.1	1.9	2.1	1.3	S	S	10	15	92	97	100
Annual	82	46	44.1	12.1	1.0	5	45	160	13	2.5	2.3	2.2	2.0	E	S	9	12	—	—	—
Years	29	10	70	70	52	3	10	10	3	30	30	5	5	5	5	3	3	74	74	74

* Computed, Halstead formula.
** At Chapman Research Station, Nabawa (28°30'S, 114°49'E, 14.6 m, *not* on the coast), Australian sunken copper tank, 13 years.
[1] At Geraldton Airport, 28°48'S, 114°42'E.

TABLE XXIII

CLIMATIC TABLE FOR KALGOORLIE, W.A., AUSTRALIA
Latitude 30°45'S, longitude 121°30'E, elevation 380 m

Month	Sta. press. (mbar) mean 09h00	mean 15h00	highest	lowest	Temp (°C) mean daily	mean daily range	extreme highest	extreme lowest	Precip. (mm) mean	highest	lowest	max. 24 h	Rel. hum. (%) daily index	09h00 highest	09h00 mean	09h00 lowest	15h00 highest	15h00 mean	15h00 lowest	Mean evapor. (mm) (*)	(**)
Jan.	971.3	967.8	995.6	959.9	25.7	16.1	45.8	8.4	24	204	0	96	46	54	36	27	43	40	19	323	328
Feb.	972.4	968.9	993.9	964.2	24.9	15.9	46.1	8.9	27	315	0	178	40	63	42	30	54	45	22	251	267
Mar.	974.2	970.8	996.9	965.2	23.0	13.7	43.9	5.3	24	166	0	71	55	72	42	34	50	50	23	191	235
Apr.	976.8	973.9	999.0	966.7	18.7	12.9	39.2	1.7	18	103	0	69	59	76	47	39	59	55	26	134	155
May	977.1	974.3	1000.4	963.7	14.7	11.8	33.3	-1.8	22	96	0	45	66	82	55	46	65	61	38	83	96
June	977.6	974.7	1000.9	963.1	12.0	10.6	27.7	-2.0	25	119	0	57	74	91	57	52	70	66	38	58	63
July	977.3	974.5	1006.7	961.4	10.8	10.9	27.2	-3.3	24	82	2	28	74	84	63	50	66	63	42	64	64
Aug.	977.7	974.1	1002.3	965.5	12.3	12.3	30.6	-2.4	23	81	0	31	66	75	55	44	61	56	34	80	91
Sept.	976.5	973.2	1003.0	966.7	15.3	14.1	35.6	-0.6	13	98	0	44	53	72	44	34	54	50	26	127	137
Oct.	974.5	970.9	997.9	958.0	18.2	14.6	40.7	-1.0	14	80	0	62	48	60	38	31	46	45	20	183	197
Nov.	973.2	969.5	995.1	963.3	21.4	15.6	43.7	3.4	15	70	0	65	44	58	31	28	44	43	19	249	264
Dec.	971.6	968.0	993.0	964.4	24.3	16.0	45.0	7.5	13	65	0	37	43	57	33	27	41	38	19	315	329
Annual	975.0	971.7	—	—	18.4	13.7	46.1	-3.3	244	458	121	178	56	62	49	37	51	42	30	2058	2228
Years	30	30	30	30	30	30	69	69	30	69	69	69	44	30	30	30	30	30	30	30	17

TABLE XXIII (continued)

Month	Number of days with								Mean cloudiness, all types		Prev. wind dir.		Visibility frequency (%)		Probability of precipitation (%)		
	precip.		max. temp.		min. temp.			thunder									
	>0.25 mm	>2.5 mm	>32.2 °C	>37.8 °C	<4.4	<2.2	<0°C		09h00	15h00	09h00	15h00	<5 km	<20 km	<12.5 mm	<25 mm	<50 mm
Jan.	3	3	18.8	7.5	0	0	0	2	2.0	2.1	SE	SE,ESE	0	58	80	87	97
Feb.	3	1	12.9	4.3	0	0	0	4	2.1	1.8	ESE	ESE	0	84	60	73	87
Mar.	4	4	10.8	2.7	0	0	0	2	2.5	2.2	SE	SE	0	97	50	70	90
Apr.	4	1	2.9	0.3	0.3	0	0	2	2.6	2.7	E	E	2	26	57	63	80
May	6	2	0.1	0	2.6	0.5	0	1	2.9	3.1	NW	WNW	2	24	23	57	80
June	7	2	0	0	5.9	2.1	0.1	<1	3.0	3.4	NW	NW	2	52	23	50	80
July	8	3	0	0	11.0	2.0	0.4	<1	2.8	3.1	NW	W	6	55	45	56	96
Aug.	7	2	0	0	7.7	2.2	0	<1	2.3	2.6	NW	NW	4	36	37	66	86
Sept.	5	<1	0.4	0	2.9	0.2	0.1	1	1.9	2.0	NW	NW	0	15	70	83	97
Oct.	4	2	2.9	0.1	0.6	0	0	2	2.1	2.2	SE	W	8	50	53	80	93
Nov.	3	2	7.4	1.3	<0.1	0	0	3	2.0	2.3	SE,ESE	SE	15	81	70	80	93
Dec.	3	1	14.8	3.9	0	0	0	3	1.9	2.0	E	E	0	70	63	77	87
Annual	57	23	71.0	20.1	31.0	7.0	0.6	19	2.3	2.5	—	—	—	—	—	—	—
Years	30	unsp.	69	69	30	30	10	13	30	30	5	5	1	1	30	30	30

* Estimated.

** Australian tank, at Coolgardie 30°54'S, 121°12'E, 427 m.

TABLE XXIV

CLIMATIC TABLE FOR MERREDIN, W.A., AUSTRALIA
Latitude 31°29'S, longitude 118°17'E, elevation 319 m

Month	M.s.l. press. (mbar)		Temperature (°C)				Precipitation (mm)									Relat. humidity (%)	
	09h00	15h00	mean	daily range	extreme max.	extreme min.	mean	highest	lowest	max. 24 h	st. dev. σ	lower quartile	median	upper quartile	daily index	09h00	15h00*
Jan.	1013	1008	25.1	16.9	45.0	7.5	8	56	0	40	13	0	5	15	42	50	29
Feb.	....	....	24.8	16.4	44.4	6.1	13	80	0	66	18	1	5	11	44	55	31
Mar.	....	....	22.2	14.7	43.1	5.1	17	161	0	83	34	2	13	28	51	64	38
Apr.	1020	1016	18.3	13.5	38.8	−1.2	22	114	0	60	27	8	15	35	58	71	42
May	....	....	13.9	12.2	34.2	−3.9	43	117	1	49	24	22	40	56	66	81	54
June	....	....	11.4	10.6	27.4	−2.8	50	131	6	41	27	32	50	65	72	85	62
July	1020	1017	10.1	11.1	24.7	−3.7	55	126	12	46	22	39	50	67	74	89	63
Aug.	....	....	10.8	12.2	27.9	−3.4	43	86	6	34	19	24	41	53	73	85	57
Sept.	....	....	13.3	14.9	33.7	−2.5	23	86	0	45	13	14	23	32	65	75	46
Oct.	1017	1013	16.1	15.7	39.2	−1.2	19	75	1	27	17	8	15	26	55	59	37
Nov.	....	....	20.5	16.4	41.7	+0.6	15	69	0	36	12	4	10	16	44	49	31
Dec.	....	....	23.3	16.9	44.4	5.0	12	92	0	49	20	1	5	22	42	47	29
Annual	—	—	17.5	15.6	45.0	−3.9	320	564	130	83	—	—	—	—	55	67	43
Years	—	—	34	34	34	34	30	41	41	41	49	49	49	49	20	20	30*

TABLE XXIV (*continued*)

Month	Evap.** (mm)	Number of days with						Cloudiness (oktas)		Sun-shine (h)***	Probability of precip. (%)		
		precip. >0.25 mm	max. temp.		min. temp.			09h00	15h00		<12.5 mm	<25 mm	<50 mm
			>32.2 °C	>37.8 °C	<2.2 °C	<0 °C							
Jan.	334	2	19.1	5.9	0	0		2.0	1.8	343	71	86	95
Feb.	274	28	15.7	4.7	0	0		2.4	2.1	295	68	82	95
Mar.	239	3	10.2	1.2	0	0		2.9	2.6	268	55	73	91
Apr.	152	5	2.0	0.1	0.2	0		3.2	2.9	230	54	64	91
May	93	8	0	0	2.1	0.2		3.5	4.0	200	14	23	64
June	59	11	0	0	4.4	1.8		4.0	4.5	170	9	18	50
July	55	13	0	0	9.2	2.3		3.5	4.4	200	0	0	50
Aug.	69	10	0.1	0	8.3	1.5		3.4	4.1	220	5	36	82
Sept.	105	7	1.5	0.1	5.3	0.4		2.8	3.4	242	14	50	91
Oct.	169	5	7.4	0.8	1.3	0		3.0	3.0	278	36	64	86
Nov.	248	3	14.0	3.3	0	0		2.5	2.3	325	72	90	100
Dec.	312	2	14.0	3.3	0	0		2.2	2.1	355	64	77	91
Annual	2109	71	70.0	16.1	30.8	6.2		3.0	3.1	3126	—	—	—
Year	41	30	41	41	41	10		unsp.	unsp.*	unsp.	22	22	22

* At Kellerberrin, 31°39'S, 117°47'E, 250 m.

** Standard Australian tank.

*** Estimated by interpolation from maps.

TABLE XXV

CLIMATIC TABLE FOR PERTH, W.A., AUSTRALIA
Latitude 31°57′S, longitude 115°51′E, elevation 60 m

Month	M.s.l. press. (mbar)	Daily ampl. (mbar)	Mean daily temp. (°C)	Mean daily temp. range (°C)	Temp. extreme (°C)				Dry and wet bulb temp. difference (°C)		
					highest		lowest		09h00	15h00	21h00
					sun	shade	shade	grass			
Jan.	1012.5	2.2	23.4	11.8	80.7	43.7	9.2	+4.2	6.2	8.1	4.2
Feb.	1012.8	2.4	23.9	12.0	78.7	44.6	8.7	+4.3	5.9	8.4	4.3
Mar.	1014.9	2.5	22.2	11.0	75.0	41.3	7.7	+2.6	5.0	7.7	3.9
Apr.	1017.5	2.3	19.2	10.5	69.4	37.6	4.1	−0.6	4.2	6.6	3.2
May	1017.7	2.1	16.1	9.0	63.3	32.4	1.3	−3.7	2.7	4.8	2.3
June	1017.9	1.9	13.7	8.1	57.5	27.6	1.6	−3.4	2.0	3.7	1.8
July	1018.5	2.0	13.1	8.2	56.2	24.7	1.2	−3.8	1.9	3.7	1.7
Aug.	1019.7	1.9	13.5	8.6	62.8	27.8	1.9	−2.9	2.4	4.1	1.8
Sept.	1019.7	1.9	14.7	9.1	67.6	32.7	2.6	−2.7	3.1	4.5	2.1
Oct.	1017.1	1.7	16.3	9.5	71.8	37.2	4.4	−1.2	4.1	5.1	2.4
Nov.	1015.2	1.9	19.2	10.8	75.0	40.3	5.6	+1.7	5.3	6.4	3.2
Dec.	1013.4	1.9	21.5	11.3	76.0	42.2	8.6	+3.3	6.1	7.1	3.4
Annual	1016.3	2.1	18.1	10.0	80.7	44.6	1.2	−3.8	4.1	5.9	2.8
Years	30	56	30	30	67	67	67	67	67	67	67

Month	Precipitation (mm)				Relat. humidity (%)*			Meridian alt. of sun	Mean daily glob. radiation (cal./cm²)	Mean evapor. (mm)
	mean	highest	lowest	max. 24 h	09h00	15h00	index			
Jan.	7	55	0	44.2	51	43	53	78°	650	263
Feb.	12	166	0	87.1	53	43	52	68°	603	219
Mar.	22	145	0	77.0	57	46	57	58°	493	191
Apr.	52	149	0	66.6	62	48	60	48°	349	117
May	125	307	20	76.2	71	58	68	38°	258	71
June	192	477	55	99.1	77	63	72	35°	219	46
July	183	425	61	76.2	77	63	73	38°	235	45
Aug.	135	318	12	73.9	72	60	71	48°	321	62
Sept.	69	199	9	46.2	67	57	64	58°	427	87
Oct.	54	200	4	43.9	61	54	64	68°	538	137
Nov.	23	71	0	39.1	53	47	57	78°	633	195
Dec.	15	81	0	46.7	51	46	54	82°	672	246
Annual	889	1250	508	99.1	63	52	62	—	449	1688
Years	30	87	87	87	67	30	30	—	11–12	30

TABLE XXV *(continued)*

Month	Number of days with										
	gale	precipit.		max. temp.		min. temp.	thunder	lightning	clear	fog	dew
		>2.5 mm	>0.25 mm	>37.8 °C	>32.2 °C	<4.4 °C					
Jan.	1	1	3	1.5	8.7	0	1	1.8	14	0	2.7
Feb.	2	2	3	1.8	8.0	0	<1	1.5	13	0	3.7
Mar.	3	2	5	0.6	5.5	0	<1	1.7	12	0	6.1
Apr.	3	4	8	0	1.2	0	<1	1.3	9	1	10.3
May	3	8	15	0	0	0.1	2	2.5	6	2	12.7
June	7	13	17	0	0	0.3	2	2.3	5	2	12.4
July	7	14	19	0	0	0.7	2	1.9	5	2	12.5
Aug.	8	12	19	0	0	0.4	1	1.5	6	1	11.7
Sept.	8	9	15	0	0	0.1	<1	1.0	8	0	10.4
Oct.	4	6	12	0	0.2	0	<1	0.9	8	0	6.4
Nov.	2	2	7	0.1	2.3	0	<1	1.3	9	0	3.6
Dec.	1	2	5	0.8	5.5	0	1	1.6	13	0	2.7
Annual	49	75	128	5.1	31.1	1.7	13	18.7	108	8	95.0
Years	5	10	30	65	65	65	42	66	30	65	unsp.

Month	Mean cloudiness (oktas)		Mean sunshine (h)	Mean vapor press. (mbar)	Wind				highest daily mean	highest gust speed	Probability of precip. (%)		
					most freq. dir.		mean speed (kt.)				< 12.5 mm	< 25 mm	< 50 mm
	all types	low			09h00	15h00	09h00	15h00					
Jan.	2.3	1.2	10.4	14.5	E	SSW	12	17	22.8	42	77	93	99
Feb.	2.5	1.0	10.0	14.7	ENE	SSW	12	17	18.5	47	75	89	97
Mar.	2.8	1.7	8.8	14.1	E	SSW	12	15	18.5	57	46	72	93
Apr.	3.4	2.5	7.3	13.0	ENE	SSW	11	13	27.3	55	18	33	59
May	4.3	2.6	5.8	11.9	NE	WSW	10	12	23.5	59	0	3	12
June	4.7	3.8	4.8	11.0	N	NW	10	12	26.1	70	0	0	0
July	4.5	4.3	5.3	10.6	NNE	W	10	12	28.8	67	0	0	0
Aug.	4.5	4.1	6.1	10.7	N	WNW	10	14	27.6	68	1	1	3
Sept.	3.9	3.8	7.1	11.4	ENE	SSW	11	15	24.9	59	3	5	20
Oct.	3.8	3.0	8.1	11.8	SE	SW	12	16	23.1	56	1	15	50
Nov.	3.1	1.8	9.7	13.0	E	SW	12	17	22.2	55	33	66	95
Dec.	2.6	1.4	10.4	13.9	E	SSW	12	18	22.1	56	55	81	97
Annual	3.5	2.6	7.8	12.5	E	SSW	11	15	28.8	70	—	—	—
Years	30	5*	67	65	30	30	10	10	63	48	76	76	76

* Maylands, 3.2 km east-northeast of Perth.

TABLE XXVI

CLIMATIC TABLE FOR EYRE, W.A., AUSTRALIA
Latitude 32°14'S, longitude 126°22'E, elevation 4.6 m

Month	Mean sta. press. (mbar)		Mean daily temp. (°C)	Mean daily temp. range (°C)	Temp. extremes (°C)		Precipitation (mm)				Relative humidity (%)		
	09h00	15h00			highest	lowest	mean	highest	lowest	max. 24 h	daily index	mean 09h00	mean 15h00
Jan.	1015.9	1014.1	20.7	10.6	47	+4	13	60	0	41	62	55	60
Feb.	1015.9	1014.0	21.2	10.1	47	+3	11	82	0	41	64	58	62
Mar.	1018.0	1016.0	19.9	10.9	43	+3	22	204	1	71	65	59	60
Apr.	1021.0	1019.0	17.9	11.4	41	+1	23	115	1	50	68	62	61
May	1020.0	1018.0	15.2	11.2	36	0	39	156	11	37	71	67	60
June	1019.0	1017.0	12.9	11.1	31	−2	43	122	6	39	71	70	60
July	1020.0	1018.0	11.9	11.4	32	−4	32	84	4	22	71	70	59
Aug.	1020.1	1017.9	12.6	12.2	32	−3	31	88	4	20	68	64	57
Sept.	1018.2	1015.8	14.6	12.6	38	−3	21	82	0	41	64	58	54
Oct.	1017.1	1015.0	16.3	12.2	41	−2	19	81	1	56	62	55	56
Nov.	1017.0	1015.0	18.1	11.7	46	+1	18	71	0	27	62	53	59
Dec.	1014.9	1013.0	19.5	10.8	46	+3	16	87	0	32	63	54	60
Annual	1018	1016	16.7	11.3	47	−4	289	506	132	71	66	60	59
Years	29	29	28	28	29	29	48	42	42	42	27	28	28

TABLE XXVI (*continued*)

Month	Mean evap. (mm)	Number of days with								Mean cloudiness (oktas)					Wind				Probability of precipitation (%)		
		precipitation		min. temp.		thunder	fog	clear	cloudy	average	highest		lowest		most freq. dir.		mean speed (kt.)		<12.5 mm	<25 mm	<50 mm
		>0.25 mm	>2.5 mm	<2.2 °C	<0 °C						09h00	15h00	09h00	15h00	09h00	15h00	09h00	15h00			
Jan.	173	13	3	0	0	1	0	7	6	3.6	5.4	4.4	2.4	1.6	SE	SE	8	11	68	75	85
Feb.	154	2	4	0	0	1	0	7	6	3.7	5.3	4.2	2.8	2.1	SE	SE	8	11	76	81	95
Mar.	142	3	5	0	0	<1	<1	7	6	3.8	6.1	5.6	2.6	1.6	SE	SE	8	10	66	78	86
Apr.	104	2	6	0.1	0	<1	<1	7	5	3.8	5.5	5.1	2.4	1.8	NE,N,SE	SE	7	9	46	71	90
May	81	5	10	0.3	0	1	<1	7	6	3.8	6.0	6.4	2.0	2.4	NW,NW	SW,SE	8	7	12	27	66
June	68	4	10	2.0	0.5	<1	<1	8	5	3.7	4.5	5.2	2.1	2.6	N,NW	SW	7	8	24	34	60
July	68	4	10	3.2	1.8	0	<1	12	3	3.2	4.4	4.6	2.1	1.2	NW	SW	8	8	17	41	90
Aug.	86	4	9	3.5	1.7	<1	<1	12	4	3.0	5.1	5.0	2.0	2.2	NW,N,W	SW,NW	8	8	15	42	88
Sept.	114	2	7	1.6	0.6	<1	<1	12	3	2.9	4.5	4.4	1.4	1.3	N	S,SE,SW	8	9	27	73	93
Oct.	145	2	6	1.1	0.4	1	0	10	2	3.0	4.5	3.8	1.7	1.6	SW	SE	8	9	42	68	93
Nov.	147	2	4	0.2	0	2	<1	9	4	3.2	5.1	4.7	2.4	1.8	SE	S.SE	8	10	54	76	93
Dec.	183	2	4	0	0	1	<1	9	5	3.2	5.2	4.6	2.2	1.6	SE	SE	7	10	73	78	90
Annual	1463	33	78	12.0	5.0	8	2	107	55	3.4	—	—	—	—	N,SE,SW	SE			—	—	—
Years	8+	10	41	10	10	10	10	10	10	28	29	29	29	29	5	5	5	5	41	41	41

* At Eucla, 31°45'S, 128°58'E, 4.5 m.

TABLE XXVII

CLIMATIC TABLE FOR ESPERANCE, W.A., AUSTRALIA
Latitude 33°50'S, longitude 121°55'E, elevation 4.3 m

Month	Mean daily sta. press. (mbar)		Temperature (°C)				Precipitation (mm)				Relative humidity (%)			Mean evaporation (mm)		Number of days precipitation	
				mean daily	extremes						daily index	mean					
	09h00	15h00	mean daily	range	highest	lowest	mean	highest	lowest	max. 24 h		09h00	15h00	(*)	(**)	>0.25 mm	>2.5 mm
Jan.	1015.2	1013.7	20.1	9.3	47	5	20	133	0	69	70	62	63	132	234	5	2
Feb.	1016.4	1014.8	20.5	9.6	44	5	18	120	0	39	69	63	64	108	185	4	1
Mar.	1017.8	1016.1	19.6	9.2	44	5	32	124	0	44	72	66	64	102	175	7	3
Apr.	1019.7	1017.6	17.4	9.9	39	3	49	176	5	126	75	70	65	86	91	9	4
May	1018.7	1017.0	14.9	9.5	33	3	92	179	20	52	77	75	64	69	69	15	8
June	1018.5	1016.9	12.8	9.3	27	0	100	273	30	72	77	78	65	54	44	15	10
July	1018.5	1017.0	12.1	9.3	26	−1	107	240	24	55	77	78	65	55	46	16	11
Aug.	1018.7	1016.9	12.6	9.9	32	0	95	284	19	59	77	74	62	65	55	15	10
Sept.	1018.7	1016.9	13.9	10.3	36	1	66	143	11	116	72	69	63	77	88	13	7
Oct.	1017.4	1015.8	15.2	10.1	40	1	52	146	14	45	73	65	64	97	128	12	5
Nov.	1016.5	1015.0	17.3	9.5	41	3	27	73	0	51	71	62	64	110	175	7	4
Dec.	1015.4	1013.8	18.9	9.4	43	4	22	83	0	71	70	61	63	125	218	6	2
Annual	1017.6	1016.0	16.3	9.7	47	−1	679	921	438	126	73	69	64	1079	1508	124	67
Years	30	30	30	30	47	47	30	60	60	60	30	30	44	—	7	30	10

TABLE XXVII (*continued*)

Month	Number of days					Monthly cloudiness						Wind				Probability precipitation (%)		
	min. temp.		thunder	clear	fog	mean		highest		lowest		most freq. dir.		mean speed (kts)		<12.5	<25	<50
	<2.2 °C	<0°C				09h00	15h00	09h00	15h00	09h00	15h00	09h00	15h00	09h00	15h00	mm	mm	mm
Jan.	0	0	2	7	0	3.5	2.6	5.6	5.6	1.0	0.5	NE,SW	SE,SW	9	13	60	81	98
Feb.	0	0	2	8	1	3.6	2.6	6.7	5.4	1.6	0.9	NE,SW	SE	8	13	57	74	88
Mar.	0	0	1	7	1	3.8	2.8	6.6	5.7	2.0	0.8	NE	SE	8	12	45	62	86
Apr.	0	0	1	6	0	3.8	3.4	6.6	5.5	2.2	1.3	SW	SW	8	11	26	50	74
May	0	0	1	6	1	3.7	3.8	5.8	5.8	2.2	1.7	SW,W	SW	10	10	0	5	26
June	0.6	0.2	<1	6	1	3.5	3.9	5.6	6.1	2.2	1.4	SW,W	SW	9	9	0	0	10
July	0.6	0	1	7	0	3.4	3.8	5.5	6.5	2.0	1.8	W,SW	SW	8	10	0	2	10
Aug.	0.5	0.1	1	7	1	3.4	3.4	5.8	5.3	1.3	1.1	W,SW	SW	12	13	0	2	17
Sept.	0.1	0	1	8	1	3.4	3.0	5.8	5.3	1.1	0.7	SW,W	SW	10	12	2	7	26
Oct.	0	0	1	7	0	3.4	3.0	6.0	5.8	2.0	1.5	SW	SW	9	11	0	14	52
Nov.	0	0	1	6	0	3.8	2.8	6.4	4.8	1.5	0.8	SW	SE	8	12	24	48	93
Dec.	0	0	1	7	0	3.5	2.6	5.5	4.5	1.0	0.5	SW	SW	8	12	45	67	86
Annual	1.8	0.3	11	82	2	3.6	3.1	6.0	5.5	1.7	1.1	SW	SW	9	12	—	—	—
Years	10	10	14	10	5	30	30	45	45	45	45	5	5	5	5	42	42	42

(*) Estimated.
(**) At Salmon Gums (32°59'S, 121°37'E, 249 m, not on the coast; standard Australian sunken tank).

TABLE XXVIII

CLIMATIC TABLE FOR ALBANY, W.A., AUSTRALIA
Latitude 35°02'S, longitude 117°55'E, elevation 12.5 m

Month	Mean sta. press. (mbar)		Temperature (°C)				Precipitation (mm)				Relative humidity (%)			Mean evap. (mm)**	Precipitation (mm)			Number of days with			
	09h00	15h00	mean daily	mean daily range	extreme highest	extreme lowest	mean	highest	lowest	max. 24 h	daily index	mean 9 h	mean 15 h		lower quartile	median	upper quartile	precipitation >0.25 mm	>2.5 mm	max. temp. >32.2 °C	>37.8 °C
Jan.	1014.2	1013.1	19.2	7.9	41.7	5.7	35	217	1	88	73	71	65	105	9	15	30	9	3	0.8	0.3
Feb.	1015.3	1014.1	19.4	7.4	44.8	5.0	26	161	0	57	73	72	66	89	7	13	31	7	2	0.3	0
Mar.	1016.4	1015.2	18.7	7.5	40.8	3.7	45	166	3	90	73	74	67	84	19	33	53	11	3	0.9	0.1
Apr.	1017.7	1016.2	16.9	8.2	37.7	4.2	74	234	5	57	75	79	68	76	38	60	98	14	7	0.6	0
May	1016.1	1014.8	14.7	7.9	35.2	1.7	135	290	44	104	77	81	69	59	84	117	149	19	12	0	0
June	1015.6	1014.4	13.1	7.3	24.6	1.7	138	292	40	72	76	83	70	47	102	131	164	20	14	0	0
July	1015.7	1014.6	12.1	7.5	23.1	0.1	152	269	52	61	76	81	70	49	112	148	179	22	16	0	0
Aug.	1016.3	1014.7	12.4	8.1	27.2	1.3	138	285	50	113	76	79	68	53	98	118	155	22	14	0	0
Sept.	1016.7	1015.4	13.4	8.2	30.6	1.1	108	202	20	79	76	77	69	60	77	93	136	18	12	0	0
Oct.	1016.0	1014.2	14.6	8.3	36.2	2.3	83	187	14	53	76	74	68	77	59	72	101	17	8	0.1	0
Nov.	1015.7	1014.5	16.4	8.2	41.1	4.8	42	170	5	78	74	71	67	87	22	38	52	12	5	0.4	0
Dec.	1014.4	1013.3	17.9	8.1	41.1	5.1	31	117	2	82	74	69	65	102	15	27	42	10	3	0.9	0.2
Annual	1015.8	1014.5	15.7	7.9	44.8	0.1	1008	1393	637	113	75	76	68	887	—	—	—	181	99	4.0	0.6
Years	29	29	29	29	50	50	30	42	42	42	42	42	42		81	81	81	30	10	30	30

TABLE XXVIII (*continued*)

Month	min. temp. <2.2 (°C)	<0 (°C)	cloudy	clear*	thunder	fog*	wet bulb temp. at 15h00 >21.1	>23.9	Cloudiness mean 09h00	15h00	highest monthly* 09h00	15h00	lowest monthly* 09h00	15h00	Sun-shine (h)	Wind prev. dir. 09h00	15h00	speed* (kt.) 09h00	15h00	Probability of precipitation (%) <12.5 mm	<25 mm	<50 mm
Jan.	0	0	10	0	1	<1	2.3	0.3	4.8	3.7	7.1	5.9	4.4	3.6	217	SE	ESE	13	13	41	72	93
Feb.	0	0	6	1	2	1	2.3	0.1	4.6	3.6	6.9	6.4	5.0	4.0	193	ESE	ESE	12	13	46	67	88
Mar.	0	0	10	0	1	1	2.5	0.2	4.6	4.0	7.4	7.0	4.6	3.4	189	ESE	ESE	12	13	18	38	72
Apr.	0	0	11	1	1	<1	0.3	0	4.2	4.2	7.2	6.4	4.3	4.5	145	NNW	ESE	12	11	2	9	36
May	0	0	10	1	1	0	0	0	4.2	4.6	7.0	6.8	4.6	4.6	140	NNW	SW	14	14	0	0	4
June	0.1	0	10	1	1	<1	0	0	4.2	4.5	6.6	6.7	4.5	4.4	97	NNW	NW	15	13	0	0	2
July	0.2	0	9	1	2	<1	0	0	4.2	4.7	6.6	7.4	4.2	4.5	121	NNW	NW	16	16	0	0	0
Aug.	0	0	11	1	1	0	0	0	4.2	4.5	6.5	6.7	4.2	4.6	140	NNW	SW	16	16	0	0	4
Sept.	0	0	8	1	1	<1	0	0	4.2	4.3	6.6	6.4	4.7	4.7	160	NNW	WSW	15	16	0	1	7
Oct.	0	0	8	1	1	1	0	0	4.6	4.2	6.6	6.8	5.2	3.9	189	NW	SW	13	13	0	3	19
Nov.	0	0	10	1	1	<1	0.3	0	4.8	3.8	7.0	6.2	4.5	3.3	193	SE	SW	13	14	10	30	71
Dec.	0	0	10	1	1	1	0.9	0.1	4.6	3.6	7.3	6.3	4.3	3.3	217	ESE	ESE	12	13	23	49	81
Annual	0.3	0	113	10	15	4	8.6	0.7	4.4	4.2	—	—	—	—	2001	—	—	14	14	—	—	—
Years	10	10	10	10	3-4	5	20	20	29	29	40	40	40	40	—	5	5	5	5	81	81	81

* On Eclipse Island, 35°12′S, 117°50′E, 118 m.
** Estimated, Australian tank, Waite formula.

TABLE XXIX

CLIMATIC TABLE FOR DARWIN, N.T., AUSTRALIA (UNTIL 1911 PALMERSTON AND PORT DARWIN, UNDER THE ADMINISTRATION OF SOUTH AUSTRALIA)
Latitude 12°28′S, longitude 130°51′E, elevation 30 m

Month	Sta. press. (mbar) mean 09h00	15h00	mean daily ampl.	Mean daily temp. (°C)	Mean daily temp. range (°C)	Temp. extremes (°C) highest sun	highest shade*	lowest shade*	Mean vapor press. (mbar)	Precipitation (mm) mean	highest	lowest	max. (24 h)	ratio mean/med.	Relative humidity (%) daily index	mean 09h00	mean 15h00	Mean Austr. tank evap. (mm)	Number of days precipitation 0 mm	>0.25 mm	>2.5 mm	>25 mm
Jan.	1004.4	1002.2	2.8	28.7	7.0	75.6	37.8	20.4	31.4	411	708	57	297	1.0	80	78	71	153	10	18	20	4
Feb.	1004.5	1002.3	2.9	28.6	7.1	73.1	38.3	17.2	31.2	314	653	11	134	1.1	80	79	72	142	8	16	18	5
Mar.	1005.5	1003.1	3.2	28.7	7.3	74.2	38.9	19.2	30.9	284	556	21	182	1.1	79	78	68	156	10	16	17	3
Apr.	1007.7	1004.9	3.4	28.8	8.9	72.8	40.0	16.0	27.2	78	603	0	168	1.4	68	69	54	165	17	6	2	2
May	1009.4	1006.5	3.6	27.4	10.2	71.1	39.0	15.1	22.1	8	356	0	56	19.0	60	63	47	185	28	1	1	<1
June	1010.8	1007.8	3.7	25.8	10.0	68.4	37.0	12.9	18.5	2	39	0	34	∞	55	61	47	177	29	1	1	0
July	1011.5	1008.1	3.9	25.1	10.4	68.9	36.7	10.4	17.7	0	65	0	43	∞	55	59	44	179	29	0	0	0
Aug.	1011.2	1007.9	4.2	26.2	10.4	69.0	36.7	13.9	20.7	1	76	0	27	∞	61	63	45	196	30	0	0	0
Sept.	1010.6	1007.0	4.2	28.1	9.5	69.4	38.9	17.2	24.8	15	69	0	51	3.2	65	65	49	205	28	2	1	0
Oct.	1009.3	1005.9	4.1	29.4	8.6	71.4	40.5	20.3	28.2	49	338	0	95	1.2	69	65	52	233	23	4	3	1
Nov.	1007.0	1004.2	3.6	29.8	8.3	76.9	39.6	19.3	29.3	110	400	10	120	1.1	79	68	58	208	19	10	6	2
Dec.	1005.4	1002.9	3.2	29.4	7.7	76.1	38.9	20.3	30.1	218	568	25	200	1.1	73	73	65	183	11	14	10	3
Annual	1008.1	1005.2	3.6	28.0	8.8	76.9	40.5	10.4	25.8	490	708	0	297	1.0	68	68	56	2183	242	88	64	20
Years	30	30	58	30	30	25	80*	80*	30	30	92	92	92	82	57	57	60	8	5	30	10	5

TABLE XXIX (*continued*)

Month	Number of days					Mean cloudiness (oktas)				Meridian altitude of sun degrees	Mean daily global radiation (cal./cm²)	Wind					Visibility freq. (%) 09h00			Probability of precipitation (%)		
	thunder	lightning	clear	fog	visibility <4 km	all types 09h00	all types 15h00	low cloud 09h00	low cloud 15h00			preval. dir. 09h00	preval. dir. 15h00	speed 09h00 (knots)	speed 15h00 (knots)	highest gust (kts)	<10 km	<20 km	<30 km	<12.5 mm	<25 mm	<50 mm
Jan.	12	16	1	0	10	4.9	5.1	1.8	3.0	82	443	NW,S	W,NW	5	8	57	6	43	100	0	0	0
Feb.	10	16	1	0	12	4.8	5.2	2.7	3.2	88	458	W,S	W,NW	7	9	47	5	29	100	0	<1	<1
Mar.	10	14	3	0	9	4.2	4.7	1.8	3.3	78	474	SE	W,NW	5	7	85	3	19	100	0	<1	<1
Apr.	4	6	11	0	3	3.2	3.8	0.7	2.2	68	444	SE	E	6	6	36	8	26	100	12	26	78
May	1	1	19	0	1	2.4	2.8	0.5	1.5	58	467	SE	E	7	8	32	0	6	100	78	85	97
June	<1	0	22	0.4	5	2.2	2.3	0.9	1.2	54	456	SE	E,SE	8	8	32	5	41	100	89	99	100
July	<1	0	23	1.1	6	1.7	1.9	0.6	0.7	58	464	SE	E,SE	7	8	31	2	31	100	99	99	99
Aug.	<1	0	23	0.7	6	1.8	1.8	0.6	0.8	68	526	SE	E,SE	4	10	30	2	16	89	97	97	99
Sept.	2	1	18	0.2	7	2.2	2.2	1.2	1.5	78	554	SE,S	NW,N	4	11	31	1	53	100	66	78	94
Oct.	7	8	10	0	8	3.0	2.6	1.8	1.7	88	561	S	NW,N	4	11	40	0	39	100	16	32	59
Nov.	14	17	4	0	6	3.8	3.9	2.6	2.2	82	512	W,S	NW,N	3	9	50	1	19	100	1	4	12
Dec.	15	17	2	0	8	4.5	4.5	1.8	2.3	79	475	NW,S	NW,N	4	8	57	0	3	99	0	1	1
Annual	75	96	137	2.4	81	3.2	3.4	1.4	1.8	—	—	SE	NW	5	9	85	3	27	99	—	—	—
Years	35	30	30	30	5	30	30	5	5	—	7-9	**	**	3	3	50	unsp.	unsp.	unsp.	74	74	74

* 80 years; records after 1941 (from Aerodrome) include all the lowest temperatures.

** Short period only; coastal location affected by breezes.

Note: for frequency distribution of monthly rainfall, see Table IV.

TABLE XXX

CLIMATIC TABLE FOR DALY WATERS, N.T., AUSTRALIA
Latitude 16°16'S, longitude 122°23'E, elevation 211 m

Month	Mean sta. level press. (mbar)		Temperature (°C)				Precipitation (mm)				Relat. humidity (%)			Mean Austr. tank evap.* (mm)	Evapor. (mm)**	Number of days with precip.	
	09h00	15h00	mean daily	mean daily range	extreme highest	extreme lowest	mean	highest	lowest	max. 24 h	daily index	09h00	15h00			>0.25 mm	>2.5 mm
Jan.	983.6	980.3	30.3	12.2	45.0	16.2	140	589	15	123	60	68	38	168	225	13	11
Feb.	983.6	980.9	29.7	12.3	43.9	11.4	129	488	23	104	62	72	43	149	188	12	9
Mar.	985.4	982.3	28.7	12.4	43.5	12.9	105	391	0	147	60	66	39	171	234	8	5
Apr.	988.6	985.2	26.7	14.5	41.2	9.4	17	256	0	137	50	55	30	169	222	2	1
May	990.7	987.2	23.6	15.0	38.3	6.1	11	48	0	43	46	52	29	173	224	1	0
June	992.1	988.8	21.3	15.1	36.6	+1.3	4	147	0	39	45	51	28	153	189	1	1
July	992.5	989.1	20.6	15.6	36.7	−1.0	1	17	0	15	40	47	25	166	202	0	0
Aug.	992.0	988.1	22.7	18.2	39.1	+3.9	0	62	0	34	56	44	21	191	238	0	0
Sept.	991.0	986.8	26.5	17.6	42.3	5.1	2	64	0	65	51	44	20	226	261	1	0
Oct.	988.8	984.5	29.9	16.5	44.4	9.1	17	88	0	61	41	47	22	254	300	3	1
Nov.	986.5	982.3	31.3	15.2	45.0	13.6	52	264	1	70	45	51	26	235	296	7	4
Dec.	984.7	981.1	31.4	14.0	46.7	14.6	92	401	2	130	53	60	31	207	279	10	4
Annual	988.3	984.7	26.9	14.9	46.7	−1.0	574	1167	228	147	49	55	30	2262	2858	58	36
Years	30	30	50	50	50	50	30	78	78	52	52	30	30	9	8	30	unsp.

TABLE XXX (*continued*)

Month	Number of days with				Cloudiness (oktas)			Sunshine (h)	Wind* most freq. dir.		Extreme gust		Visibility frequ. (%)		Probability of precipitation (%)		
	thunder	fog	visi-bility <4 km	cloud base <300 m	all types 09h00	15h00	low cloud		09h00	15h00	dir.	speed	<5 km	<20 km	<12.5 mm	<25 mm	<50 mm
Jan.	8	0	1	3	3.9	4.5	2.6	8.1	C,NW,N	N,C	SE	18	15	50	0	<3	12
Feb.	7	0	2	3	4.1	4.5	2.5	8.5	C,NW,N	S,C	SE	30	5	25	0	1	14
Mar.	5	0	4	5	3.4	4.2	2.6	8.7	C,SE	E,SE	SE	22	0	15	9	14	29
Apr.	0	0	0	0	2.3	3.0	1.0	9.1	SE	SE	SSE,E	22	0	10	60	72	87
May	0	0	1	2	1.8	2.2	0.7	9.9	SE	SE	SE	22	4	0	86	93	100
June	0	0	1	1	1.8	2.0	0.7	10.1	SE	SE	ESE,SE	22	0	15	86	94	99
July	0	0	1	1	1.4	1.5	0.3	10.2	SE	SE	ESE	22	2	10	96	100	100
Aug.	0	1	2	1	1.3	1.4	0.3	10.9	SE,N	SE,E	E	22	0	15	95	96	98
Sept.	0	0	2	0	1.3	1.8	0.7	10.5	N,SE	SE,E	NNE	22	0	45	90	95	99
Oct.	3	0	3	0	1.7	2.8	1.2	10.0	N	NE,N	ENE,SE	22	0	45	55	68	88
Nov.	5	0	2	0	2.6	3.8	1.8	9.3	N	N,E	SE	22	0	40	15	26	55
Dec.	7	0	2	1	3.6	4.3	2.3	8.9	N,C	N,E,C	NNE	22	0	20	3	9	18
Annual	35	1	21	17	2.4	3.0	1.4	9.5	SE	SE,E	SE	30	—	—	—	—	—
Years	unsp.	6	5	5	30	30	6	13	5	5	8	8	unsp.	unsp.	78	78	78

* At Katherine, 141°30'S, 132°40'E, 120 m. C = calms.

** Estimated by interpolation from maps.

TABLE XXXI

CLIMATIC TABLE FOR TENNANT CREEK, N.T., AUSTRALIA
Latitude 19°34'S, longitude 134°13'E, elevation 327 m

Month	Mean sta. press. (mbar)		Mean daily temp. (°C)	Mean daily temp. range (°C)	Precipitation (mm)					Relative humidity (%)				Mean evap. (mm)	Number of days precipitation		Temp. extremes	
	09h00	15h00			mean	highest	lowest	max. 24 h		daily index	mean 09h00	mean 15h00			>0.25 mm	>2.5 mm	highest (°C)	lowest (°C)
Jan.	965.9	962.9	30.7	12.6	103	282	0	187		41	42	27		305	7	4	46.1	15.6
Feb.	965.8	962.9	29.8	12.3	90	379	0	234		45	50	29		248	5	2	43.7	11.1
Mar.	968.4	965.3	28.5	12.4	53	430	0	117		40	39	27		285	3	1	43.9	12.1
Apr.	971.9	968.7	25.2	12.6	9	196	0	58		32	33	22		264	1	1	40.0	10.0
May	973.7	970.7	21.3	12.6	5	99	0	66		36	36	24		196	1	0	37.8	5.1
June	975.2	972.3	18.1	12.6	9	70	0	43		40	39	27		158	1	2	34.2	2.8
July	975.5	972.5	17.4	13.5	6	94	0	34		36	37	25		155	1	1	33.3	2.4
Aug.	974.7	971.5	19.9	15.1	2	83	0	15		31	32	21		191	0	0	35.6	2.8
Sept.	973.9	970.3	23.7	15.5	3	71	0	28		29	29	19		252	1	0	38.9	5.6
Oct.	971.0	967.6	27.5	15.1	10	78	0	36		29	28	19		312	2	2	42.8	10.6
Nov.	968.9	965.5	29.7	14.2	27	106	0	75		32	31	22		342	3	2	44.2	13.3
Dec.	966.9	963.6	30.9	13.5	35	219	0	90		36	38	24		325	5	1	44.9	11.7
Annual	971.0	967.8	25.2	13.6	351	864	94	234		34	36	34		3033	30	16	46.1	2.4
Years	25	25	30	30	30	77	77	90		31	30	30		30	30	unsp.	90	90

TABLE XXXI (*continued*)

Month	Frequency of wet periods			Frequency of wet periods with rainfall			Rainfall variability (%)	Mean cloudiness (oktas)		Most freq. wind dir.		Probability of precipitation (%)		
	>0 days	>2 days	>5 days	>12.7 mm	>25.4 mm	>50.8 mm		09h00	15h00	09h00	15h00	<12.5 mm	<25 mm	<50 mm
Jan.	2.58	1.16	0.34	1.26	0.92	0.58	70	2.7	3.5	NE	E	21	29	43
Feb.	1.94	0.98	0.44	1.14	0.80	0.56	88	2.5	3.4	E	E	21	31	47
Mar.	1.24	0.42	0.24	0.50	0.34	0.16	107	2.0	3.0	E	E	40	55	70
Apr.	0.72	0.18	0	0.22	0.12	0	122	1.4	1.9	E	E	72	82	96
May	0.36	0.12	0	0.18	0.08	0	154	1.5	1.8	E	SE	83	87	92
June	0.66	0.16	0	0.16	0.14	0	130	1.3	1.3	E	SE	83	87	95
July	0.48	0.12	0	0.20	0.06	0	145	1.0	0.9	E	SE	87	90	93
Aug.	0.24	0	0	0.08	0	0	162	0.6	0.6	E	E	90	96	99
Sept.	0.52	0	0	0.10	0.06	0	148	0.7	0.9	E	E	82	92	97
Oct.	1.24	0.43	0.10	0.35	0.10	0.06	86	1.4	2.0	E	SE	62	84	93
Nov.	1.96	0.65	0.14	0.82	0.35		63	1.8	2.8	E		35	58	80
Dec.	2.39	1.02	0.35	1.12	0.64	0.21	77	2.3	3.1		NE,E	21	44	64
Annual	14.33	5.24	1.61	6.13	3.61	1.57	36	1.6	2.1	—	—	—	—	—
Years	>30	>30	>30	>30	>30	>30	>30	30	30	5	5	77	77	77

TABLE XXXII

CLIMATIC TABLE FOR ALICE SPRINGS, N.T., AUSTRALIA (INITIALLY STUART)
Latitude 23°38'S, longitude 132°35'E, elevation 579 m

Month	Mean sta. press. (mbar) 09h00	Mean sta. press. (mbar) 15h00	Mean daily temp. (°C)	Mean daily temp. range (°C)	Precipitation (mm) mean	Precipitation (mm) highest	Precipitation (mm) lowest	Precipitation (mm) max. 24 h	Relative humidity (%) daily index	Relative humidity (%) mean 09h00	Relative humidity (%) mean 15h00	Mean Austr. tank evap. (mm)	Number of days precipitation 0 mm	Number of days precipitation >0.25 mm	Number of days precipitation >2.5 mm	Number of days min. temp. <2.2 (°C)	Number of days min. temp. <0 (°C)	Temp. extremes highest (°C)	Temp. extremes lowest (°C)
Jan.	947.9	945.1	28.1	14.2	44	281	0	98	33	32	20	308	28	4	5	0	0	46.7	10.0
Feb.	948.4	945.5	27.5	14.7	34	236	0	84	36	38	22	260	25	3	3	0	0	45.6	8.5
Mar.	951.2	948.2	24.7	15.2	28	227	0	147	38	39	22	243	26	3	1	0	0	45.0	3.9
Apr.	954.7	951.7	19.8	15.3	10	117	0	72	41	46	26	176	27	2	1	0	0	39.3	1.9
May	956.0	953.2	15.3	15.2	15	109	0	39	49	57	30	122	27	2	1	1.6	0.4	38.3	−2.8
June	956.8	954.1	12.3	14.4	13	74	0	51	54	63	35	86	26	2	2	>4	3.4	30.6	−5.6
July	957.0	954.2	11.6	15.6	7	106	0	50	49	59	29	92	25	1	1	>7	6.2	31.1	−7.2
Aug.	956.3	953.3	14.3	16.3	8	158	0	63	40	47	24	128	27	2	1	3.3	2.0	35.8	−3.9
Sept.	955.3	952.0	18.2	17.2	7	90	0	31	34	34	18	182	28	1	0	1.3	0	37.6	−1.1
Oct.	952.6	948.7	22.8	16.2	18	115	0	58	30	36	19	236	25	3	1	0	0	45.1	2.4
Nov.	950.5	947.6	25.5	15.4	29	139	0	68	31	29	19	267	25	4	2	0	0	46.1	4.4
Dec.	948.5	945.9	27.4	14.7	39	288	0	118	32	30	20	295	24	4	1	0	0	47.2	7.8
Annual	952.9	950.9	20.6	15.3	252	726	60	147	37	38	22	2388	313	31	19	>17	12.0	47.2	−7.2
Years	30	30	30	30	30	77	77	90	60	30	30	32	5	30	unsp.	10	10	90	90

TABLE XXXII (*continued*)

Month	Frequency of wet periods			Frequency of wet periods with rainfall			Mean cloudiness		Me-ridian altitude of sun (21st of month) degrees	Sun-shine (h)	Mean daily global radi-ation (cal./cm²)	Wind			Probability of precipitation (%)		
	>0 days	>2 days	>5 days	>12.7 mm	>25.4 mm	>50.8 mm	09h00	15h00				most freq. dir.		speed (knots)	<12.5 mm	<25 mm	<50 mm
												09h00	15h00				
Jan.	1.70	0.57	0.17	0.62	0.40	0.19	1.8	2.6	86	10.3	622	E	SE	4.4	42	53	70
Feb.	1.28	0.76	0.26	0.70	0.56	0.41	1.8	2.5	76	10.4	615	SE	SE	4.7	43	57	65
Mar.	1.20	0.42	0.17	0.57	0.39	0.20	1.6	2.1	66	9.3	567	E	ESE	3.7	51	60	74
Apr.	0.87	0.17	0.06	0.31	0.22	0.08	1.5	1.8	56	9.2	463	SE	ESE,SE	3.5	64	78	91
May	0.67	0.26	0.11	0.28	0.17	0	2.0	2.2	46	8.0	382	SE	SE	2.7	66	74	92
June	1.07	0.26	0.06	0.37	0.15	0.09	1.9	1.8	43	8.0	359	SE	SE	2.5	65	82	92
July	0.74	0.22	0.09	0.26	0.07	0	1.6	1.4	46	8.9	389	WSW	SE	2.3	75	89	96
Aug.	0.74	0.28	0.06	0.24	0.11	0	1.0	1.0	56	9.8	468	E	SE	2.4	82	88	93
Sept.	0.74	0.20	0	0.15	0.09	0	1.0	1.1	66	10.0	549	E	SE	3.2	79	88	95
Oct.	1.44	0.54	0.11	0.57	0.28	0.09	1.7	1.9	76	9.7	594	E	SE	3.7	45	74	92
Nov.	1.83	0.72	0.24	0.81	0.41	0.20	1.8	2.3	86	10.1	626	E	SE	4.0	42	58	87
Dec.	1.68	0.70	0.31	0.59	0.42	0.17	2.2	2.5	90	10.0	631	SE	SE	4.3	40	57	78
Annual	13.96	5.10	1.64	5.47	3.27	1.43	1.7	1.9	—	9.5	522	—	—	3.5	—	—	—
Years	>30	>30	>30	>30	>30	>30	30	30	—	16	10–11	5	5	5	77	77	77

TABLE XXXIII

CLIMATIC TABLE FOR OODNADATTA, S.A., AUSTRALIA
Latitude 27°30′S, longitude 135°23′E, elevation 117 m

Month	Temperature (°C)				Precipitation (mm)					Relative humidity (%)	
	mean	daily range	extreme		mean	highest	lowest	max. 24h*	ratio mean/ median*	09h00	15h00
			max.	min.							
Jan.	29.3	14.8	46.1	11.7	18	132	0	36	∞	33	22
Feb.	28.3	13.9	45.1	8.4	34	155	0	64	2.7	34	23
Mar.	26.2	14.5	44.9	9.6	18	110	0	62	3.5	37	25
Apr.	20.5	13.7	41.8	5.1	6	45	0	30	∞	46	31
May	16.0	13.1	35.0	+0.9	16	88	0	43	2.7	51	35
June	12.7	13.3	30.4	−0.9	10	70	0	38	3.2	58	42
July	12.4	13.8	31.2	−1.7	12	46	0	30	25.0	54	35
Aug.	14.2	14.9	36.5	−0.2	10	39	0	18	17.0	49	31
Sept.	18.7	15.4	39.1	2.2	7	56	0	19	21.0	41	26
Oct.	22.1	15.1	45.1	3.4	12	59	0	50	2.9	34	23
Nov.	25.5	15.2	47.1	10.1	8	62	0	77	47.0	34	23
Dec.	28.7	15.1	48.3	11.3	15	83	0	34	2.8	33	22
Annual	21.2	14.4	48.3	−1.7	167	295	29	77	1.3	42	28
Years	50	50	50	50	50	55	55	38	38	50	50

Month	Evapor. (mm)	Number of days with				Probability of precipitation (%)			Mean monthly sun-shine*** (h)
		precipitation		min. temp.**		<12.5 mm	<25 mm	<50 mm	
		>0.25 mm	>2.5 mm	<2.2 °C	<0 °C				
Jan.	447	3	1	0	0	62	77	87	325
Feb.	373	3	1	0	0	69	78	86	290
Mar.	356	2	0	0	0	76	85	98	295
Apr.	254	2	0	0	0	87	95	100	270
May	170	3	0	0.9	0	82	94	96	240
June	130	3	1	4.3	0.7	69	76	91	225
July	117	2	0	6.4	2.1	82	93	100	255
Aug.	163	3	1	2.2	0.3	87	96	100	280
Sept.	239	2	0	0	0	86	96	98	285
Oct.	345	4	1	0	0	73	89	94	300
Nov.	391	4	0	0	0	82	91	98	315
Dec.	394	4	1	0	0	71	84	98	330
Annual	3368	35	6	13.8	3.1	–	–	–	3410
Years	3	50	unsp.	10	10	55	55	55	–

* At Mulka, S.A., 28°24′S, 138°42′E, 30 m; mean annual rainfall 109 mm.
** At Charlotte Waters, N.T., 26°56′S, 134°55′W, 197 m.
*** Estimated by interpolation from maps.

TABLE XXXIV

CLIMATIC TABLE FOR FARINA, S.A., AUSTRALIA (INITIALLY FARINA TOWN)
Latitude 30°05′S, longitude 138°08′E, elevation 93 m

Month	Mean daily temp. (°C)	Mean daily temp. range (°C)	Extreme temp.* highest	Extreme temp.* lowest	Precipitation (mm) mean	Precipitation (mm) highest	Precipitation (mm) lowest	max. 24h*	Relative humidity daily index	Relative humidity mean 15h00	Relative humidity mean 08h30
Jan.	27.6	15.6	45.3	3.9	14	102	0	66	34	22	32
Feb.	27.7	15.5	42.1	5.9	15	114	0	127	35	23	34
Mar.	24.5	14.9	40.4	5.6	13	102	0	74	40	26	39
Apr.	19.3	14.4	36.2	2.2	10	77	0	41	49	32	48
May	14.8	13.5	30.0	−0.3	15	76	0	39	57	39	58
June	11.6	12.2	25.8	−2.8	15	132	0	49	65	47	70
July	10.7	13.2	24.2	−3.6	9	52	0	24	62	43	67
Aug.	12.7	14.1	31.4	−1.8	7	52	0	48	55	35	56
Sept.	16.1	15.2	35.6	−0.9	8	42	0	54	47	28	45
Oct.	20.3	15.4	36.7	1.4	14	84	0	23	39	24	36
Nov.	24.0	15.8	42.5	3.9	13	74	0	41	36	23	32
Dec.	26.6	15.7	43.9	2.2	13	88	0	42	34	22	32
Annual	19.7	14.6	45.3	−3.6	146	365	47	127	44	30	46
Years	50	50	21	21	30	71	71	21	50	50	50

Month	Evap. (mm)**	Number of days precip. >0.25 mm	Number of days min. temp. <2.2°C	Number of days min. temp. <0°C	Cloudiness* (oktas) 09h00	Cloudiness* (oktas) 15h00	Mean monthly sunshine (h)**	Prev. wind dir.* 09h00	Prev. wind dir.* 15h00	Probability of precipitation (%) <12.5 mm	Probability of precipitation (%) <25 mm	Probability of precipitation (%) <50 mm
Jan.	395	2	0	0	1.9	2.3	330	SE	SE	72	80	87
Feb.	315	2	0	0	2.1	2.8	285	SE	SE	62	76	89
Mar.	295	2	0	0	2.2	2.6	280	SE	SE	65	76	93
Apr.	195	2	0	0	2.4	2.9	255	SE	SE	72	89	96
May	115	3	0.3	0	3.2	3.6	230	SE	SW	71	76	94
June	85	3	4.5	1.4	3.8	3.8	205	NNW	NNW	53	76	91
July	80	2	7.1	1.4	3.4	3.3	230	N	N	79	92	99
Aug.	115	2	3.8	0.8	2.8	3.3	250	NW	NW	72	89	99
Sept.	180	2	0	0	2.4	2.7	270	NW, SE	SW	65	84	100
Oct.	235	3	0	0	2.7	2.7	295	SE	SW	72	87	97
Nov.	305	3	0	0	2.7	3.0	310	SE	NW, SW	66	90	98
Dec.	365	2	0	0	2.2	2.6	340	SE	SW	63	84	94
Annual	2680	28	15.7	3.6	2.7	3.0	3270	SE	SE, SW	−	−	−
Years	−	30	10	10	15	15	−	5	5	71	71	71

* At Angorichina, 31°07′S, 138°33′E, 616 m.
** Estimated by interpolation from maps.

TABLE XXXV

CLIMATIC TABLE FOR COOK, S.A., AUSTRALIA
Latitude 30°37′S, longitude 130°27′E, elevation 123 m

Month	Mean sta. press. (mbar)		Mean daily temp. (°C)	Mean daily temp. range (°C)	Extreme temp.		Precip. (mm)					Relative humidity (%)		
	09h00	15h00			highest (°C)	lowest (°C)	mean	highest	lowest	ratio mean/ median*	max. 24 h	daily index	mean 09h00	mean 15h00
Jan.	999.5	997.2	23.9	17.5	47.2	+4.4	13	82	0	4.0	78	49	54	31
Feb.	1000.3	998.1	23.1	16.2	46.1	+5.6	11	55	0	1.9	32	52	61	34
Mar.	1002.6	1000.5	21.8	15.9	45.6	+2.8	18	72	0	1.7	59	55	62	36
Apr.	1005.5	1003.2	18.0	14.4	40.6	+0.6	14	56	0	1.6	32	63	68	42
May	1006.0	1003.9	14.9	13.7	34.6	−1.1	13	46	0	1.3	25	68	74	45
June	1006.6	1004.6	12.1	12.8	29.4	−3.3	22	45	0	1.9	54	72	80	53
July	1006.4	1004.5	11.3	13.4	29.6	−2.4	15	32	0	1.4	26	66	79	50
Aug.	1005.9	1003.6	12.7	14.4	36.7	−3.4	15	63	0	1.2	27	62	73	44
Sept.	1005.0	1002.6	15.4	16.1	39.3	−1.1	7	45	0	1.6	25	54	61	36
Oct.	1002.4	1000.2	17.7	16.3	42.5	+0.6	18	70	0	1.6	35	53	57	34
Nov.	1001.1	998.8	20.3	16.7	45.6	+1.1	18	45	0	1.7	33	50	55	33
Dec.	999.5	997.2	22.5	17.3	45.0	+5.0	13	90	0	2.0	32	50	55	32
Annual	1003.4	1001.2	17.8	15.4	47.2	−3.3	176	300	61	1.1	78	57	54	37
Years	27	27	30	30	50	50	30	31	31	30	50	15	29	29

Month	Mean Austr. tank evap. (mm)	Number of days				Mean cloudiness (oktas)		Mean monthly sunshine (h)***	Most freq. wind dir.		Probability of precipitation (%)		
		precip. <0.25 mm	min. temp. (°C)		thunder**	09h00	15h00		09h00	15h00	<12.5 mm	<25 mm	<50 mm
			<2.2	<0									
Jan.	382	2	0	0	2	2.3	2.0	310	SE	S	78	84	94
Feb.	300	3	0	0	1	2.6	2.2	265	SE	SE	67	76	97
Mar.	271	3	0	0	1	2.6	2.2	260	SE	SE	55	73	94
Apr.	177	4	0	0	1	2.8	3.0	235	NE	SE	67	85	97
May	139	4	0.6	0	<1	2.7	3.3	230	NE	SW	61	79	100
June	90	6	4.8	2.2	<1	2.8	3.3	195	N	NW	61	85	100
July	102	6	<5	3.9	0	2.3	3.1	225	NE	NW	73	97	100
Aug.	135	5	3.6	1.6	<1	2.2	3.0	235	N	NW	52	85	97
Sept.	199	3	2.0	0.4	2	2.1	2.5	260	N, NE	S	85	97	100
Oct.	256	4	0.8	0	1	2.6	2.7	280	NE	S	64	79	97
Nov.	263	4	0	0	1	2.9	2.9	290	NE	S	55	73	100
Dec.	329	3	0	0	4	2.6	2.3	325	SE	SE	58	82	97
Annual	2642	47	<16	8.1	15	2.6	2.7	3110	–	–	–	–	–
Years	7	30	10	10	3	30	30	–	5	5	33	33	33

Note: for frequency distribution of monthly rainfall at Deakin, W.A., see Table II.
  * At Deakin, W.A., 30°48′S, 129°00′E, 152 m.
 ** At Forrest, W.A., 30°55′S, 128°07′E, 150 m.
*** Estimated by interpolation from maps.

TABLE XXXVI

CLIMATIC TABLE FOR TARCOOLA, S.A., AUSTRALIA

Latitude 30°42′S, longitude 134°34′E, elevation 120 m

Month	Mean sta. press. (mbar)		Mean daily temp. (°C)	Mean daily temp. range (°C)	Temperatures extremes (°C)		Precipitation (mm)				Relative humidity (%)			Evap. (mm)*
	09h00	15h00			max.	min.	mean	max.	min.	max. 24h	daily index	mean 09h00	15h00	
Jan.	999.8	997.4	25.9	17.0	47.7	5.6	10	55	0	44	40	42	24	365
Feb.	999.7	997.2	25.5	16.6	46.1	6.7	21	167	0	141	42	47	24	305
Mar.	1003.3	1000.3	23.3	16.4	45.0	4.4	12	48	0	38	44	49	27	280
Apr.	1006.4	1003.3	18.3	15.5	38.9	1.7	7	62	0	53	53	58	32	180
May	1006.7	1004.6	14.6	14.5	33.3	—1.7	12	90	0	72	60	65	39	130
June	1006.6	1005.0	11.2	13.8	28.1	—4.7	14	73	0	37	65	74	46	90
July	1007.6	1005.7	10.8	14.4	28.9	—4.4	11	46	0	23	64	73	43	85
Aug.	1006.1	1003.7	12.7	15.3	34.4	—3.3	15	54	0	27	57	64	38	115
Sept.	1005.6	1003.1	15.8	16.6	36.1	—1.7	11	61	0	45	50	53	33	175
Oct.	1002.4	998.4	18.8	16.9	42.2	+0.6	18	68	0	36	46	48	31	230
Nov.	1000.0	997.6	22.2	17.2	45.0	2.2	11	56	0	30	41	42	24	280
Dec.	999.0	996.7	24.7	17.5	48.9	5.0	14	97	0	55	40	40	25	330
Annual	1003.6	1001.1	18.7	15.9	48.9	—4.7	156	350	68	141	48	52	32	2565
Years	11	9	28	28	28	28	20	46	46	28	25	25	20	–

Month	Number of days			Cloudiness all types		Mean monthly sun-shine* (h)	Wind prev. dir.		Probability of precipitation (%)		
	precip. >0.25 mm	min. temp. (°C) < 2.2	< 0	09h00	15h00		09h00	15h00	< 12.5 mm	< 25 mm	< 50 mm
Jan.	2	0	0	1.7	1.8	330	E	SW	71	89	98
Feb.	2	0	0	1.7	1.9	280	NE, E	SW	54	76	92
Mar.	2	0.1	0	1.7	2.0	275	E	SW	66	76	100
Apr.	2	0.4	0	2.1	2.3	250	E	E	77	91	98
May	3	1.4	0.5	2.4	2.7	230	NE	W	66	83	92
June	4	3.2	3.4	2.6	2.7	195	NE	NE	51	74	94
July	4	2.5	3.9	2.1	2.4	230	NE	NE	70	83	100
Aug.	4	2.6	2.3	1.9	2.3	250	NE, W	SW	45	79	96
Sept.	2	1.5	0	1.7	1.8	260	SW	SW, W	68	85	98
Oct.	3	0.5	0	2.2	2.2	300	SW	SW	55	72	89
Nov.	3	0.1	0	2.2	2.2	310	NE	SW	62	83	95
Dec.	2	0	0	2.1	2.0	330	SW	SW	62	85	91
Annual	33	12.3	10.1	2.0	2.2	3240	NE	SW	–	–	–
Years	27	10	10	25	22	–	3	3	47	47	47

* Estimated by interpolation from maps.

TABLE XXXVII

CLIMATIC TABLE FOR PORT AUGUSTA, S.A., AUSTRALIA
Latitude 32°29'S, longitude 137°45'E, elevation 5.5 m

Month	Mean sta. press. (mbar)		Mean daily temp. (°C)	Mean daily temp. range (°C)	Temp. extremes (°C)		Precipitation (mm)				Relative humidity (%)			Mean evapor. (mm)*
	09h00	15h00			highest	lowest	mean	highest	lowest	max. 24 h	daily index	mean 09h00	mean 15h00	
Jan.	1013.0	1010.6	25.2	13.4	48	10	15	81	0	52	50	45	33	371
Feb.	1013.8	1011.2	25.4	12.5	47	9	18	114	0	54	47	47	35	309
Mar.	1016.8	1014.1	23.2	12.8	44	6	16	143	0	118	49	51	36	269
Apr.	1020.0	1017.4	19.2	12.1	38	5	14	101	0	48	55	57	41	173
May	1020.6	1018.4	15.6	11.3	33	+1	23	97	0	44	61	65	46	109
June	1021.5	1019.2	12.6	9.8	27	0	26	95	1	64	68	75	54	71
July	1021.2	1018.9	11.8	10.5	28	−1	20	59	0	19	66	74	52	74
Aug.	1020.2	1017.6	13.4	11.5	32	0	26	87	0	32	62	66	46	104
Sept.	1019.4	1016.5	16.1	12.7	35	+2	21	104	1	52	54	55	39	165
Oct.	1016.7	1013.8	19.2	13.1	41	4	22	104	0	64	49	48	36	231
Nov.	1015.0	1012.1	22.1	13.4	43	6	18	141	0	53	46	44	35	289
Dec.	1013.1	1010.4	24.2	13.4	46	8	17	82	0	32	45	44	34	338
Annual	1017.6	1015.0	19.0	12.3	48	−1	236	469	56	118	53	56	41	2507
Years	30	30	62	62	60	60	30	84	84	42	49	58	49	8

TABLE XXXVII (*continued*)

Month	Number of days with							Mean cloudiness			Wind			Probability of precipitation (%)		
	precipitation		min. temp.		thunder	fog	clear	all types		low clouds	most freq. dir.		mean speed (kt.)	<12.5 mm	<25 mm	<50 mm
	>0.25 mm	>2.5 mm	<2.2 °C	<0 °C				09h00	15h00		09h00	15h00				
Jan.	3	1	0	0	1	0	17	2.0	1.9	1.2	S	S	3	64	77	94
Feb.	3	1	0	0	1	0	14	2.0	1.9	0.9	S	S	4	61	78	91
Mar.	3	2	0	0	1	0	15	1.9	1.9	0.7	S	S	4	61	80	91
Apr.	5	2	0	0	1	1	15	2.6	2.6	1.0	calm,SE	S	3	39	72	94
May	7	2	0	0	1	1	10	3.0	3.0	1.0	calm	S,N,calm	2	34	61	85
June	7	2	0.5	0.1	0	1	9	3.4	3.4	—	calm	N	2	24	56	85
July	10	3	0.9	0	1	1	11	3.1	3.3	1.2	N,calm	N	2	36	73	98
Aug.	9	3	0.1	0	1	1	10	2.7	3.1	0.9	N	N,S	3	35	67	94
Sept.	6	2	0	0	1	0	12	2.4	2.6	1.3	N,S	N,S	4	34	73	89
Oct.	6	3	0	0	2	0	11	2.8	1.9	1.0	S	S	4	42	62	91
Nov.	6	2	0	0	1	0	11	2.9	2.6	1.3	S	S	4	57	76	91
Dec.	4	2	0	0	2	0	13	2.5	2.3	0.7	S	S	4	55	77	94
Annual	69	25	1.5	0.1	10	2	148	2.6	2.6	1.0	S	S	3	—	—	—
Years	30	10	10	10	7	7	10	30	30	3	5	5	5	91	91	91

* At Yudnapinna, 32°06′S, 137°06′E, Australian tank.

TABLE XXXVIII

CLIMATIC TABLE FOR PORT LINCOLN, S.A., AUSTRALIA
Latitude 34°43'S, longitude 135°52'E, elevation 4 m

Month	Mean sta. press. (mbar)		Mean daily temp. (°C)	Mean daily temp. range (°C)	Temp. extremes (°C)		Precipitation (mm)				Relative humidity (%)			Evapor. (mm)*	Number of days precipitation	
	09h00	15h00			highest	lowest	mean	highest	lowest	max. 24 h	daily index	mean 09h00	mean 15h00		>0.25 mm	>2.5 mm
Jan.	1014.1	1013.0	19.9	10.5	45.6	7.9	13	121	0	42	64	61	52	175	4	2
Feb.	1015.1	1013.8	20.3	10.2	42.2	7.1	19	96	0	71	67	64	55	132	5	1
Mar.	1017.6	1016.1	18.9	9.6	41.1	6.1	15	126	0	104	68	66	57	117	5	1
Apr.	1019.3	1018.1	16.8	9.0	37.8	5.0	35	115	1	45	75	70	63	91	11	3
May	1019.4	1017.7	14.6	8.3	32.2	0.0	54	145	13	35	74	74	66	64	15	5
June	1019.6	1018.0	12.6	7.3	27.2	0.0	65	197	10	38	76	77	70	48	17	7
July	1019.0	1017.5	11.8	7.6	25.0	0.0	75	185	19	50	76	77	69	51	19	8
Aug.	1018.0	1016.4	12.2	8.4	29.0	0.8	63	186	14	24	68	74	67	58	19	8
Sept.	1017.9	1016.2	13.4	9.3	33.3	2.2	47	123	15	32	72	69	64	74	13	5
Oct.	1016.2	1014.8	15.1	10.1	38.9	1.1	37	106	4	36	70	64	60	117	12	4
Nov.	1015.3	1014.0	17.0	10.2	41.4	4.4	27	79	1	36	67	62	56	140	8	2
Dec.	1014.1	1012.8	18.8	10.3	43.9	5.9	18	74	0	48	64	61	53	165	6	1
Annual	1017.1	1015.7	15.9	9.3	45.6	0.0	468	762	302	104	70	68	61	1232	134	47
Years	29	29	59	59	63	63	30	78	78	42	56	62	62	–	30	10

TABLE XXXVIII (*continued*)

Month	Number of days							Mean cloudiness			Wind				Probability of precipitation (%)		
	max. temp.		min. temp.	thunder	fog	clear	cloudy	all types		low clouds	most freq. dir.		mean speed (kt.)		<12.5 mm	<25 mm	<50 mm
	>32.2 °C	>37.8 °C	<2.2 °C					09h00	15h00		09h00	15h00	09h00	15h00			
Jan.	3	1	0	1	1	8	4	3.8	2.8	2.2	S	S	4	6	60	86	96
Feb.	2	1	0	1	0	6	3	4.0	3.0	2.2	SE,S	S	4	5	55	79	96
Mar.	2	0	0	1	0	5	5	4.0	3.3	2.2	SE	E,S	4	4	40	75	96
Apr.	0	0	0	1	1	4	6	4.3	4.2	2.9	S	E	4	5	19	43	74
May	0	0	0	1	0	3	6	4.1	4.2	4.2	N,NW	NE,N,S	3	5	0	9	48
June	0	0	0	0	0	1	6	4.1	4.3	4.2	N	NE,N,S	3	5	1	7	27
July	0	0	2	1	0	2	6	3.9	4.2	4.0	N	N,NW,S	4	5	0	5	20
Aug.	0	0	0	1	1	3	6	3.8	4.0	3.6	N	N	4	5	1	5	34
Sept.	0	0	0	1	1	4	6	3.8	3.8	3.6	S	S	5	6	0	18	54
Oct.	1	0	0	1	0	3	6	4.2	3.8	3.6	S	S	5	5	9	35	78
Nov.	1	0	0	1	0	5	4	4.2	3.5	3.4	S	S	5	5	41	64	89
Dec.	2	1	0	1	0	4	4	4.2	3.2	3.4	S	S	4	6	55	78	95
Annual	11	3	2	10	1	48	62	4.1	3.6	3.3	S	S	4	5	—	—	—
Years	65	65	13	6	6	10	10	30	30	3	5	5	5	5	85	85	85

* Calculated by Waite Institute formula.

TABLE XXXIX

CLIMATIC TABLE FOR ADELAIDE, S.A., AUSTRALIA
Latitude 34°56'S, longitude 138°35'E, elevation 42.7 m

Month	M.s.l. press. (mbar)		Mean daily temp. (°C)	Mean daily temp. range (°C)	Temp. extremes (°C)				Extreme rainfall (mm)		Precipitation (mm)			Relat. humidity (%)			Probability of precip. (%)			Vapor press. (mbar)
	mean daily	mean ampl.			highest		lowest		highest	lowest	mean	max.		09h00	15h00	daily index	<12.5 mm	<25 mm	<50 mm	
					day	night	day	night				24 h	48 h							
Jan.	1013.4	2.1	22.6	13.2	47.6	35.1	16.1	7.3	84	0	23	58	78	39	31	41	50	71	90	11.1
Feb.	1014.4	2.3	21.0	13.3	45.3	32.7	15.8	7.5	155	0	23	141	141	43	33	43	56	73	85	11.9
Mar.	1017.2	2.5	20.9	12.3	43.6	28.9	14.7	6.6	117	0	21	89	91	46	37	45	34	65	86	11.3
Apr.	1019.6	2.3	17.2	10.3	37.0	26.4	11.4	4.2	148	0	50	80	83	57	46	56	12	31	68	11.1
May	1020.1	2.2	14.6	8.9	31.9	21.0	11.1	2.7	196	3	66	70	96	67	55	63	3	9	37	10.6
June	1020.6	2.1	12.1	8.0	25.6	17.4	8.9	+0.3	218	6	61	54	76	75	64	71	1	4	33	10.0
July	1019.9	2.0	11.2	8.1	23.3	14.9	8.4	0.0	138	10	61	44	60	76	63	71	1	8	29	9.5
Aug.	1018.7	2.1	12.0	8.9	29.4	18.4	9.3	+0.2	157	8	59	57	58	70	58	67	2	4	37	9.5
Sept.	1018.3	2.2	13.4	9.2	35.1	22.9	10.6	+0.4	148	7	49	40	57	60	50	61	1	16	53	9.8
Oct.	1016.1	2.3	16.0	11.6	39.4	25.0	12.2	2.3	133	4	47	57	79	52	43	52	5	24	62	9.7
Nov.	1014.9	2.1	18.5	12.7	45.2	30.2	13.4	4.9	113	2	36	75	85	44	36	45	21	52	81	9.9
Dec.	1012.7	1.9	20.7	13.2	45.9	31.2	15.4	6.1	101	0	27	61	82	41	33	43	30	58	88	10.9
Annual	1017.2	2.2	16.7	10.9	47.6	35.1	8.4	0.0	784	289	523	141	141	53	43	53	—	—	—	10.3
Years	30		30	30	106	106	106	106	124	124	30	124	124	95	95	30	76	76	76	30

TABLE XXXIX (*continued*)

Month	Mean evapor. (mm)	Number of days							Mean cloudiness (oktas)	Mean sunshine (h)	Wind					Days with			
		precip. >0.25 mm	hail	thunder	light-ning	dew	fog	clear			most freq. dir.		speed (kt.)			max. temp.		min. temp.	ground frost*
											09h00	15h00	mean	highest daily	highest gust	>37.8 C°	>32.2 °C	<4.4 °C	<-0.9 °C
Jan.	236	4	0.1	1.5	2.2	3.5	0	8	2.9	10.0	SW	SW	8.6	28	63	4.2	10.4	0	0
Feb.	191	4	0.1	1.1	1.7	5.5	0	8	3.0	9.3	NE	SW	7.7	25	57	2.8	9.1	0	0
Mar.	158	5	0.1	0.9	1.8	9.9	0	7	3.2	7.8	S	SW	7.1	23	68	1.0	5.7	0	0
Apr.	96	10	0.3	1.0	1.5	12.9	0	5	4.2	6.0	NE	SW	6.9	28	70	0	0.7	0	0
May	58	13	0.5	1.0	1.6	15.3	0.5	2	4.6	4.8	NE	NW	7.0	27	61	0	0	0.4	0.5
June	37	15	0.7	0.9	1.5	15.1	1.1	2	4.9	4.1	NE	N	7.1	27	58	0	0	2.5	2.2
July	37	16	0.8	0.8	1.5	16.2	1.3	2	4.8	4.3	NE	NW	7.4	24	52	0	0	4.4	3.2
Aug.	53	16	0.9	1.1	1.8	15.7	0.6	3	4.4	5.2	NE	SW	8.2	28	55	0	0	3.0	2.0
Sept.	80	13	0.8	1.4	1.9	14.5	0.1	4	4.2	6.1	NNE	SW	8.2	26	60	0.1	0.1	1.3	0.8
Oct.	127	11	0.6	2.0	2.8	11.7	0	4	4.7	7.1	NNE	SW	8.5	28	65	0.8	1.3	0.4	0.2
Nov.	172	8	0.3	2.1	3.0	6.0	0	5	3.9	8.5	SW	SW	8.6	28	69	2.5	4.7	0	0
Dec.	219	6	0.1	1.5	2.2	4.3	0	7	3.4	9.4	SW	SW	8.6	24	65	11.4	8.0	0	0
Annual	1463	121	5.3	15.3	23.5	130.6	3.6	57	4.0	6.9	NE	SW	7.8	28	70	11.4	40.0	12.0	8.9
Years	91	124	91	91	91	91	91	59	30	81	30	30	82		43	106	106	106	100

* At the Adelaide Weather Bureau there are only 0.3 days per year with a minimum <2.2°C, but in the Trinity Gardens the total is 3.3 days (of which 1.5 in July).

TABLE XL

CLIMATIC TABLE FOR MOUNT GAMBIER, S.A., AUSTRALIA
Latitude 37°30'S, longitude 140°30'E, elevation 65 m.

Month	Mean sta. level press. (mbar)		Temperature (°C)				Precipitation (mm)				Relat. humidity (%)			Evapor. (mm)	Number of days with precip.	
	09h00*	15h00*	mean	daily range	extreme max.	extreme min.	mean	highest	lowest	max. 24 h**	daily index	09h00	15h00		>0.25 mm	>2.5 mm
Jan.	991.4	990.3	18.1	12.4	44.8	+0.6	34	158	2	51	65	58	51	159	7	3
Feb.	992.5	991.3	18.5	12.6	43.1	+1.1	29	119	0	45	65	60	51	125	7	2
Mar.	995.0	993.6	17.6	10.8	41.3	+0.6	38	153	2	36	70	65	53	112	9	3
Apr.	996.6	995.2	14.4	10.3	36.8	−1.2	60	177	<1	46	76	75	61	63	13	8
May	996.5	995.3	12.1	8.8	28.1	−3.2	85	206	21	60	80	82	68	39	17	7
June	996.9	995.6	10.3	7.8	22.2	−4.8	97	201	15	37	81	85	74	30	19	9
July	995.7	994.4	9.6	8.2	21.4	−4.6	105	226	21	33	79	85	73	32	20	9
Aug.	994.5	993.0	10.3	8.8	25.0	−2.5	98	177	14	43	79	81	67	41	20	12
Sept.	994.5	993.1	11.6	9.7	31.7	−2.0	77	164	24	33	77	74	60	63	17	8
Oct.	992.7	991.5	13.3	10.6	35.0	−0.9	65	148	15	33	74	68	61	83	15	6
Nov.	991.9	990.7	14.9	11.4	40.0	−0.6	45	130	9	56	71	64	57	102	12	5
Dec.	990.8	989.7	16.6	12.1	42.0	+1.1	41	277	1	44	67	61	54	144	9	4
Annual	994.1	992.8	13.9	10.3	44.8	−4.8	774	1414	457	60	73	72	61	993	165	76
Years	30	30	50	50	50	50	50	89	89	42	30	50	50	5	50	unsp.

TABLE XL (*continued*)

Month	Number of days — min. temp. C° <2.2	<0	gale	thunder **	fog **	clear **	cloudy **	Cloudiness (oktas) — all types 09h00	15h00	low cloud 09h00	15h00	Wind** — most freq. dir. 09h00	15h00	mean speed (m/sec.) 09h00	15h00	Probability of precip. (%) <12.5 mm	<25 mm	<50 mm
Jan.	0		0	1	1	3	7	3.6	3.2	3.3	2.7	SE,SW,W	SE	13	13	24	52	79
Feb.	0		0	<1	2	3	7	3.8	3.4	3.0	2.7	SE	SE	12	13	31	57	81
Mar.	0		0	1	1	1	10	3.7	3.6	2.5	2.4	SE	SE	12	13	19	44	78
Apr.	0.4		0	1	<1	2	11	4.2	4.3	3.8	4.2	N,SW	W,NW	13	14	3	12	46
May	0.8		0	1	2	1	11	4.3	4.6	4.2	4.6	N,NW	W,NW	12	13	0	2	23
June	3.8	1.5	0	1	1	1	9	4.5	4.7	4.3	4.3	N	E,N	11	12	0	2	19
July	3.1	1.4	0	1	2	1	10	4.6	4.9	3.8	4.1	N,NW	NW,W	12	13	0	1	7
Aug.	2.8	0.5	0	1	1	1	9	4.2	4.8	3.3	4.3	NW,N	W,NW	14	16	0	2	9
Sept.	1.7	0.2	0	1	<1	1	8	4.0	4.6	3.3	4.1	NW	W	14	16	0	1	17
Oct.	0.9		0	1	<1	1	9	4.3	4.7	3.8	4.6	W,NW	W	13	15	0	8	32
Nov.	0		<1	<1	1	2	9	4.6	4.6	3.7	3.5	SE,SW,W	SE,W	13	14	6	29	68
Dec.	0		0	1	1	2	8	4.1	3.8	3.8	2.9	SE,SW	SE	13	14	12	35	73
Annual	13.5	3.6	<1	11	13	19	108	4.2	4.2	3.6	3.7	N,NW	W,SE	13	14	—	—	—
Years	10	10	3	7	7	10	10	30	30	unsp.	unsp.	5	5	5	5	90	90	90

* At Hamilton, Vic., 37°44'S, 142°01'E, 187 m.
** At Cape Northumberland, S.A., 38°05'S, 140°40'E, 187 m.
*** At Mount Burr, 37°33'S, 140°24'E, 64 m.

TABLE XLI

CLIMATIC TABLE FOR THURSDAY ISLAND, QLD., AUSTRALIA
Latitude 10°34'S, longitude 142°12'E, elevation 5 m

Month	Mean daily sta. press. (mbar)	Mean daily temp. (°C)	Mean daily temp. range (°C)	Temp. extremes (°C)		Precipitation (mm)				Relative humidity (%)			Number of days		
				highest	lowest	mean	highest	lowest	max. 24 h	daily index	mean 09h00	mean 15h00	precipitation >0.25 mm	>2.5 mm	thunder
Jan.	1007.3	28.0	5.4	36.1	21.1	441	895	164	178	83	81	77	20	16	5
Feb.	1006.9	27.7	5.3	34.4	21.1	378	791	130	173	84	83	77	20	14	5
Mar.	1007.5	27.7	5.3	33.9	21.1	350	649	86	148	83	82	75	20	14	6
Apr.	1008.7	27.6	5.1	34.4	21.1	203	721	13	217	81	80	74	14	9	5
May	1009.8	26.9	4.9	32.8	18.9	41	209	1	75	80	79	71	10	3	1
June	1010.0	26.1	5.1	31.7	17.8	16	68	1	28	80	79	71	8	2	0
July	1011.8	25.3	5.2	32.2	17.8	13	51	0	35	79	79	70	7	1	0
Aug.	1011.6	25.3	5.4	31.7	20.0	5	48	0	6	78	76	68	4	1	0
Sept.	1011.6	26.0	5.6	32.8	20.0	3	18	0	6	77	74	67	3	1	0
Oct.	1010.8	27.2	5.8	33.9	21.1	5	78	0	48	75	72	67	2	1	1
Nov.	1008.8	28.3	6.0	35.6	21.7	36	190	0	92	75	71	67	4	1	1
Dec.	1007.8	28.8	6.0	36.7	21.1	198	526	1	120	77	74	69	12	3	7
Annual	1009.5	27.1	5.4	36.7	17.8	1687	2520	815	217	79	77	71	124	72	29
Years	10	30	30	49	49	30	53	53	49	30	32	32	30	10	10

TABLE XLI (*continued*)

Month	Number of days									fog	Mean cloudiness (oktas)	Wind		mean speed (kt.)	Probability of precipitation (%)		
	cloud cover											most freq. dir.			<12.5 mm	<25 mm	<50 mm
	09h00			15h00			21h00					09h00	15h00				
	0–2.5 clear	2.6–5.6	5.7–8 cloudy	0–2.5 clear	2.6–5.6	5.7–8 cloudy	0–2.5 clear	2.6–5.6	5.7–8 cloudy								
Jan.	3	9	19	2	8	21	11	10	10	0	5.3	NW	NW	4	0	0	0
Feb.	3	9	16	3	9	16	11	8	9	0	5.3	NW	NW	4	0	0	0
Mar.	5	9	17	3	10	18	12	11	8	0	5.1	SE	NW	4	0	0	0
Apr.	6	10	14	6	11	13	14	9	7	0	4.4	SE	SE	6	0	6	12
May	7	12	12	8	13	10	14	11	6	0	3.9	SE	SE	9	35	55	75
June	7	11	12	9	10	11	14	9	7	0	3.8	SE	SE	8	59	84	98
July	10	8	13	10	10	11	12	11	8	0.1	3.6	SE	SE	8	80	92	98
Aug.	11	9	11	12	12	7	18	8	5	0	3.5	SE	SE	9	94	96	100
Sept.	7	11	12	13	10	7	18	7	5	0	3.4	SE	SE	8	98	100	100
Oct.	6	15	10	14	12	5	18	9	4	0	3.4	SE	SE	9	87	92	96
Nov.	5	13	12	11	11	8	17	8	5	0	3.6	SE	NW	6	55	67	72
Dec.	6	11	14	7	10	14	13	9	9	0	4.5	SE	SE	4	11	14	24
Annual	76	127	162	98	126	141	172	110	83	0.1	4.2	SE	SE	7	—	—	—
Years	unsp.	unsp.	unsp.	unsp.	unsp.	unsp.	unsp.	unsp.	unsp.	10	10	5	5	5	49	49	49

TABLE XLII

CLIMATIC TABLE FOR COEN, QLD., AUSTRALIA
Latitude 13°57'S, longitude 143°12'E, elevation 600 m

Month	Mean sta. level press. (mbar)*		Temperature (°C)		extreme*		Precipitation (mm)				Relat. humidity (%)		
	09h00	15h00	mean	daily range	max.	min.	mean	highest	lowest	max. 24 h	daily index	09h00	14h30*
Jan.	1007.4	1004.8	27.0	9.6	40.0	19.4	247	579	23	356	81	71	85
Feb.	1007.3	1004.5	26.4	8.9	38.9	17.2	272	782	36	170	84	72	84
Mar.	1008.7	1006.2	26.0	8.9	36.7	19.4	272	746	48	343	83	74	86
Apr.	1011.5	1008.9	25.1	9.3	33.3	16.1	80	563	2	282	80	72	85
May	1013.4	1010.9	23.6	10.1	32.2	12.2	11	78	0	104	79	72	85
June	1014.7	1012.3	22.5	10.7	33.3	10.6	13	54	0	94	77	71	83
July	1015.3	1012.8	21.7	11.6	30.6	8.3	7	44	0	23	76	69	80
Aug.	1015.2	1012.7	22.3	11.8	30.6	10.6	3	44	0	43	72	65	77
Sept.	1015.3	1012.5	23.9	12.1	31.7	11.1	1	11	0	18	69	65	75
Oct.	1013.9	1010.8	25.8	11.8	34.4	14.4	8	106	0	66	66	64	73
Nov.	1011.2	1008.3	27.3	11.3	40.0	16.1	44	400	0	69	67	66	72
Dec.	1008.8	1006.2	27.7	10.8	40.6	16.1	187	604	6	196	74	68	77
Annual	1011.9	1009.2	24.9	10.6	40.6	8.3	1145	1818	482	356	76	69	80
Years	30	28	26	26	31	31	30	51	51	29	unsp.	29	24

TABLE XLII (*continued*)

Month	Evapor. (mm)**	Number of days with						Mean cloudiness (oktas)	Mean monthly sunshine** (h)	Preval. wind dir.*		Probability of precipit. (%)		
		precipitation		thunder**	cloudiness 09h00					09h00	15h00	<12.5 mm	<25 mm	<50 mm
		>0.25 mm	>2.5 mm		0–2.5 oktas	2.6–5.5 oktas	5.6–8 oktas							
Jan.	130	15	13	10	6	13	12	5.1	180	SE	SE	0	2	2
Feb.	110	15	14	6	4	11	13	5.2	160	SE	SE	0	0	2
Mar.	95	15	10	2	6	13	12	5.0	180	SE	SE	0	2	8
Apr.	105	7	5	1	11	9	10	4.2	180	SE	SE	20	35	51
May	100	1	1	0	14	12	5	3.6	190	SE	SE	66	81	96
June	95	3	1	0	12	12	6	3.5	180	SE	SE	67	88	94
July	80	1	1	0	16	10	5	3.3	220	SE	SE	86	96	100
Aug.	100	1	1	0	20	7	4	3.3	215	SE	SE	96	98	100
Sept.	125	1	0	0	20	7	3	3.0	235	SE	SE	100	100	100
Oct.	160	2	1	2	19	9	3	3.3	235	SE	SE	82	86	92
Nov.	180	4	4	2	12	10	8	3.7	230	SE	NE	33	49	61
Dec.	175	9	8	8	12	12	7	4.4	205	SE	SE	6	10	18
Annual	1455	74	59	31	152	125	88	4.0	2410	—	—	—	—	—
Years	—	unsp.	unsp.	—	unsp.	unsp.	unsp.	unsp.	—	5	5	51	51	51

* At Cooktown, 15°28'S, 145°17'E, 5 m (on the coast).
** Estimated from maps, standard Australian tank.
*** At Musgrave, 14°47'S 143°31'E, 100 m.

TABLE XLIII

CLIMATIC TABLE FOR WILLIS ISLAND, QLD., AUSTRALIA
Latitude 16°18'S, longitude 149°59'E, elevation 8.5 m

Month	Sta. press. (mbar)		Mean daily temp. (°C)	Mean daily temp. range (°C)	Temp. extremes (°C)		Precipitation (mm)				Relative humidity (%)				Number of days with precip. >0.25 mm	
	mean daily	mean daily ampl.			highest	lowest	highest	mean	lowest	max. 24 h	daily index	mean 09h00	mean 15h00	lowest	max.	min.
Jan.	1007	2.1	28.1	4.7	34	22	476	167	10	118	82	81	78	58	22	7
Feb.	1007	2.3	27.8	4.5	33	22	684	283	23	329	80	81	81	51	23	5
Mar.	1009	2.5	27.4	4.6	33	21	326	175	7	150	81	81	80	56	24	10
Apr.	1011	2.6	26.6	4.2	32	20	510	177	20	230	80	80	79	55	23	4
May	1013	2.7	25.4	4.1	31	20	313	67	4	246	80	78	77	53	20	2
June	1014	2.6	24.3	3.8	29	17	320	72	3	100	77	77	74	49	20	4
July	1015	2.6	23.6	3.9	28	18	186	49	4	161	74	73	73	44	20	1
Aug.	1015	2.8	23.8	4.4	28	19	109	18	0	46	73	73	73	33	18	0
Sept.	1015	2.9	24.6	4.8	30	19	207	22	0	73	72	74	70	42	15	0
Oct.	1013	2.8	26.8	4.8	31	19	102	17	0	62	72	74	74	45	10	0
Nov.	1011	2.5	26.9	5.2	32	21	171	36	1	120	75	74	75	56	14	1
Dec.	1008	2.3	27.8	5.1	33	21	274	92	0	171	79	80	78	56	16	0
Annual	1011	2.6	26.0	4.5	34	17	1596	1175	739	329	78	78	76	33	194	82
Years	20	20	16	16	20	20	20	20	20	20	20	20	20	3	20	20

TABLE XLIII (*continued*)

Month	Number of days			thunder	clear	cloudy	fog	gusts >34 kt.	Mean cloudiness			Wind				Days with			Mean daily solar radiat. (cal./cm²)
	precip. >2.5 mm								all types	low		most freq. dir.		mean speed (kt.)		calm	light wind	gale	
	max.	mean	min.							09h00	15h00	09h00	15h00	09h00	15h00				
Jan.	16	8	1	2	1	9	0	2	5.1	2.0	2.3	E,SE	E,SE	12	11	0.06	1.17	2	489
Feb.	15	10	3	1	1	10	0	3	5.4	2.6	2.2	SE	SE	11	11	0.11	1.05	2	519
Mar.	19	10	6	1	2	7	0	1	5.0	2.3	2.6	SE,E	SE	13	13	0.11	0.78	2	441
Apr.	15	8	1	1	2	8	0	3	5.0	2.7	3.4	SE	SE	15	14	0.11	0.39	3	464
May	9	4	0	1	2	4	0	2	4.3	2.6	2.6	SE	SE	16	15	0	0.17	3	353
June	9	4	0	<1	2	7	0	1	4.3	3.1	3.2	SE	SE	15	15	0	0.11	2	358
July	6	3	0	0	4	6	0	1	4.4	2.8	3.2	SE	SE	14	13	0	0.44	1	392
Aug.	8	2	0	0	4	7	0	<1	3.8	2.0	2.2	SE	SE	15	15	0	0.22	1	465
Sept.	8	2	0	0	4	4	0	<1	3.7	2.2	2.1	SE	SE	14	12	0.06	0.22	1	546
Oct.	4	1	0	<1	5	3	0	<1	3.3	1.8	1.9	SE,E	SE	14	12	0	0.33	1	584
Nov.	11	3	0	<1	3	3	0	<1	3.8	1.5	1.9	E,SE	SE	12	11	0	0.28	1	606
Dec.	11	4	0	2	2	4	0	<1	4.3	1.5	1.8	SE	SE	10	9	0	0.44	1	581
Annual	82	59	38	9	32	72	0	15	4.4	2.3	2.4	SE	SE	13	13	0.44	5.61	19	483
Years	20	20	20	20	20	20	20	13	20	3	3	20	20	5	5	18	18	20	1

TABLE XLIV

CLIMATIC TABLE FOR INNISFAIL, QLD., AUSTRALIA
Latitude 17°32′S, longitude 146°03′E, elevation 6.7 m

Month	Mean sta. press.* (mbar)	Mean daily temp. (°C)	Mean daily temp. range (°C)	Temp. extremes (°C) highest	Temp. extremes (°C) lowest	Precipitation (mm) mean	Precipitation (mm) highest	Precipitation (mm) lowest	Precipitation (mm) max. 24h	Relative humidity (%) daily index	Relative humidity (%) mean 09h00	Relative humidity (%) mean* 14h30
Jan.	1007.5	26.7	8.7	40	16	490	1581	32	531	84	80	69
Feb.	1006.9	26.4	8.4	41	17	602	1515	60	435	85	84	68
Mar.	1008.5	25.6	8.4	37	16	686	1591	87	350	87	86	69
Apr.	1011.9	24.0	8.7	34	11	467	1653	31	521	87	86	68
May	1013.9	21.9	9.0	32	6	323	1063	28	171	85	86	68
June	1015.0	20.2	9.3	30	5	188	491	10	111	86	86	67
July	1015.9	19.2	10.0	30	4	119	397	14	86	84	85	63
Aug.	1015.4	19.6	10.9	32	6	106	450	0	235	83	83	61
Sept.	1015.3	21.4	10.9	33	7	81	426	0	59	85	80	61
Oct.	1013.7	23.3	10.7	35	10	94	359	0	154	87	76	63
Nov.	1011.1	24.8	10.1	36	13	134	716	0	239	84	76	62
Dec.	1008.8	26.1	9.9	41	13	244	1304	10	539	84	77	66
Annual	1012.0	23.3	9.6	41	4	3535	5365	1775	58	84	82	65
Years	20	29	29	27	27	30	61	61		29	27	27

Month	Mean evapor. (mm)**	Number of days with precipitation 0 mm*	Number of days with precipitation >0.25 mm	Number of days with precipitation >2.5 mm	Number of days with precipitation >25 mm*	clear (*)	cloudy (*)	Mean cloudiness (oktas)	Wind (09h00) most freq. dir.	Wind (09h00) mean wind speed (kt.)	Probability of precipitation (%) <12.5 mm	Probability of precipitation (%) <25 mm	Probability of precipitation (%) <50 mm
Jan.	112	9	15	13	5	2	8	3.9	NW	2	0	0	4
Feb.	94	6	16	12	6	2	7	4.0	NW, S	2	0	0	0
Mar.	99	12	18	15	6	2	7	4.6	S, NW	2	0	0	0
Apr.	96	9	17	11	2	4	5	4.2	S, NW	2	0	0	2
May	33	16	16	15	1	5	6	4.1	SW	2	0	0	2
June	72	17	13	10	<1	8	5	4.2	SW, S	2	2	5	18
July	76	21	11	9	0	9	4	3.7	S	3	0	3	19
Aug.	94	22	9	7	0	10	4	3.3	S	3	7	12	21
Sept.	107	22	9	6	0	9	2	2.9	SE, S, NW	3	16	26	40
Oct.	140	21	7	6	<1	10	2	2.8	NW	2	21	38	44
Nov.	140	22	8	7	<1	5	3	3.1	E	2	9	14	32
Dec.	144	17	10	7	1	6	3	3.2	NW	2	4	7	26
Annual	1255	194	149	118	23	72	56	3.7	NW, S	2	–	–	–
Years	10	5	30	10	5	25	25	27	5	5	57	57	57

* At Cairns (16°55′S, 145°47′E, 4.9 m). There are no days with fog, and about 7 with thunder.
** At South Johnstone (17°36′S, 146°00′E).
Note: for frequency distribution of monthly rainfall at Tully (17°54′S, 145°54′E, 15 m) see Table VI.

TABLE XLV

CLIMATIC TABLE FOR NORMANTON, QLD., AUSTRALIA
Latitude 17°39′S; longitude 141°05′E; elevation 9.5 m

Month	Station press. (mbar)		Mean daily temp. (°C)	Mean daily temp. range (°C)	Temp. extremes (°C)		Precipitation (mm)				Relative humidity (%)			Evap.* (mm)
	09h00	15h00			max.	min.	mean	max.	min.	max. 24h	daily index	mean 09h00	mean 15h00	
Jan.	1006.7	1003.6	29.8	9.6	43	17	295	670	2	272	70	73	55	182
Feb.	1006.5	1003.7	29.2	8.9	41	18	249	770	18	280	72	78	58	160
Mar.	1008.3	1005.2	29.2	10.1	42	16	170	546	0	299	65	74	51	182
Apr.	1011.4	1007.8	27.9	11.9	40	12	35	305	0	174	50	57	38	175
May	1013.5	1010.0	25.2	13.0	38	9	7	198	0	48	49	53	56	157
June	1015.1	1011.5	22.8	13.1	36	7	16	106	0	78	51	56	37	152
July	1015.5	1011.9	21.8	14.1	36	6	3	85	0	60	48	52	34	145
Aug.	1015.2	1011.4	23.4	15.1	37	7	1	34	0	25	42	48	32	165
Sept	1014.4	1010.0	26.6	14.5	41	11	2	51	0	50	43	47	31	200
Oct.	1012.4	1008.0	29.1	13.9	44	11	6	99	0	72	48	50	35	248
Nov.	1010.0	1006.0	30.6	12.4	43	12	46	213	0	137	52	55	40	256
Dec.	1007.9	1004.3	30.6	11.1	43	13	122	619	0	173	59	52	47	228
Annual	1011.4	1007.8	27.2	11.7	44	6	954	1530	354	299	54	59	41	2250
Years	82	26	23	23	32	32	30	66	66	50	18	29	25	–

Month	Number of days with precip.		Mean cloudiness (oktas)		Mean monthly sunshine* (h)	Wind				Probability of precipitation (%)		
	>0.25 mm	>2.5 mm	09h00	15h00		most freq. dir.		mean speed (kt)		<12.5 mm	<25 mm	<50 mm
						09h00	15h00	09h00	15h00			
Jan.	14	8	3.6	4.2	198	N, NW	NW	4	6	2	3	5
Feb.	13	9	3.9	4.4	180	N	NW	5	5	0	2	7
Mar.	9	6	2.9	3.5	220	E	E	6	6	6	14	15
Apr.	3	2	1.6	2.4	250	SE	SE, E	8	6	39	51	71
May	1	1	1.3	2.0	255	E, SE	SE	8	6	88	88	94
June	1	1	1.2	1.6	240	SE	SE	8	5	77	83	91
July	1	1	1.0	1.0	280	SE	SE	7	4	94	97	98
Aug.	0	1	0.8	1.0	285	E, SE	SE	7	4	94	94	100
Sept.	0	1	0.7	1.4	290	SE, N, E	SE	6	4	97	98	98
Oct.	1	1	1.2	1.8	302	SE	NW	6	6	76	84	91
Nov.	4	2	2.3	2.8	275	N, NW	NW	6	6	18	33	71
Dec.	9	6	3.1	3.5	250	NW	NW	4	5	3	3	17
Annual	56	36	2.0	2.5	3025	SE, E	SE, NW	6	5	–	–	–
Years	30	10	29	28	–	5	5	5	5	66	66	66

* Estimated by interpolation from maps.

TABLE XLVI

CLIMATIC TABLE FOR TOWNSVILLE, QLD., AUSTRALIA
Latitude 19°14'S, longitude 146°51'E, elevation 22.2 m

Month	Mean sta. press. (mbar)		Mean daily temp. (°C)	Mean daily temp. range (°C)	Temp. extremes (°C)		Precipitation (mm)				Rainfall varia- bility (%)	Relative humidity (%)				Evap.* (mm)	Number of days with precipitation	
	09h00	15h00			highest	lowest	mean	highest	lowest	max. 24 h		daily index	mean 09h00	mean 15h00	min.		>0.25 mm	>2.5 mm
Jan.	1008.5	1006.1	27.5	6.8	39.6	18.7	284	831	15	347	56	75	74	69	39	168	14	8
Feb.	1008.1	1005.6	27.2	7.0	43.4	17.9	346	871	2	291	57	76	77	68	38	135	16	7
Mar.	1010.2	1007.6	26.6	7.6	36.7	16.7	230	601	1	366	67	73	75	67	47	137	13	6
Apr.	1013.6	1010.6	24.9	8.8	36.1	11.7	72	592	0	237	84	70	71	63	45	109	7	2
May	1015.9	1012.8	22.5	9.9	31.7	6.2	29	206	0	122	92	67	62	60	16	102	6	2
June	1017.3	1014.2	20.1	10.3	30.5	4.5	31	181	0	110	92	66	69	56	28	89	5	2
July	1017.9	1014.7	19.4	10.4	31.6	3.7	21	137	0	90	117	64	70	57	23	99	3	1
Aug.	1018.1	1014.5	20.3	10.8	33.3	1.1	11	113	0	59	122	66	65	55	14	117	2	1
Sept.	1017.2	1013.7	22.3	10.0	35.4	8.7	8	178	0	66	125	67	63	56	10	145	2	1
Oct.	1015.2	1011.7	24.7	8.3	37.1	8.2	21	274	0	229	104	68	63	59	17	175	4	1
Nov.	1012.4	1009.3	26.5	7.5	37.9	14.9	62	335	0	152	93	69	66	60	16	198	6	4
Dec.	1009.9	1007.0	27.5	7.2	39.6	18.7	102	616	0	167	84	73	69	64	37	190	8	5
Annual	1013.7	1010.7	24.2	8.7	43.4	1.1	1,215	2,482	268	—	—	70	70	61	10	1664	86	40
Years	29	27	30	30	34	34	30	72	72	72	—	23	30	27	1½	—	30	10

TABLE XLVI (continued)

Month	Number of days					Mean cloudiness				Wind				Sunshine (h)	Meridian altitude of sun (21st of month) (degrees)	Mean daily global radiation cal./cm² (°C)	Probability of precipitation (%)		
	thunder	cloudy	clear	fog	gusts >34 kt.	all types		low		most freq. dir.		mean speed (kt.)					<12.5 mm	<25 mm	<50 mm
						09h00	15h00	09h00	15h00	09h00	15h00	09h00	15h00						
Jan.	4	6	1	0	0	4.2	3.6	3.4	2.3	E,SE	NE	5	9	7.2	89	514	0	1	13
Feb.	3	5	1	0	1	4.4	4.1	3.4	3.3	SE	NE	6	10	7.1	81	484	3	6	13
Mar.	1	4	1	0	1	3.8	3.8	4.0	2.9	SE	NE,E	6	10	6.9	71	481	6	9	22
Apr.	1	3	2	1	1	3.0	3.2	2.6	3.2	SE,S	E,NE	7	10	8.5	61	451	30	36	52
May	1	2	6	1	0	3.0	3.2	1.7	2.4	S	NE	4	7	7.3	51	386	43	63	73
June	1	3	8	1	0	2.7	2.9	1.8	2.0	S	NE	6	8	8.0	47	372	43	60	75
July	0	2	10	1	0	2.6	2.4	1.0	1.2	S	NE	5	8	8.8	51	401	68	83	91
Aug.	0	2	7	2	0	2.2	2.1	0.6	1.1	S,SE	NE	5	9	9.2	61	471	76	84	94
Sept.	1	2	4	1	0	2.5	1.8	1.8	1.0	SE	NE	6	10	9.4	71	559	68	76	89
Oct.	1	1	3	0	0	3.0	1.8	3.4	1.3	E	NE	6	10	9.5	81	595	50	61	76
Nov.	1	1	2	0	0	3.4	2.3	3.8	1.4	E	NE	7	10	9.2	89	610	39	52	70
Dec.	5	4	1	0	1	3.6	2.8	3.0	1.4	E	NE	7	10	8.4	86	600	10	22	33
Annual	16	35	46	6	3	3.2	2.8	2.6	2.0	SE,S	NE	6	9	8.3	—	494	—	—	—
Years	10	10	10	10	1½	30	28	3	3	5	5	5	5	6	—	10-12	67	67	67

* Estimated by interpolation from maps.

TABLE XLVII

CLIMATIC TABLE FOR CLONCURRY, QLD., AUSTRALIA
Latitude 20°43'S, longitude 140°30'E, elevation 193 m

Month	M.s.l. press. (mbar)	Mean daily temp. (°C)	Mean daily temp. range (°C)	Temp. extr. (°C) min.	Temp. extr. (°C) max.	Precipitation (mm) mean	Precipitation (mm) min.	Precipitation (mm) max.	Precipitation (mm) max. 24h	Relative humidity (%) daily index	Relative humidity (%) mean 09h00	Relative humidity (%) mean 15h00	Evap.* (mm)
Jan.	1006.1	30.9	12.3	52.8	15.0	120	498	4	63	40	44	30	298
Feb.	1006.3	29.9	11.6	46.1	15.0	101	444	0	90	49	48	34	248
Mar.	1008.1	28.8	11.3	43.9	11.7	47	279	0	122	46	47	33	236
Apr.	1012.6	25.8	12.8	39.4	8.9	16	90	0	13	37	37	27	212
May	1015.0	21.8	12.9	37.2	5.0	12	122	0	34	39	39	27	180
June	1016.7	18.7	12.9	37.2	2.2	20	179	0	26	43	44	30	155
July	1017.3	17.8	13.8	35.6	1.7	6	83	0	33	40	41	27	132
Aug.	1012.6	19.9	15.1	39.4	1.7	3	57	0	–	32	32	19	180
Sept.	1014.7	23.7	15.2	41.1	5.0	4	112	0	17	30	30	18	230
Oct.	1011.6	27.6	14.9	44.9	10.6	11	92	0	28	31	29	19	285
Nov.	1008.9	30.1	13.9	48.3	12.2	40	130	0	19	34	34	23	295
Dec.	1006.7	31.3	13.4	51.7	10.0	48	515	0	38	38	39	25	292
Annual	1011.7	25.5	13.4	52.8	1.7	429	1,047	134	–	39	39	26	2743
Years	20	30	30	32	32	30	54	54	7	28	30	30	—

Month	Number of days precipitation >0.25 mm	Number of days thunder	Number of days fog	cloud base below 300 m	visibility <4 km	Mean cloudiness all types	Mean cloudiness low	Wind most freq. dir. 09h00	Wind most freq. dir. 15h00	Wind extreme gust dir.	Wind extreme gust speed (kt.)	Probability of precipitation (%) <12.5 mm	Probability of precipitation (%) <25 mm	Probability of precipitation (%) <50 mm
Jan.	7	6	0	1	4	3.6	2.0	NE, SE	SE	W	57	2	11	33
Feb.	7	1	0	1	4	3.6	2.8	SE	SE	SE	57	22	28	41
Mar.	4	2	0	3	4	2.8	1.5	SE	SE	NE	50	26	40	57
Apr.	2	0	0	1	1	1.8	1.1	SE	SE	SW	42	59	70	84
May	1	0	0	1	2	1.8	0.7	SE	SE	NNE	35	72	87	91
June	2	0	0	0	1	1.7	0.7	SE	SE	ESE	35	72	82	89
July	1	0	0	0	0	1.1	0.6	SE, S	SE	SSW	34	85	89	95
Aug.	0	0	0	0	0	0.8	0.1	SE	SE	SSW	36	89	96	96
Sept.	1	0	0	0	1	1.0	0.6	SE	SE, S	SSE, SW	32	81	91	98
Oct.	2	1	0	0	3	1.7	0.6	SE	SE, S	S	46	70	89	94
Nov.	3	4	0	0	3	2.5	1.0	SE, S	S, SE	W, ENE	52	33	50	74
Dec.	5	3	0	1	5	3.0	2.0	SE, S	SE, S	SE	68	18	33	59
Annual	35	17	0	8	28	2.1	1.1	SE	SE	SE	68	–	–	–
Years	30	6	6	5	5	32	4	5	5	5	5	54	54	54

* Estimated by interpolation from maps.

TABLE XLVIII

CLIMATIC TABLE FOR HUGHENDEN, QLD., AUSTRALIA
Latitude 20°51'S, longitude 144°13'E, elevation 327 m

Month	Mean sta. press. (mbar)		Mean daily temp. (°C)	Mean daily temp. range (°C)	Temp. extremes (°C)		Precipitation (mm)			
	09h00	15h00			highest	lowest	mean	highest	lowest	max. 24 h
Jan.	971.7	967.7	29.1	14.4	44.6	11.1	101	549	6	124
Feb.	972.7	969.1	28.2	13.8	42.8	11.1	112	335	0	161
Mar.	974.7	970.7	26.8	13.9	41.6	10.8	66	215	0	94
Apr.	977.8	973.1	24.0	15.4	39.4	4.1	16	212	0	139
May	980.2	975.9	20.3	14.6	36.7	1.8	21	139	0	83
June	981.4	977.0	17.4	15.3	33.8	−2.2	19	169	0	77
July	981.6	977.6	16.6	17.0	35.8	−2.8	16	62	0	69
Aug.	981.3	977.2	18.7	17.9	36.8	−2.3	6	57	0	39
Sept.	980.1	975.5	22.3	18.0	39.7	1.7	7	70	0	57
Oct.	977.7	973.2	26.1	17.6	43.3	2.9	21	145	0	121
Nov.	975.5	970.7	28.2	16.2	45.2	8.9	32	132	0	53
Dec.	972.8	968.5	29.5	15.9	45.6	8.6	61	294	4	98
Annual	977.3	973.0	23.9	15.9	45.6	−2.8	478	1085	150	161
Years	10	6	28	28	35	35	30	54	54	35

Month	Relative humidity (%)			Number of days			Mean cloudiness		Probability of precipitation (%)		
	daily index	mean 09h00	mean 15h00	precip. >0.25 mm	min. temp. (°C) <2.2	<0	09h00	15h00	<12.5 mm	<25 mm	<50 mm
Jan.	54	59	35	7	0	0	2.2	2.6	2	9	28
Feb.	57	64	37	8	0	0	2.4	3.3	11	15	35
Mar.	57	62	36	4	0	0	2.0	2.3	26	42	61
Apr.	54	56	31	2	0	0	1.2	1.6	63	72	83
May	54	60	35	1	0	0	1.3	1.6	68	77	91
June	58	63	42	2	0.5	0	1.4	1.7	50	72	83
July	53	63	38	2	0.6	0	1.1	1.1	72	78	94
Aug.	47	53	30	1	0.2	0	0.7	0.9	83	89	96
Sept.	44	51	30	1	0.1	0	0.6	1.4	76	85	97
Oct.	39	43	28	3	0	0	0.9	1.7	59	72	91
Nov.	40	44	25	4	0	0	1.3	2.2	39	54	70
Dec.	45	46	29	5	0	0	1.5	2.2	15	33	57
Annual	50	56	33	40	1.4	0	1.3	1.8	—	—	—
Years	21	13	5	30	10	10	21	10	54	54	54

TABLE XLIX

CLIMATIC TABLE FOR ROCKHAMPTON, QLD., AUSTRALIA
Latitude 23°24′S, longitude 150°30′E, elevation 11 m

Month	Mean sta. press. (mbar)		Mean daily temp. (°C)	Mean daily temp. range (°C)	Temp. extremes (°C)		Precipitation (mm)				Relative humidity (%)			Mean Austr. tank evapor. (mm)	Number of days with precip.			>2.5 mm
															>0.25 mm			
	09h00	15h00			highest	lowest	mean	highest	lowest	max. 24 h	daily index	mean 09h00	mean 15h00		max.	mean	min.	
Jan.	1009.6	1006.9	27.3	9.8	41.6	15.6	154	873	9	267	68	69	55	162	23	11	5	7
Feb.	1009.5	1006.9	26.8	9.2	42.4	15.6	187	924	6	389	69	70	55	131	22	12	2	6
Mar.	1011.8	1008.9	25.8	9.7	40.4	10.1	118	637	1	162	69	72	54	135	19	11	2	5
Apr.	1015.0	1011.8	23.6	10.8	36.7	6.3	44	551	0	357	67	70	50	117	15	7	3	3
May	1017.1	1013.8	20.4	11.7	34.6	4.3	44	246	0	108	67	71	49	94	12	6	0	3
June	1018.3	1015.0	17.9	11.3	31.2	0.4	41	266	0	174	68	73	50	78	13	5	0	3
July	1019.0	1015.6	16.9	12.5	33.2	1.4	50	496	0	293	65	71	45	75	13	6	0	4
Aug.	1018.4	1014.6	18.2	13.2	34.6	2.4	19	95	0	77	64	68	42	95	9	4	1	1
Sept.	1018.1	1013.9	21.2	13.0	37.9	4.4	20	107	0	88	64	64	42	119	11	4	0	1
Oct.	1015.5	1011.6	23.8	12.3	39.3	6.1	50	171	0	86	63	61	44	151	12	6	1	3
Nov.	1012.9	1009.4	25.7	11.4	42.2	12.2	68	188	0	74	64	61	46	164	15	7	1	4
Dec.	1010.4	1007.4	26.9	10.6	44.2	15.1	93	493	5	136	66	64	49	178	16	9	4	6
Annual	1014.6	1011.3	22.9	11.3	44.2	0.4	887	2081	399	389	66	68	48	1499	—	88	—	46
Years	29	29	30	30	48	49	30	71	71	68	29	41	44	24	47	30	47	10

TABLE XLIX *(continued)*

Month	Number of days — max.temp.(°C) >32.2	>37.8	thunder	clear	cloudy	fog	gusts (kt.)* >34	>32 (a)	>32 (b)	Cloudiness — all types 09h00	all types 15h00	low cloud 09h00	low cloud 15h00	Wind — most freq. dir. 09h00	most freq. dir. 15h00	mean speed (kt.)	Probability of precipitation (%) <12.5 mm	<25 mm	<50 mm
Jan.	19.9	1.1	4	7	4	0	1	0.1	1.4	4.0	3.8	3.8	3.3	E	E	7	5	8	15
Feb.	11.4	0.4	2	7	3	0	2	0.4	0.7	4.0	4.0	4.3	4.9	E,SE	E	7	4	6	17
Mar.	9.4	0.1	1	11	3	1	1	0.8	0.7	3.4	3.8	3.1	3.9	SE	E	8	10	14	34
Apr.	3.0	0	1	16	1	1	2	0.1	0.5	2.4	3.2	2.6	3.6	SE,E	E	5	24	41	63
May	0	0	1	16	2	2	0	0	0.1	2.4	3.0	1.8	3.2	SE	E,SE	5	27	46	71
June	0	0	1	12	4	2	<1	0	0.1	2.4	2.9	2.2	2.8	SE	E	4	45	43	63
July	0	0	1	16	2	3	0	0	0.1	2.2	2.5	1.9	2.4	SE	E	6	51	52	68
Aug.	0	0	0	18	1	2	<1	0.1	0.2	1.8	2.1	1.1	1.8	E,SE	E	6	40	70	87
Sept.	1.7	0	1	17	1	2	0	0	0.2	2.1	2.3	1.5	2.0	SE,E	E	6	20	54	76
Oct.	6.1	0	2	14	2	0	<1	0	0.7	2.9	2.6	3.2	2.8	E,SE	E,NE	7	12	38	62
Nov.	13.1	0.8	3	10	2	0	1	0.1	1.0	3.2	2.8	3.8	2.9	E	E,NE	6	2	25	47
Dec.	17.5	1.2	4	10	2	0	1	0.2	1.6	3.6	3.1	3.8	2.9	E	E,NE	6	—	7	22
Annual	82.1	3.6	19	154	27	11	9	1.8	7.3	2.9	3.0	2.8	3.0	E,SE	E		—	—	—
Years	10	10	10	10	10	10	3	18	18	30	30	3	3	5	5	5	83	83	83

* (a) = from tropical cyclones; (b) = from thunderstorms.

TABLE L

CLIMATIC TABLE FOR TAMBO, QLD., AUSTRALIA
Latitude 24°52'S, longitude 146°16'E, elevation 294 m

Month	Mean sta. press. (mbar)		Mean daily temp. (°C)	Mean daily temp. range (°C)	Temp. extremes (°C)		Precipitation (mm)			
	09h00	15h00			highest	lowest	mean	highest	lowest	max. 24 h
Jan.	965.2	962.1	28.1	15.2	46.6	9.4	74	307	0	113
Feb.	965.4	962.0	27.2	14.9	43.3	10.0	77	279	3	101
Mar.	968.1	964.6	25.1	15.6	41.9	4.0	50	361	0	173
Apr.	971.2	967.6	21.0	17.5	38.7	−1.0	41	159	0	117
May	973.1	969.8	16.5	17.9	34.6	−7.1	30	164	0	65
June	974.2	970.8	13.6	16.5	31.7	−5.6	27	121	0	64
July	974.3	971.1	12.6	17.8	32.2	−7.2	34	157	0	59
Aug.	973.7	970.2	14.4	19.0	33.6	−6.7	15	151	0	68
Sept.	972.9	969.2	18.3	18.8	37.7	−2.2	16	131	0	92
Oct.	970.4	966.4	22.6	17.9	41.1	−0.6	38	180	0	90
Nov.	967.9	963.9	25.5	16.7	43.6	4.3	46	174	0	97
Dec.	965.8	961.8	27.2	16.6	45.8	5.5	71	215	4	76
Annual	970.2	966.6	21.0	17.0	46.6	−7.2	521	1360	207	173
Years	27	23	30	30	55	55	30	57	57	55

Month	Relative humidity (%)			Number of days			Mean cloudiness		Prev. wind dir.		Probability of precipitation (%)		
	daily index	mean 09h00	mean 15h00	precip. >0.25 mm	min. temp. (°C) <2.2	<0	09h00	15h00	09h00	15h00	<12.5 mm	<25 mm	<50 mm
Jan.	50	52	31	7	0	0	2.0	3.4	NE	E	12	16	38
Feb.	56	62	40	7	0	0	2.4	3.4	NE	NE,E	12	21	46
Mar.	56	59	38	6	0	0	1.9	3.2	NE	NE	25	33	51
Apr.	57	56	34	3	0.3	0	1.5	2.5	NE	SE	47	61	76
May	61	64	37	3	1.5	0.4	1.7	2.2	E	SE	41	54	74
June	65	71	39	4	> 5	5.2	1.8	2.2	E	NE	39	52	77
July	62	68	38	3	> 5	4.8	1.6	1.9	E	SE	40	60	77
Aug.	55	56	31	2	> 4	3.7	1.0	1.5	E	SW	54	72	95
Sept.	49	48	27	3	1.4	0.7	1.0	1.8	E	SW	58	70	86
Oct.	45	45	29	5	0.2	0	1.5	2.5	NE	W	32	56	74
Nov.	48	44	27	5	0	0	1.7	3.0	NE	NE	23	42	70
Dec.	48	45	28	6	0	0	1.7	3.2	NE	NW	13	23	49
Annual	53	55	33	54	>18	14.8	1.7	2.6	—	—	—	—	—
Years	29	26	22	30	10	10	29	26	5	5	57	57	57

TABLE LI

CLIMATIC TABLE FOR WINDORAH, QLD., AUSTRALIA
Latitude 25°26′S, longitude 142°36′E, elevation 119 m

Month	Mean sta. press. (mbar)		Mean daily temp. (°C)	Mean daily temp. range (°C)	Extreme temp. (°C)		Precipitation (mm)			
	09h00	15h00			highest	lowest	mean	highest	lowest	max. 24 h
Jan.	994.9	991.3	30.9	14.6	46.7	10.6	35	140	0	130
Feb.	995.1	992.0	30.1	13.6	45.6	12.8	60	188	0	186
Mar.	998.0	994.7	27.6	13.9	43.2	8.3	47	378	0	103
Apr.	1002.3	998.5	22.7	14.7	39.7	1.6	14	167	0	89
May	1004.5	1001.2	17.9	14.7	35.6	1.0	13	111	0	86
June	1006.0	1002.8	14.5	13.8	35.0	−3.3	16	78	0	52
July	1006.0	1002.9	13.5	14.8	32.0	−3.3	14	141	0	42
Aug.	1005.3	1001.8	15.9	16.3	35.6	−0.7	10	125	0	47
Sept.	1003.8	999.8	19.8	16.8	40.0	−1.1	10	96	0	64
Oct.	1000.7	996.9	24.4	16.5	44.4	3.9	21	89	0	63
Nov.	997.8	994.1	27.7	15.8	45.0	7.0	15	102	0	50
Dec.	995.2	992.1	29.8	15.3	46.1	9.9	22	186	0	96
Annual	1000.8	997.3	22.9	15.1	46.7	−3.3	279	751	91	186
Years	27	21	32	32	35	35	30	50	50	80

Month	Relative humidity (%)			Number of days			Mean cloudiness		Prev. wind dir.		Probability of precipitation (%)		
	daily index	mean 09h00	mean 15h00	precip. >0.25 mm	min. temp. (°C) <2.2	<0	09h00	15h00	09h00	15h00	<12.5 mm	<25 mm	<50 mm
Jan.	34	31	20	4	0	0	1.6	2.8	N	NE	36	56	74
Feb.	38	40	24	5	0	0	2.0	2.7	N	SE	34	54	70
Mar.	39	40	25	4	0	0	1.6	2.5	NE	SE	38	58	70
Apr.	40	42	27	2	0	0	1.4	2.1	E	SE	60	72	84
May	47	49	31	2	0	0	1.6	2.0	E	SE	64	74	84
June	54	57	34	2	2.5	0.1	1.6	1.9	E	SE	48	64	84
July	51	53	29	2	2.0	0	1.4	1.6	E	SE	70	82	94
Aug.	41	41	23	2	1.5	0	1.0	1.0	E	SE,SW	70	92	98
Sept.	34	31	17	2	0	0	1.0	1.2	N	SE	70	84	96
Oct.	32	28	16	3	0	0	1.4	1.9	N	SE	62	78	88
Nov.	34	26	17	3	0	0	1.4	2.4	N	SE	52	68	84
Dec.	33	27	17	3	0	0	1.4	2.2	N,NE	NE	36	60	78
Annual	38	36	23	34	6.0	0.1	1.4	2.0	—	—	—	—	—
Years	28	29	23	30	10	10	29	23	5	5	50	50	50

TABLE LII

CLIMATIC TABLE FOR CHARLEVILLE, QLD., AUSTRALIA
Latitude 26°25'S, longitude 146°13'E, elevation 294 m

Month	Mean sta. press. (mbar)		Mean daily temp. (°C)	Mean daily temp. range (°C)	Extreme temp. (°C)		Precipitation (mm)				Relative humidity (%)			Mean Austr. tank evapor. (mm)	Number of days precip. >0.25 mm
	09h00	15h00			highest	lowest	mean	highest	lowest	max. 24 h	daily index	mean 09h00	mean 15h00		
Jan.	975.8	972.8	29.0	14.9	47.0	11.1	75	307	3	109	44	45	26	274	7
Feb.	976.8	973.5	28.4	14.4	46.1	9.4	70	399	0	130	47	52	32	237	6
Mar.	979.4	975.8	25.8	14.2	43.3	5.0	67	382	0	159	50	53	33	211	5
Apr.	982.8	979.3	21.2	13.8	38.8	1.0	33	198	0	81	52	54	33	148	3
May	984.7	981.4	16.6	16.2	33.3	−2.8	28	171	0	69	59	61	36	107	4
June	985.7	982.7	13.2	15.0	31.0	−5.0	27	128	0	70	65	71	40	77	4
July	985.9	982.5	12.3	15.7	30.7	−5.0	31	220	0	86	61	67	36	81	4
Aug.	984.9	981.7	14.2	17.1	34.3	−4.4	20	125	0	44	53	51	29	113	3
Sept.	984.1	980.3	18.2	17.5	38.9	−1.7	18	127	0	57	47	43	24	162	3
Oct.	981.4	977.6	22.7	16.9	43.2	+0.9	41	130	0	86	42	40	23	212	5
Nov.	978.9	975.2	26.1	16.2	47.2	4.4	40	190	0	83	42	38	22	264	5
Dec.	976.7	973.0	28.1	15.5	47.8	6.7	49	235	3	161	43	38	22	270	6
Annual	981.4	978.0	21.3	15.8	47.8	−5.0	498	1202	202	161	49	49	29	2,156	55
Years	30	30	30	30	35	35	30	60	60	35	29	30	30	11	30

TABLE LII (*continued*)

Month	Number of days						Mean cloudiness			Wind				Most freq. visibility (km)	Probability of precipitation (%)		
	min. temp.		thunder	fog	visibility <4 km	cloud base <300 m	all types		low	prev. direction		extreme gusts			<12.5 mm	<25 mm	<50 mm
	<2.2	<0°C					09h00	15h00		09h00	15h00	direction	speed (kt.)				
Jan.	0	0	1	0	6	0	2.0	3.2	1.2	NNE	E	NNW	26	>50	19	25	60
Feb.	0	0	1	0	6	2	2.3	3.4	1.6	NNE	E,SE	WNW	19	20–50	27	38	55
Mar.	0	0	0	1	5	1	1.8	3.0	1.2	N	SE	E,SW,SSW	9	20–50	25	48	61
Apr.	0	0	0	0	2	0	1.7	2.5	0.9	NE	SE	NNW,SSE	19	>50	40	52	73
May	1.1	0.1	0	0	4	1	2.0	2.4	0.9	E	SE	N,NNE,SSE,SSW	9	>50	42	58	75
June	5.9	2.8	0	2	3	3	2.1	2.4	1.5	SE	S	E	9	>50	32	48	75
July	7.1	2.5	0	0	3	3	1.9	2.0	1.4	NNE	S	NW	12	>50	42	59	75
Aug.	4.5	1.7	0	0	3	3	1.4	1.6	1.2	NNE	SW,W	WNW,NW	9	>50	57	68	90
Sept.	1.0	0.1	1	0	3	0	1.0	1.8	0.9	NNE	SW	various	9	20–50	56	77	85
Oct.	0	0	1	0	3	0	1.6	2.4	1.0	N	NW	N	19	20–50	40	55	77
Nov.	0	0	3	0	7	0	1.6	2.6	1.0	N	SSW	NW,NNW	9	20–50	25	55	70
Dec.	0	0	3	0	6	1	1.7	3.0	1.5	N	SE	ESE,NW	26	20–50	13	22	50
Annual	19.6	7.2	10	3	51	14	1.8	2.5	1.2	—	—	—	—	—	—	—	—
Years	10	10	3	3	3	3	30	30	3	5	5	5	5	3	60	60	60

TABLE LIII

CLIMATIC TABLE FOR BRISBANE, QLD., AUSTRALIA
Latitude 27°28'S, longitude 153°02'E, elevation 41.8 m

Month	M.s.l. press. (mbar)	Mean daily temp. (°C)	Mean daily temp. range (°C)	Temp. extremes (°C)				Mean vapor press. (mbar)	Precipitation (mm)				Relat. humidity (%)			Probability of precip. (%)		
				highest sun	shade	lowest shade	grass		highest	mean	lowest	max. 24 h	index	09h00	15h00	<12.5 mm	<25 mm	<50 mm
Jan.	1013.3	25.0	9.1	76.1	43.2	14.9	+9.9	21.6	704	143	8	465	69	66	60	1	2	14
Feb.	1012.4	24.7	8.8	74.0	40.9	14.7	+9.5	21.8	1026	183	15	269	72	69	61	0	6	18
Mar.	1014.8	23.6	8.9	72.5	37.4	11.3	+7.4	20.5	864	147	0	284	72	72	60	2	13	26
Apr.	1017.1	21.2	9.8	67.7	35.1	6.9	+2.6	17.4	388	78	0	139	71	71	56	11	21	32
May	1018.4	18.2	10.1	63.9	32.4	4.8	−1.2	14.3	351	57	0	142	69	71	55	17	29	51
June	1019.0	15.8	9.9	57.8	31.6	2.4	−3.7	12.1	356	56	0	163	67	73	55	21	35	45
July	1019.6	15.0	10.7	63.4	29.1	2.3	−4.9	11.2	218	49	0	89	66	71	51	20	40	59
Aug.	1018.9	16.1	11.7	61.1	32.8	3.0	−2.7	11.5	372	30	0	124	64	67	48	26	46	72
Sept.	1018.7	18.1	11.5	68.6	38.3	4.8	−0.9	13.4	137	45	3	63	64	62	50	9	29	61
Oct.	1016.2	20.7	10.5	69.7	40.7	6.3	+1.6	15.6	290	77	1	135	64	59	54	10	19	47
Nov.	1014.3	22.5	9.8	72.4	41.2	9.2	+3.8	18.1	315	92	0	113	66	61	58	2	9	24
Dec.	1012.2	24.3	9.4	74.4	41.1	13.5	+9.5	19.9	441	136	9	168	67	62	58	2	7	20
Annual	1016.2	24.0	10.1	76.1	43.2	2.3	−4.9	16.4	2240	1092	411	465	68	67	55	—	—	88
Years	30	30	30	50	74	74	74	30	109	30	109	109	30	30	30	88	88	88

TABLE LIII (*continued*)

Month	Mean evapor. (mm)	Number of days with						Mean cloudiness (oktas)	Mean sunshine (h)	Wind						Mean terrestr. min. temp. (°C)
		precip.		thunder-storm	light-ning	clear	fog			most freq. dir.		speed (kt.)		highest daily mean	highest gust	
		>2.5 mm	>0.25 mm							09h00	15h00	09h00	15h00			
Jan.	171	8	12	5	9.8	3.5	0.6	4.6	7.6	SE	NE	5	10	17	50	18.6
Feb.	140	7	12	4	6.5	2.4	0.9	4.5	7.4	SE	NE	6	10	20	58	18.6
Mar.	128	7	14	3	5.9	5.4	1.6	4.1	7.0	S	E	6	9	17	56	17.3
Apr.	103	7	11	2	5.0	7.8	4.0	3.4	7.1	S	E	5	7	14	56	14.0
May	79	4	9	<1	4.1	8.3	5.4	3.4	6.6	SW	SE	6	7	15	43	10.3
June	63	4	8	<1	2.9	9.2	4.5	3.5	6.3	SW	W,SW	5	6	16	50	7.7
July	68	4	8	<1	2.8	12.4	4.9	3.0	6.8	SW	W,SW	6	7	19	58	6.2
Aug.	89	2	7	2	3.8	13.1	5.9	2.5	7.9	SW	NE	5	8	13	49	6.9
Sept.	114	3	7	4	5.8	13.0	2.8	2.6	8.2	S	NE	5	9	14	55	9.6
Oct.	147	4	8	5	7.1	8.5	1.6	3.4	8.4	SE,N	NE	5	10	13	54	12.9
Nov.	161	7	10	7	9.5	5.9	0.7	3.9	8.2	SE	NE	6	10	13	54	15.7
Dec.	178	7	11	7	10.6	3.8	0.4	4.2	8.2	SW	NE	5	11	17	69	17.7
Annual	1441	64	117	40	73.8	93.3	33.3	3.6	7.5				9	20	69	12.9
Years	30	21	30	53	30	30	30	30	30	30	30	5	5	46	46	75

345

TABLE LIV

CLIMATIC TABLE FOR TOOWOOMBA, QLD., AUSTRALIA
Latitude 27°33'S, longitude 151°58'S, elevation 585 m

Month	Mean sta. press. (mbar)		Mean daily temp. (°C)	Mean daily temp. range (°C)	Extreme temp. (°C)		Precipitation (mm)			
	09h00	15h00			highest	lowest	mean	highest	lowest	max. 24 h
Jan.	945.9	943.8	22.2	11.9	39.8	7.5	153	349	12	157
Feb.	945.4	943.2	21.7	11.1	38.2	7.8	143	736	2	164
Mar.	947.9	945.7	20.3	10.8	37.2	+0.6	112	389	0	132
Apr.	950.4	947.9	17.2	11.7	32.6	−0.6	62	243	0	97
May	949.7	947.5	13.8	11.4	29.0	−1.8	49	269	<1	80
June	951.1	948.9	11.2	10.9	26.9	−5.6	54	350	0	97
July	949.7	947.5	10.5	11.3	25.7	−5.3	58	194	1	145
Aug.	950.3	947.9	11.8	12.6	30.0	−4.2	32	267	0	51
Sept.	950.2	947.4	14.8	13.2	31.9	−1.1	42	168	8	61
Oct.	948.7	945.7	17.9	13.3	36.2	+0.2	79	234	6	89
Nov.	947.6	944.7	20.2	13.1	38.3	4.1	89	228	5	72
Dec.	945.7	943.1	21.6	12.6	40.6	4.5	131	292	18	127
Annual	948.5	946.1	16.9	12.0	40.6	−5.6	1,005	1,735	432	164
Years	10	10	30	30	58	58	30	66	66	80

Month	Relative humidity (%)			Number of days			Mean cloudiness		Prev. wind dir.		Probability of precipitation (%)		
	daily index	mean 09h00	mean 15h00	precip. >0.25 mm	min. temp. (°C) <2.2	<0	09h00	15h00	09h00	15h00	<12.5 mm	<25 mm	<50 mm
Jan.	73	73	51	11	0	0	3.6	3.8	E	E	3	3	17
Feb.	76	77	59	12	0	0	3.8	4.3	E	E	6	9	31
Mar.	74	77	58	11	0	0	3.5	3.8	E	E	9	18	35
Apr.	77	73	54	9	0.2	0.1	2.6	3.4	E	E	12	24	45
May	78	72	49	7	0.7	0	2.3	3.0	E	E	23	32	59
June	80	76	54	7	4.5	1.7	2.6	3.1	E	E	18	33	59
July	79	73	50	7	5.9	1.6	2.2	2.8	W	W	17	35	56
Aug.	72	67	42	6	4.6	1.2	1.9	2.2	E	E,W	24	48	74
Sept.	71	65	48	7	1.1	0.2	2.0	2.6	E	E	9	26	56
Oct.	67	64	43	8	0.1	0	2.7	3.2	E	E	9	15	44
Nov.	67	65	40	10	0	0	3.0	3.0	E	E	5	12	33
Dec.	70	66	41	11	0	0	3.1	3.4	E	E	0	3	14
Annual	74	69	48	106	17.1	4.8	2.8	3.3	—	—	—	—	—
Years	29	30	11	30	10	10	30	11	5	5	66	66	66

TABLE LV

CLIMATIC TABLE FOR BOURKE, N.S.W., AUSTRALIA
Latitude 30°13′S, longitude 145°58′E, elevation 110 m

Month	Mean sta. press. (mbar)		Mean daily temp. (°C)	Mean daily temp. range (°C)	Temp. extremes (°C)		Precipitation (mm)			
	09h00	15h00			highest	lowest	mean	highest	lowest	max. 24 h
Jan.	999.1	996.8	28.7	15.9	52.8	8.8	33	311	0	115
Feb.	999.8	997.5	28.2	15.4	48.8	9.4	45	284	0	76
Mar.	1003.0	1000.7	25.2	15.3	47.1	1.9	38	211	0	93
Apr.	1006.3	1003.8	19.9	14.6	41.6	1.9	26	183	0	114
May	1008.1	1005.8	15.4	13.4	34.9	−2.9	27	134	0	45
June	1009.0	1007.1	11.9	12.3	30.0	−3.9	31	128	0	54
July	1009.1	1006.8	11.3	12.8	28.9	−3.4	23	158	0	36
Aug.	1008.1	1005.7	13.4	14.5	34.4	−2.9	18	94	0	32
Sept.	1007.1	1004.4	17.1	15.7	37.8	−1.8	17	98	0	48
Oct.	1004.0	1001.3	21.3	16.0	44.4	+1.9	31	109	0	50
Nov.	1001.6	998.8	24.9	16.1	46.0	3.2	26	159	0	72
Dec.	999.1	996.5	27.2	15.8	49.3	5.3	33	157	0	73
Annual	1004.5	1002.1	20.4	14.8	52.8	−3.9	348	754	102	115
Years	29	29	30	30	90	90	30	72	72	90

Month	Relative humidity (%)			Number of days			Mean cloudiness		Prev. wind direction		Probability of precipitation (%)		
	daily index	mean 09h00	mean 15h00	precip. >0.25 mm	min. temp. (°C) <2.2	<0	09h00	15h00	09h00	15h00	<12.5 mm	<25 mm	<50 mm
Jan.	37	40	25	4	0	0	1.5	2.2	NE	SE	40	58	78
Feb.	41	48	29	5	0	0	1.8	2.6	NE,SE	SE	43	55	75
Mar.	44	50	31	4	0	0	1.5	2.1	NE	SE	53	67	83
Apr.	57	57	36	3	0	0	1.6	2.1	SE	SE	46	65	83
May	58	67	44	4	0.1	0	1.8	2.0	SE	SW	39	61	81
June	66	74	47	4	4.2	1.0	2.2	2.4	SE	SE	36	55	78
July	64	74	46	5	5.5	0.3	1.9	2.3	SW	SW	40	68	93
Aug.	56	62	38	4	3.1	0.3	1.6	1.8	SW	SW	50	73	91
Sept.	47	49	31	3	0	0	1.2	1.8	NE	SW	54	71	90
Oct.	42	42	25	5	0	0	1.6	2.1	NE	SW	40	64	89
Nov.	40	37	24	4	0	0	1.5	2.2	NE	SW	39	54	74
Dec.	41	37	23	4	0	0	1.5	2.1	NE,SE	SW	29	47	75
Annual	40	49	32	49	12.9	1.6	1.7	2.1	—	—	—	—	—
Years	30	30	28	30	10	10	30	29	5	5	72	72	72

TABLE LVI

CLIMATIC TABLE FOR ARMIDALE, N.S.W., AUSTRALIA
Latitude 30°39′S, longitude 151°38′E, elevation 1015 m

Month	Mean sta. press. (mbar)		Mean daily temp. (°C)	Mean daily temp. range (°C)	Extreme temp. (°C)		Precipitation (mm)				Relative humidity (%)		
	09h00	15h00			highest	lowest	mean	highest	lowest	max. 24 h	daily index	mean 09h00	15h00
Jan.	904.1	902.4	20.4	13.5	39.7	4.4	107	274	8	96	60	61	40
Feb.	904.1	902.5	19.8	13.2	37.8	3.3	105	279	3	75	63	68	46
Mar.	906.5	904.7	17.6	12.9	34.4	− 0.6	67	235	1	86	67	70	46
Apr.	907.4	905.5	13.9	13.8	30.1	− 3.9	42	236	4	64	69	75	46
May	907.3	905.5	10.1	12.2	26.7	− 6.7	36	140	3	80	69	80	50
June	907.2	905.5	7.3	11.3	24.4	− 8.3	55	190	5	94	68	82	53
July	907.0	905.1	6.6	11.2	21.1	−10.0	52	152	1	59	61	78	51
Aug.	906.8	904.8	7.7	12.7	25.7	− 8.1	56	209	4	155	64	72	43
Sept.	906.9	904.7	10.8	13.8	28.3	− 5.6	56	146	<1	76	62	64	39
Oct.	905.1	903.1	14.3	14.1	32.5	− 3.3	72	183	1	64	59	54	38
Nov.	904.0	902.1	17.3	14.3	36.4	0.0	78	232	4	65	58	51	37
Dec.	902.8	900.9	19.3	13.3	37.7	2.2	88	289	<1	61	60	55	39
Annual	905.8	903.9	13.8	12.9	39.7	−10.0	815	1507	422	155	68	66	44
Years	24	24	30	30	80	80	30	79	79	100	30	29	29

Month	Evapor. (mm)*	Number of days					Mean cloudiness		Prev. wind dir.		Probability of precipitation (%)		
		precip. >0.25 mm	min. temp. (°C) <2.2	<0	fog	snow	09h00	15h00	09h00	15h00	<12.5 mm	<25 mm	<50 mm
Jan.	188	10	0	0	1	0	3.0	3.4	SE	SE	4	11	23
Feb.	132	11	0	0	2	0	3.4	3.7	SE	W	7	16	41
Mar.	130	11	0.1	0	4	0	3.3	3.4	SE	SE	8	19	44
Apr.	95	9	2.5	1.6	4	0	3.0	3.2	SE	SE	18	41	58
May	71	9	> 6	6.0	4	0	3.1	3.1	W	W	13	41	65
June	48	11	>12	11.8	4	0.5	3.7	3.3	W	W	4	19	41
July	49	10	>13	12.7	3	1.2	3.3	3.0	W	W	12	24	54
Aug.	66	9	>13	12.2	2	0.2	2.8	2.8	W	W	8	24	64
Sept.	84	9	> 6	5.0	3	0.5	2.4	2.6	W	W	17	24	49
Oct.	110	10	1.9	0.8	1	0.2	2.6	3.0	W	W	7	10	40
Nov.	156	9	0.1	0	1	0	2.6	3.4	W	W	5	12	28
Dec.	192	11	0	0	1	0	2.6	3.4	E	W	2	8	29
Annual	1319	119	>55	50.1	30	2.6	3.0	3.2	—	—	—	—	—
Years	10	30	10	10	20	30	30	30	5	5	79	79	79

* At Inverell, 29°47′S, 151°10′E, 604 m; Australian tank.

TABLE LVII

CLIMATIC TABLE FOR LORD HOWE ISLAND, N.S.W., AUSTRALIA
Latitude 31°31′S, longitude 159°07′E, elevation 4.5 m

Month	Mean sta. level press. (mbar)	Temperature (°C)				Precipitation (mm)				Relative humidity (%)		
		mean		extreme		mean	highest	lowest	max. 24h	daily index	09h00	15h00
		max.	min.	max.	min.							
Jan.	1012	25.6	19.3	30.6	11.7	125	291	10	140	77	71	71
Febr.	1013	25.6	19.6	31.7	12.2	106	337	24	161	75	70	68
Mar.	1016	25.0	18.9	29.4	12.8	127	375	51	111	75	69	69
Apr.	1016	23.3	17.2	28.3	10.6	171	702	38	305	77	71	70
May	1018	21.1	15.3	25.6	8.3	158	376	45	117	77	72	70
June	1016	19.4	13.8	23.9	7.8	195	387	82	148	77	70	70
July	1016	18.6	13.0	23.9	6.1	196	496	80	131	77	73	70
Aug.	1017	18.8	12.6	22.8	6.1	135	269	13	79	75	71	68
Sept.	1017	19.9	13.4	25.0	6.7	134	322	27	247	76	71	69
Oct.	1017	21.4	14.8	27.2	7.8	131	337	36	170	78	71	71
Nov.	1015	22.9	16.3	27.8	9.4	114	292	14	127	78	71	72
Dec.	1013	24.4	18.0	29.4	10.6	123	339	17	106	78	71	72
Annual	1015	22.2	16.1	31.7	7.2	1715	2866	996	305	78	71	70
Years	9	27	27	27	27	45	42	42	30	28	28	9

Month	Number of days with							Mean cloudiness (oktas)			Wind				
	precipitation		thunder	clear	cloudy	fog	gale	all types	low cloud		most freq. dir.			mean speed (kt.)	
	>0.25 mm	>2.5 mm							09h00	15h00	09h00	15h00		09h00	15h00
Jan.	11	7	1	2	10	0	0	5.4	4.5	4.2	SE	SE		9	10
Feb.	11	6	<1	1	9	0	0	5.2	4.4	4.2	SE, NE	SE, NE		10	11
Mar.	14	8	2	1	11	0	<1	5.1	4.2	3.1	SE	SE		9	10
Apr.	18	12	2	1	10	0	1	5.4	4.4	4.2	S, SE, SW	SE, SW, S		11	11
May	20	13	2	1	9	0	1	5.4	4.3	4.2	SE, S, SW	SE, S, SW		11	11
June	21	13	3	1	11	0	2	5.8	4.7	4.6	S, SE, SW	SW, S, SE		12	12
July	22	15	5	1	7	0	1	5.6	4.3	4.4	S, SW, SE	SW, SE, NW		11	12
Aug.	19	13	8	1	9	0	1	5.3	4.3	4.2	SW	S, NW		12	13
Sept.	14	8	5	2	8	0	2	5.0	4.2	4.2	S, NW	NW, N		11	11
Oct.	13	8	3	2	10	0	1	5.4	4.7	4.2	NW, N	S, SE, NW		11	12
Nov.	13	7	4	1	11	0	<1	5.5	4.0	3.9	SE, S	S, SE, NW		11	11
Dec.	11	8	2	1	11	0	1	5.5	4.1	3.8	NE	NE, NW		10	10
Annual	187	118	37	15	116	0	11	5.4	4.2	4.1	SE, S	SE, S		11	11
Years	15	10	4	10	10	4	4	4	3	3	5	5		4	4

349

TABLE LVIII

CLIMATIC TABLE FOR PORT MACQUARIE, N.S.W., AUSTRALIA
Latitude 31°35'S, longitude 152°54'E, elevation 13.4 m

Month	Mean sta. press (mbar)		Mean daily temp. (°C)	Mean daily temp. range (°C)	Extreme temp. (°C)		Precipitation (mm)				Relative humidity (%)		
	09h00	15h00			highest	lowest	mean	highest	lowest	max. 24h	daily index	mean 09h00	15h00
Jan.	1011.1	1009.5	21.9	7.9	40.0	10.0	137	1,387	6	274	85	80	79
Feb.	1011.6	1010.1	21.9	7.4	41.0	8.9	172	844	2	203	84	81	79
Mar.	1014.7	1012.8	20.8	8.5	36.1	6.4	176	494	18	146	85	83	78
Apr.	1016.3	1014.0	18.3	9.1	33.8	4.4	153	505	14	298	82	79	72
May	1017.2	1014.9	15.4	10.0	28.9	1.4	112	506	11	180	80	79	69
June	1017.2	1014.9	13.1	10.3	26.4	−0.8	125	652	6	189	77	80	70
July	1017.7	1015.3	12.4	10.7	28.9	−1.4	112	361	6	141	74	80	66
Aug.	1016.8	1014.2	13.1	11.3	32.9	−0.6	94	775	0	142	73	74	65
Sept.	1016.9	1014.1	13.9	10.7	34.4	+0.3	85	326	0	124	78	72	70
Oct.	1014.2	1011.6	17.2	9.1	37.2	3.3	92	286	9	151	81	72	73
Nov.	1012.3	1010.1	19.2	8.3	40.0	5.1	91	462	10	178	84	73	75
Dec.	1010.5	1008.8	20.8	7.7	37.8	8.9	112	637	8	104	85	77	74
Annual	1014.7	1012.5	17.4	9.3	41.0	−1.4	1,463	3,211	915	298	81	77	74
Years	30	30	30	30	100	100	30	68	68	103	30	30	30

Month	Number of days							Mean cloudiness		Wind				Probability of precipitation (%)		
	precipitation		min. temp.		clear	cloudy	fog	09h00	15h00	prev. direction		speed (kt.)		<12.5 mm	<25 mm	<50 mm
	>0.25 mm	>2.5 mm	<2.2 °C	<0 °C						09h00	15h00	09h00	15h00			
Jan.	13	6	0	0	5	3	0.3	3.6	3.5	NE, SW	SE	8	11	0	8	20
Feb.	13	8	0	0	5	4	0.6	3.8	3.8	NE	SE	7	10	<1	6	14
Mar.	15	10	0	0	5	4	0.3	3.6	3.4	SW	SE	7	9	0	3	10
Apr.	13	8	0	0	6	3	0	3.1	3.5	SW	SE	8	10	0	2	13
May	11	9	0	0	6	3	0.6	2.8	3.1	W	SE	7	10	4	9	20
June	10	7	1.2	0.2	8	3	0	3.0	3.2	W	SW	8	10	7	20	36
July	9	6	1.4	0.1	7	3	0.3	2.7	2.9	W	SW	8	10	12	21	34
Aug.	9	5	0.7	0.1	7	1	0.3	2.5	2.8	W	NE	8	13	11	25	47
Sept.	10	5	0.2	0	8	3	0.3	2.6	3.0	W	NE	9	16	10	22	41
Oct.	12	6	0	0	7	2	0	3.0	3.3	SW	NE	9	14	1	12	39
Nov.	11	6	0	0	4	3	0.3	3.1	3.4	NE	NE	9	13	0	11	32
Dec.	12	7	0	0	4	2	0	3.4	3.4	NE	NE	9	13	1	8	23
Annual	138	83	3.5	0.4	72	34	5	3.1	3.3	−	−	8	12	−	−	−
Years	30	10	10	10	10	10	3	30	30	5	5	5	5	74	74	74

TABLE LIX

CLIMATIC TABLE FOR BROKEN HILL, N.S.W., AUSTRALIA
Latitude 31°57′S, longitude 141°28′E, elevation 305 m

Month	Mean sta. press. (mbar)		Mean daily temp. (°C)	Mean daily temp. range (°C)	Extreme temp. (°C)		Precipitation (mm)				Relative humidity (%)		
	09h00	15h00			highest	lowest	mean	highest	lowest	max. 24 h	daily index	mean 09h00	15h00
Jan.	978.1	976.4	25.3	14.4	46.1	7.2	18	90	0	60	36	39	25
Feb.	978.9	977.2	25.2	14.2	46.6	5.6	28	109	0	67	40	46	29
Mar.	981.8	980.0	22.4	13.6	45.5	4.4	21	71	0	60	44	50	31
Apr.	984.4	982.6	17.6	12.3	37.7	1.1	16	114	0	93	53	57	36
May	985.3	983.6	13.8	10.9	31.0	−0.8	17	93	0	62	63	65	43
June	985.8	984.3	10.7	9.7	26.1	−2.8	15	148	0	48	70	72	51
July	985.5	983.7	10.2	10.2	26.7	−1.9	17	76	0	33	67	71	47
Aug.	984.8	982.7	11.9	11.3	28.9	−1.7	15	91	0	46	57	61	39
Sept.	983.9	981.7	14.9	12.5	34.4	+0.4	13	73	0	48	48	50	31
Oct.	981.5	979.3	18.4	13.6	39.9	1.8	25	102	0	55	41	45	30
Nov.	979.8	977.6	21.5	14.0	43.8	4.4	23	122	0	103	38	41	26
Dec.	977.9	975.9	24.2	14.4	45.5	5.0	14	115	0	60	38	40	25
Annual	982.3	980.4	18.0	12.6	46.6	−2.8	224	447	57	103	47	51	33
Years	29	29	30	30	85	85	30	56	56	85	30	29	29

Month	Mean Austr. tank evapor. (mm)*	Number of days				Mean cloudiness		Prev. wind dir.		Probability of precipitation (%)		
		precipitation		min. temp. °C		09h00	15h00	09h00	15h00	<12.5 mm	<25 mm	<50 mm
		>2.5 mm	>0.25 mm	<2.2	<0							
Jan.	323	1	2	0	0	1.5	1.8	S	S	68	75	92
Feb.	274	2	3	0	0	1.7	2.1	S	S	56	76	89
Mar.	236	0	3	0	0	1.6	2.0	S	S	61	81	94
Apr.	151	1	2	0	0	1.8	2.2	S	S	58	79	92
May	105	1	3	0.2	0	2.4	2.6	W	SW	52	68	85
June	72	1	4	1.9	0.7	2.6	3.0	W	W	43	63	86
July	73	1	5	4.7	1.0	2.4	2.9	W	W	47	73	97
Aug.	99	2	4	2.9	0.6	2.1	2.6	S	W	36	74	95
Sept.	147	1	3	0.2	0	1.8	2.1	N	W	59	79	94
Oct.	216	3	4	0	0	2.2	2.4	S	S	49	67	86
Nov.	258	1	3	0	0	2.0	2.4	S	S	61	78	92
Dec.	309	2	2	0	0	1.7	2.2	S	S	63	74	92
Annual	2263	16	38	9.9	2.3	2.0	2.4	—	—	—	—	—
Years	31	unsp.	30	10	10	30	30	5	5	56	56	56

* At Umberumberka, 31°48′S, 141°12′E, 244 m.

TABLE LX

CLIMATIC TABLE FOR DUBBO, N.S.W., AUSTRALIA
Latitude 32°18'S, longitude 148°35'E, elevation 265 m

Month	Mean sta. level press. (mbar)		Temperature (°C)				Precipitation (mm)				Relat. humidity (%)		
	09h00	15h00	mean	daily range	extreme max.	extreme min.	mean	highest	lowest	max. 24 h.	daily index.	09h00	15h00
Jan.	983	981	25.5	15.7	45.2	6.1	65	232	0	131	48	48	32
Feb.	985	983	25.3	15.3	43.8	6.1	73	219	0	51	50	51	34
Mar.	988	986	22.4	14.9	40.3	3.9	47	287	0	114	55	57	37
Apr.	990	988	17.7	14.5	36.6	−1.1	50	181	0	60	62	67	44
May	993	990	13.2	13.6	32.7	−3.3	44	140	0	79	68	79	50
June	991	989	10.1	12.1	26.6	−5.6	52	194	1	51	74	87	57
July	992	989	9.2	12.3	23.8	−6.7	45	148	2	42	74	85	56
Aug.	991	988	10.5	14.0	30.6	−5.0	45	162	1	51	70	77	50
Sept.	990	987	13.6	15.3	33.8	−2.8	37	146	3	48	64	64	41
Oct.	988	985	17.7	16.2	38.8	−2.3	52	155	0	48	56	55	36
Nov.	986	984	21.6	16.1	43.3	+1.4	52	190	0	81	51	50	34
Dec.	984	981	24.1	15.7	45.8	3.3	36	179	0	62	50	46	32
Annual	988	985	17.6	14.6	45.8	−6.7	598	961	287	131	58	64	42
Years	15	15	30	30	42	42	30	72	72	76	30	10	15

TABLE LX (*continued*)

Month	Mean Austr. tank evapor. (mm)	Number of days with			Cloudiness (oktas)		Sun-shine (h)**	Preval. wind dir.		Visibility (% of hours)		Probability of precipitation (%)		
		precip. >0.25 mm	thunder	fog	09h00	15h00		09h00	15h00	<10 km	<4 km	<12.5 mm	<25 mm	<50 mm
Jan.	279	6	1.1	0	2.2	2.9	324	C*,E	C,E	1.8	0.3	15	28	56
Feb.	234	6	0.4	0	2.2	3.2	277	C,E	C	2.1	0.3	30	52	64
Mar.	185	5	0.2	0	2.2	3.0	274	C,NE,E	C	1.1	0.1	21	41	65
Apr.	109	5	0.2	0.2	2.6	3.2	242	C,NE,E	C	0.8	0	28	44	64
May	57	6	0.3	0.5	2.4	3.0	220	C	C	2.0	0.4	24	33	64
June	39	8	0.0	1.6	3.3	3.8	190	C	C	3.1	1.0	8	27	55
July	38	8	0.1	2.2	3.3	3.8	210	C	C	3.0	0.9	15	33	65
Aug.	51	8	0.1	1.8	2.8	3.4	235	C	C,SW	2.2	0.4	9	25	69
Sept.	77	6	0.3	0.5	2.3	3.1	254	C	C,S	2.0	0.2	19	44	75
Oct.	130	7	0.4	0.1	2.4	3.4	292	C	NW,C	0.8	0.2	16	39	73
Nov.	193	6	0.5	0	2.7	3.4	306	NE,C,E	SW,C,S	1.4	0.2	17	37	58
Dec.	274	5	0.6	0	2.2	3.3	330	N,C,NE	C,SW	2.1	0.9	27	39	64
Annual	1669	76	4.2	6.9	2.6	3.3	3154	C	C	22.4	4.9	—	—	—
Years	23	30	17	17	10	15	—	5	5	9	9	72	72	72

* C = calms
** Estimated by interpolation from maps.

TABLE LXI

CLIMATIC TABLE FOR SYDNEY, N.S.W., AUSTRALIA
Latitude 33°51′S, longitude 151°13′E, elevation 42.1 m

Month	M.s.l. press. (mbar)		Mean daily temp. (°C)	Mean daily temp. range (°C)	Temp. extremes (°C)				Mean vapor. press. (mbar)	Precip. (mm)			
					highest		lowest			highest	mean	lowest	max. 24 h
	daily mean	daily ampl.			sun	shade	shade	grass					
Jan.	1012.6	2.4	22.0	7.3	73.9	45.3	10.6	6.5	18.8	388	104	6	180
Feb.	1013.9	2.0	21.9	7.1	75.7	42.1	9.6	6.0	19.4	564	125	3	226
Mar.	1016.4	2.3	20.8	7.2	70.2	39.2	9.3	4.4	18.3	521	129	11	281
Apr.	1018.2	2.5	18.3	7.6	62.3	33.0	7.0	+0.7	15.1	622	101	2	191
May	1018.6	1.9	15.1	7.8	54.3	30.0	4.6	−1.5	12.1	585	115	4	212
June	1018.7	1.6	12.8	7.4	51.9	26.9	2.1	−2.2	10.1	643	141	4	132
July	1018.4	2.7	11.8	7.9	51.5	25.7	2.2	−4.4	9.7	336	94	3	198
Aug.	1018.0	2.7	13.0	8.7	69	30.4	2.7	−3.3	9.6	378	83	1	135
Sept.	1017.1	2.3	15.2	8.8	61.2	33.5	4.9	−1.1	11.2	356	72	2	145
Oct.	1014.8	2.3	17.6	8.6	66.8	37.4	5.7	+0.4	13.0	282	80	5	162
Nov.	1013.4	2.2	19.5	8.2	70.3	40.3	7.7	2.2	15.0	518	77	2	133
Dec.	1011.9	2.2	21.1	7.8	73.6	42.2	9.1	5.2	17.1	402	86	6	121
Annual	1016.0	2.3	17.4	7.8	75.7	45.3	2.1	−4.4	14.1	2.102	1.205	546	281
Years	53	30	104	104	84	104	104	104	87	104	30	104	104

Month	Relat. humidity (%)				Probability of precip. (%)			Mean Austr. tank evap. (mm)*		Number of days with			
	09h00	15h00	mean	daily index	<12.5 mm	<25 mm	<50 mm			precipitation (mm)			thunder
										>2.5	>25	>0.25	
Jan.	67	64	68	68	3	12	35	128	134	6	0.9	13	5
Feb.	70	66	71	71	9	17	40	96	106	7	0.9	13	3
Mar.	73	65	74	72	1	6	30	90	92	7	1.1	14	2
Apr.	76	65	75	71	5	9	24	72	69	8	1.4	13	2
May	77	64	77	70	5	13	31	54	49	6	1.3	14	0.5
June	77	63	76	68	2	15	36	40	38	7	1.4	12	0.5
July	76	59	74	67	13	17	28	39	39	6	1.2	12	1
Aug.	70	56	69	64	13	24	44	53	51	5	0.8	11	2
Sept.	66	55	64	62	8	19	44	68	69	5	0.7	11	2
Oct.	62	57	62	63	3	17	48	102	99	5	0.6	12	4
Nov.	63	60	63	65	12	23	39	124	119	5	0.6	12	4
Dec.	64	62	65	68	5	21	42	135	136	6	0.6	13	5
Annual	70	61	69	66	—	—	—	1000	1003	73	11.5	150	32
Years	65	65	86	30	—	—	—	16	48	81	87	104	20

TABLE LXI (*continued*)

Month	Number of days with				Mean cloudiness (oktas)					Mean sun-shine (h)	Merid. alt. sun** (°)	Wind most freq. dir.	
	light-ning	clear	fog	gale gusts	all types				low				
					mean	09h	15h	21h				09h00	15h00
Jan.	4.7	5.1	0.3	4	4.7	5.0	4.4	4.3	3.6	7.3	77	S	ENE
Feb.	4.1	4.7	0.8	3	4.8	5.0	4.3	4.0	3.6	6.6	67	NE,W	ENE
Mar.	3.7	5.9	1.7	2	4.4	4.7	4.2	3.7	3.5	6.3	57	W	ENE
Apr.	3.4	7.0	2.4	3	4.1	4.3	4.3	3.3	3.2	6.1	47	W	NE,E
May	2.7	7.9	3.4	1	3.9	4.2	4.5	3.4	2.9	5.8	37	W	S,E
June	2.0	8.3	2.8	1	4.0	4.2	4.4	3.3	3.1	6.3	34	W	W
July	2.0	10.4	2.5	1	3.5	3.6	4.1	3.0	2.6	6.1	37	W	W,S
Aug.	2.8	10.8	2.0	3	3.3	3.3	3.8	2.7	2.1	6.9	47	W	NE,W
Sept.	3.6	9.1	1.0	3	3.5	3.7	3.7	3.0	3.0	7.2	57	W,NW	ENE
Oct.	4.4	6.9	0.6	2	4.0	4.5	4.2	3.4	3.0	7.4	67	W,E	ENE
Nov.	5.2	5.5	0.5	4	4.5	4.9	4.5	4.0	3.4	7.6	77	S,NE	ENE
Dec.	5.6	4.9	0.4	4	4.6	5.0	4.5	4.3	3.7	7.4	81	S,NE	ENE
Annual	44.4	86.2	18.4	31	4.2	4.3	4.2	3.5	3.7	6.7	—	W	NE,E
Years	42	48	42	5	101	38	38	38	5	42	—	48	48

Month	Mean wind speed (kt.)					Mean daily glob. radiat.** (cal./cm²)	Fine spells		Wet spells	
	09h00	15h00	daily	highest daily	highest gust		freq. of 7 or more consecut. fine days	highest number of consecut. fine days	freq. of 7 or more consecut. rain days	highest number of consecut. rain days
Jan.	7	12	7.9	22	81	536	39	22	12	12
Feb.	6	11	7.7	17	55	454	45	17	15	17
Mar.	6	10	6.6	18	50	438	42	17	16	15
Apr.	6	9	6.3	20	61	317	47	22	24	15
May	6	7	6.2	18	55	256	52	26	24	16
June	6	7	6.9	19	73	215	52	19	23	13
July	7	8	6.9	23	59	251	63	26	20	12
Aug.	7	9	6.8	21	59	315	63	21	15	15
Sept.	6	9	7.2	19	61	394	50	21	12	12
Oct.	7	10	7.6	21	83	514	47	13	12	11
Nov.	7	11	7.6	19	62	584	46	19	11	13
Dec.	7	12	7.8	22	65	554	49	23	8	11
Annual	7	10	7.1	23	83	—	595	32	192	21
Years	6	8	86	46	44	—	87	87	87	87

* First column: period of strictly uniform observations.
** At Williamtown Airport (meridian alt. of sun on 21st of month).

Note : only 0.2 days in June have a minimum <2.2°C.

TABLE LXII

CLIMATIC TABLE FOR MILDURA, N.S.W., AUSTRALIA
Latitude 34°11′S, longitude 142°12′E, elevation 54 m

Month	Mean sta. press. (mbar)*		Mean daily temp. (°C)	Mean daily temp. range (°C)	Extreme temp. (°C)		Precipitation (mm)				Ratio mean–median	Relative humidity (%)		
	09h00	15h00			highest	lowest	mean	highest	lowest	max. 24h**		daily index	mean* 09h00	15h00
Jan.	1005.1	1002.7	24.1	16.0	50.8	4.4	19	48	0	40	3.5	48	49	28
Feb.	1006.1	1003.5	24.4	15.7	47.8	6.1	23	145	0	103	2.3	51	56	32
Mar.	1008.9	1006.6	21.6	15.1	46.4	2.8	18	109	0	58	2.3	55	61	35
Apr.	1011.8	1009.7	16.9	13.3	37.2	+1.1	14	59	0	26	1.5	61	71	45
May	1012.5	1010.2	13.4	11.8	32.2	−2.8	26	77	0	33	1.8	67	82	54
June	1012.8	1010.9	10.4	10.6	26.7	−3.3	27	101	0	31	1.2	70	88	62
July	1012.4	1010.5	10.1	10.6	25.6	−4.4	23	63	2	20	1.1	71	89	63
Aug.	1010.9	1008.8	11.8	11.9	30.5	−1.7	26	75	1	27	1.3	65	80	54
Sept.	1011.0	1008.5	14.4	13.2	35.0	−1.7	24	108	1	24	1.3	60	70	46
Oct.	1007.9	1005.4	17.6	14.2	40.0	+1.1	25	107	0	48	1.5	56	62	40
Nov.	1006.5	1003.7	20.7	15.4	45.0	−1.1	21	72	0	39	1.4	50	52	33
Dec.	1004.9	1002.6	23.3	15.9	49.7	+4.4	18	133	0	55	1.5	43	48	28
Annual	1009.2	1006.9	17.4	13.7	50.8	−4.4	264	556	120	103	1.0	57	64	41
Years	28	27	30	30	60	60	30	46	46	29	29	28	30*	28*

Month	Number of days				Mean evapor.*** (mm)	Mean cloudiness*		Prev. wind dir.*		Probability of precipitation (%)		
	precip. (mm)		min. temp. (°C)			09h00	15h00	09h00	15h00	<12.5 mm	<25 mm	<50 mm
	>0.25	>2.5	<2.2	<0								
Jan	3	3	0	0	235	1.8	2.2	SE	SW	58	80	93
Feb.	3	1	0	0	184	1.9	2.3	SE	SW	59	73	87
Mar.	3	2	0	0	156	1.8	2.3	SE	SW	55	75	90
Apr.	4	2	0	0	99	2.4	3.1	SE	SE	52	78	95
May	6	1	0.6	0.1	62	3.1	3.8	SW	SE	30	59	86
June	8	3	3.1	0.5	44	3.4	4.0	NW	NW	25	48	81
July	8	4	3.8	0.4	42	3.4	4.1	SW	NW	23	72	94
Aug.	8	3	1.9	0.1	60	2.7	3.8	SW	SW	26	52	87
Sept.	6	2	0.1	0	94	2.6	3.6	SW	SW	40	71	90
Oct.	5	2	0.1	0	134	3.0	3.6	SE	SW	28	57	87
Nov.	4	1	0	0	179	3.0	3.2	SW	SW	50	71	89
Dec.	3	2	0	0	219	2.3	2.6	SE	SE	47	71	95
Annual	61	26	9.6	1.1	1,511	2.6	3.2	–	–	–	–	–
Years	30	unsp.	<10	10	24	30	28	5	5	46	46	46

*   At Swan Hill, Vic., 35°22′S, 143°35′E, 70 m.
**   At Neds Corner, Vic., 34°12′S, 141°30′E, 30 m.
*** At Merbein, Vic., 34°10′S, 142°04′E, 56 m.

TABLE LXIII

CLIMATIC TABLE FOR HAY, N.S.W., AUSTRALIA
Latitude 34°30′S, longitude 144°56′E, elevation 94 m

Month	Mean sta. press. (mbar)		Mean daily temp. (°C)	Mean daily temp. range (°C)	Extreme temp. (°C)		Precipitation (mm)				Relative humidity (%)		
	09h00	15h00			highest	lowest	mean	highest	lowest	max. 24 h	daily index	mean	
												09h00	15h00
Jan.	1001.9	1099.2	23.9	15.8	47.9	5.1	29	99	0	81	44	44	27
Feb.	1002.8	1000.4	24.4	15.7	46.9	5.0	34	123	0	64	45	51	31
Mar.	1006.2	1003.5	21.2	15.2	42.5	1.7	31	108	0	66	51	55	32
Apr.	1008.7	1006.1	16.6	15.6	36.9	+0.6	28	141	0	59	60	70	42
May	1009.8	1007.6	13.0	12.3	30.3	−3.4	32	120	0	45	67	79	52
June	1010.2	1008.2	9.8	10.7	26.9	−5.1	35	116	2	46	74	84	60
July	1010.0	1007.9	9.2	11.2	25.0	−4.4	29	99	1	51	73	85	58
Aug.	1008.8	1006.3	10.8	12.2	29.2	−4.9	26	107	0	35	69	77	51
Sept.	1008.3	1005.3	13.5	13.6	34.9	−3.4	26	106	3	46	63	65	41
Oct.	1005.2	1002.9	16.8	14.6	40.8	−0.6	40	93	1	49	55	54	36
Nov.	1003.4	1000.7	20.2	15.3	43.1	+0.6	23	152	0	115	49	47	32
Dec.	1001.5	998.9	22.8	15.7	46.3	2.2	19	115	0	55	47	46	31
Annual	1006.4	1003.9	16.8	13.8	47.9	−5.1	352	683	157	115	56	60	39
Years	29	26	39	29	84	84	30	64	64	84	30	28	25

Month	Mean evapor. (mm)	Number of days				Mean cloudiness		Prev. wind dir.		Probability of precipitation (%)		
		precip. (mm)		min. temp. (°C)		09h00	15h00	09h00	15h00	<12.5 mm	<25 mm	<50 mm
		>0.25	>2.5	<2.2	<0							
Jan.	199	3	3	0	0	1.4	1.9	SW	SW	55	67	80
Feb.	161	3	1	0	0	1.6	2.0	SW	SW	47	61	84
Mar.	128	4	1	0	0	1.6	1.9	NE	SW	44	61	85
Apr.	74	5	4	0.3	0	1.9	2.5	NE	SW	36	57	76
May	43	6	1	2.1	0.2	2.6	2.9	SW	SW	18	44	74
June	29	7	5	5.4	2.3	3.0	3.2	SW	SW	12	22	66
July	29	8	4	6.7	2.8	2.9	3.4	SW	SW	17	47	85
Aug.	38	8	4	6.3	2.2	2.5	3.2	SW	SW	15	37	78
Sept.	62	6	2	3.1	0.1	2.0	2.7	SW	SW	25	47	82
Oct.	105	7	3	0.3	0	2.4	2.8	SW	SW	24	51	78
Nov.	152	5	1	0	0	2.2	2.5	SW	SW	48	67	77
Dec.	181	3	2	0	0	1.8	2.2	SW	SW	45	65	74
Annual	1201	65	31	24.2	7.6	2.1	2.6	—	—	—	—	—
Years	49	30	unsp.	10	10	30	27	5	5	64	64	64

TABLE LXIV

CLIMATIC TABLE FOR CANBERRA, A.C.T., AUSTRALIA
Latitude 35°17′S, longitude 149°08′E, elevation 559 m

Month	M.s.l. press. (mbar)	Mean daily temp. (°C)	Mean daily temp. range (°C)	Temp. extremes (°C)			Mean vapor. press. (mbar)
				highest shade	lowest shade	grass	
Jan.	1012.0	20.7	14.7	41.9	3.3	− 1.1	12.5
Feb.	1012.4	20.2	13.7	37.7	1.7	− 3.1	13.2
Mar.	1015.8	18.0	13.1	37.3	+1.6	− 3.1	12.8
Apr.	1018.4	13.4	11.8	32.1	−1.7	− 7.2	10.7
May	1018.9	9.6	11.2	22.6	−5.3	− 9.1	8.6
June	1020.6	6.7	9.4	18.3	−7.7	−12.8	7.3
July	1020.3	6.0	10.0	17.5	−6.7	−11.8	6.6
Aug.	1018.6	7.4	10.9	21.7	−6.1	−12.2	7.5
Sept.	1017.9	10.1	12.5	27.5	−3.8	−10.6	8.1
Oct.	1013.9	13.1	12.7	32.2	−1.7	− 7.7	9.5
Nov.	1011.8	16.0	13.4	38.6	+0.1	− 5.1	10.3
Dec.	1010.3	19.1	14.6	39.7	2.2	− 1.6	11.5
Annual	1015.9	13.4	12.3	41.9	−7.7	−12.8	9.7
Years	20	20	29	33	33	33	26

Month	Precipitation (mm)				Relat. humidity (%)			
	highest	mean	lowest	max. 24 h	mean	09h	15h00	daily index
Jan.	170	54	1	82	53	56	35	55
Feb.	153	55	0	82	59	62	38	59
Mar.	322	63	0	65	66	68	41	65
Apr.	132	55	2	64	71	75	51	70
May	156	52	2	99	79	81	57	72
June	155	49	5	59	81	85	64	75
July	129	41	7	51	81	85	62	75
Aug.	120	50	9	53	75	81	57	74
Sept.	115	41	3	44	66	72	50	70
Oct.	177	70	9	132	60	63	43	66
Nov.	113	54	7	62	55	58	40	61
Dec.	223	47	4	58	51	57	35	58
Annual	322	632	0	132	66	70	48	65
Years	33	20	33	33	28	16	16	20

TABLE LXIV *(continued)*

Month	Probability of precipitation (%)			Mean Austr. tank evap. (mm)	Number of days					
					precip.		light-ning	fog	clear	min. temp. <0°C
	<12.5 mm	<25 mm	<50 mm		>0.25 mm	>2.5 mm				
Jan.	15	35	28	221	7	1.6	1.5	0.1	7.3	0.0
Feb.	23	38	54	182	7	1.4	2.3	0.2	6.3	0.0
Mar.	11	35	46	146	7	1.5	0.2	1.0	6.9	0.1
Apr.	19	27	65	86	8	1.7	0.3	1.4	4.7	1.0
May	23	50	73	50	8	1.2	0.2	4.8	5.8	8.9
June	8	26	52	32	9	0.9	0.1	5.8	4.5	12.4
July	7	19	66	33	10	0.7	0.0	5.3	5.6	14.8
Aug.	7	15	44	48	11	1.0	0.1	2.4	5.7	10.9
Sept.	4	31	65	78	9	0.7	0.4	1.4	6.1	5.9
Oct.	0	15	61	123	11	1.6	1.0	0.4	5.2	1.1
Nov.	19	27	62	160	8	1.3	1.1	0.1	4.5	0.2
Dec.	11	27	50	206	8	0.8	0.7	0.0	6.3	0.0
Annual	—	—	—	1366	103	14.4	7.9	22.9	68.9	55.3
Years	26	26	26	17	29	29	20	25	27	27

Month	Mean cloudiness (oktas)			Mean daily sun-shine (h)	Wind				
					most freq. dir.		speed (kt.)		
	mean	09h	15h		09h00	15h00	mean	highest daily	highest gust
Jan.	3.9	3.4	3.8	8.4	NW	NW	4.1	13	56
Feb.	4.1	3.5	4.1	7.3	E	NW	3.5	13	56
Mar.	4.1	3.7	3.8	7.2	E	NW	3.2	16	45
Apr.	4.3	3.8	3.7	6.7	NW	NW	3.1	16	45
May	4.5	3.9	3.8	5.2	NW	NW	2.6	11	56
June	4.8	4.3	4.0	4.2	NW	NW	3.1	14	52
July	4.5	4.2	3.8	4.8	NW	NW	2.9	20	54
Aug.	4.3	4.0	3.9	5.8	NW	NW	3.5	14	51
Sept.	4.1	3.2	3.5	7.2	NW	NW	3.6	15	53
Oct.	4.4	3.7	4.2	7.8	NW	NW	3.7	13	64
Nov.	4.4	3.7	4.1	8.2	NW	NW	4.1	15	57
Dec.	4.0	3.8	4.2	8.5	NW	NW	4.1	14	57
Annual	4.2	3.8	3.9	6.8	NW	NW	3.5	20	64
Years	26	21	14	27	27	27	27	31	22

TABLE LXV

CLIMATIC TABLE FOR KOSCIUSKO, N.S.W., AUSTRALIA
Latitude 36°26′S, longitude 148°16′E, elevation 1,529 m

Month	Mean daily temp*. (°C)	Mean daily temp. range (°C)	Precipitation (mm)					Relative humidity (%)	
			mean	highest	lowest	max. 24h**	ratio mean/ median**	daily index	09h00
Jan.	12.1	12.7	84	260	8	162	1.2	62	55
Feb.	12.5	12.7	74	172	1	134	1.5	65	57
Mar.	10.2	11.6	86	240	5	174	1.2	67	61
Apr.	6.2	9.9	87	242	3	190	1.4	71	66
May	3.3	9.1	111	317	13	155	1.3	70	66
June	+0.8	7.7	106	226	12	217	1.2	77	75
July	−0.1	7.7	107	216	26	140	1.0	79	79
Aug.	+0.6	7.9	130	391	7	93	1.0	77	72
Sept.	3.4	8.8	129	296	23	145	1.0	66	63
Oct.	6.4	10.3	132	288	25	130	1.1	61	55
Nov.	8.8	11.7	92	200	13	97	1.3	61	53
Dec.	11.1	12.2	100	273	6	110	1.1	62	55
Annual	6.3	10.2	1238	1810	676	217	1.0	67	63
Years	27	27	33	33	33	35	35	29	29

Month	Temp. <0°C (days)	Probability of precipitation (%)			
		<12.5 mm	<25 mm	<50 mm	<100 mm
Jan.	1.9	7	15	30	67
Feb.	1.8	10	20	40	64
Mar.	3.1	6	13	34	67
Apr.	10.7	5	11	26	72
May	19.2	3	8	28	63
June	25.5	1	5	15	49
July	27.4	0	1	13	49
Aug.	25.6	3	6	12	38
Sept.	18.9	1	3	8	40
Oct.	12.8	2	4	8	34
Nov.	7.3	3	7	20	58
Dec.	3.3	3	10	24	62
Annual	157.5	–	–	–	–
Years	10	31	31	31	31

* Extreme maximum 32.2°C; extreme minimum −14.4°C.
** On Mount Buffalo, Vic., 36°48′S, 146°48′E, 1,332 m.

TABLE LXVI

CLIMATIC TABLE FOR BENDIGO, VIC., AUSTRALIA
Latitude 36°46′S, longitude 144°17′E, elevation 223 m

Month	Mean sta. press. (mbar)		Mean daily temp. (°C)	Mean daily temp. range (°C)	Extreme temp. (°C)		Precipitation (mm)				Relative humidity (%)		
	09h00	15h00			highest	lowest	mean	highest	lowest	max. 24h	daily index	mean	
												09h00	15h00
Jan.	987.3	985.1	21.2	15.0	47.4	2.8	32	110	0	74	49	50	31
Feb.	988.3	986.1	20.8	14.1	44.2	4.4	42	214	0	87	50	57	35
Mar.	991.8	989.4	18.5	13.3	40.4	3.3	32	155	1	88	58	62	40
Apr.	993.4	991.2	14.3	11.2	34.8	+0.8	45	142	0	70	65	70	48
May	993.9	991.9	11.2	9.7	26.7	−2.6	52	156	0	43	74	82	58
June	994.3	992.5	8.7	8.3	25.2	−4.5	54	159	7	69	79	85	64
July	993.6	991.9	7.8	8.1	22.8	−4.7	63	133	4	46	77	88	66
Aug.	992.2	990.2	8.9	9.3	24.3	−3.3	53	169	7	51	76	82	59
Sept.	992.1	989.9	11.1	10.8	32.2	−1.8	49	153	9	64	73	71	50
Oct.	989.7	987.6	13.7	11.7	37.6	0.0	58	194	1	58	61	62	44
Nov.	988.3	986.1	16.6	13.1	41.4	+0.8	43	164	1	52	54	56	38
Dec.	986.8	984.6	19.4	14.3	44.2	2.8	31	136	<1	70	50	50	31
Annual	991.0	988.9	14.4	11.6	47.4	−4.7	553	975	276	88	62	66	44
Years	29	29	29	29	105	105	30	74	74	105	24	29	29

Month	Evapor. Austr. tank* (mm)	Number of days			Mean cloudiness		Prev. wind dir.		Probability of precipitation (%)		
		precip. >0.25 mm	min. temp. (°C)		09h00	15h00	09h00	15h00	<12.5 mm	<25 mm	<50 mm
			<2.2	<0							
Jan.	209	6	0	0	2.2	2.6	N	W	30	49	79
Feb.	163	5	0	0	2.3	2.6	S	S	37	55	80
Mar.	138	6	0	0	2.5	2.9	S	S	35	55	70
Apr.	80	9	0.1	0	3.0	3.5	S	S, W	21	40	70
May	45	11	3.2	0.2	3.5	3.9	S	S	9	22	50
June	32	14	5.0	2.6	3.8	4.2	S	W	3	13	37
July	29	16	>6	3.8	4.0	4.5	S	W	6	18	52
Aug.	44	16	5.7	1.6	3.8	4.2	S	W	6	18	51
Sept.	62	12	4.2	0.4	3.2	3.0	S, W	S	4	19	57
Oct.	89	12	0.9	0.1	3.5	3.9	S	S	12	26	60
Nov.	118	9	0	0	3.4	3.7	S	S	20	45	74
Dec.	167	7	0	0	2.8	3.1	S	S	26	54	80
Annual	1.175	123	>25	8.7	3.2	3.6	–	–	–	–	–
Years	11	29	10	10	30	30	5	5	74	74	74

* At Eildon, 37°12′S, 145°54′E, 229 m.

TABLE LXVII

CLIMATIC TABLE FOR GABO ISLAND, VIC., AUSTRALIA
Latitude 37°34'S, longitude 149°55'E, elevation 15 m

Month	Mean sta. press. (mbar)		Temperature (°C)				Precipitation (mm)				Relat. humidity (%)	
	09h00	15h00	mean	daily range	extreme max.	extreme min.	mean	highest	lowest	max. 24 h	09h00	15h00
Jan.	1012.6	1011.4	18.3	5.4	39.4	5.0	71	256	3	105	81	81
Feb.	1014.6	1013.4	18.7	5.3	38.9	6.1	68	236	3	62	82	82
Mar.	1016.7	1015.3	18.1	5.5	38.3	3.3	73	381	1	102	83	81
Apr.	1018.8	1017.2	16.3	6.8	31.7	2.8	84	208	5	99	82	80
May	1018.8	1017.2	13.9	5.8	28.9	0.6	100	434	17	116	83	80
June	1017.8	1016.3	11.9	5.7	28.3	0.0	107	314	11	55	83	80
July	1017.7	1016.3	11.1	6.1	26.7	-0.6	68	315	23	60	82	79
Aug.	1016.8	1015.1	11.7	6.3	26.7	0.0	72	251	14	47	80	78
Sept.	1015.8	1014.1	12.8	6.3	29.4	0.0	76	239	17	70	80	79
Oct.	1014.9	1013.2	14.1	6.1	33.9	2.2	79	197	10	56	79	80
Nov.	1013.8	1012.1	15.6	5.6	35.0	1.1	65	247	0	114	81	81
Dec.	1011.5	1010.5	17.1	5.5	38.3	3.9	62	267	7	65	82	81
Annual	1015.7	1014.3	14.9	5.8	39.4	-0.6	943	1455	508	116	82	80
Years	32	32	50	50	66	66	78	84	84	29	33	33

TABLE LXVII (*continued*)

Month	Evapor. (mm)*	Number of days with — precip. >0.25 mm	>2.5 mm	thunder	fog	clear	cloudy	Cloudiness (oktas) all types	low clouds 09h00	low clouds 15h00	Mean monthly sunshine* (h)	Wind prevalent dir. 09h00	prevalent dir. 15h00	mean speed (kt.) 09h00	15h00	Probability of precip. (%) <12.5 mm	<25 mm	<50 mm
Jan.	124	11	5	1	1	2	9	4.9	4.2	4.1	248	SW,NE	SW	11	14	8	9	41
Feb.	99	9	5	2	1	3	7	4.8	4.2	3.2	222	NE	NE,SW	10	12	14	28	51
Mar.	89	12	6	2	0	3	9	4.8	5.0	4.0	218	W,NE,SW	NE,SW	9	12	9	21	46
Apr.	64	14	8	1	0	3	7	5.0	4.9	3.8	173	W	SW,NE	10	13	3	11	29
May	43	15	6	<1	0	3	6	5.0	4.0	3.9	170	W	SW,NE	8	10	1	8	32
June	32	16	9	1	<1	1	9	5.2	4.2	3.8	147	W	SW,W	11	12	1	4	20
July	36	16	7	1	0	3	7	5.0	2.9	3.9	148	W	SW	9	12	1	4	21
Aug.	44	14	6	1	0	2	6	4.6	3.6	3.8	180	W,SW	SW	9	13	2	11	36
Sept.	58	14	5	1	<1	2	6	4.7	3.4	3.0	192	W,SW	SW	9	13	1	9	38
Oct.	83	14	6	<1	1	3	9	4.8	3.4	2.8	210	SW,NE	SW,NE	10	13	1	7	34
Nov.	110	13	5	2	1	2	9	5.0	4.6	2.9	220	SW,NE	SW,NE	12	15	5	17	48
Dec.	116	12	6	2	1	2	10	5.0	3.3	3.2	245	SW,NE	SW,NE	12	15	5	17	47
Annual	898	160	74	14	5	29	94	5.0	4.0	3.5	2373	W,SW NE	SW,NE	10	13	—	—	—
Years	—	30	10	6	6	10	10	33	3	3	—	5	5	5	5	84	84	84

* Estimated by interpolation from maps.

TABLE LXVIII

CLIMATIC TABLE FOR MELBOURNE, VIC., AUSTRALIA
Latitude 37°49'S, longitude 144°58'E, elevation 35 m

Month	M.s.l. press. (mbar)	Mean daily ampl. (mbar)	Mean daily temp. (°C)	Mean daily temp. range (°C)	Temp. extreme (°C)				Mean vapor press. (mbar)	Precipitation (mm)				Relat. humidity (%)			Probability of precipitation (%)		
					highest sun	highest shade	lowest shade	lowest grass		highest	09h00	lowest	max. 24 h	09h00	15h00	daily index	<12.5 mm	<25 mm	<50 mm
Jan.	1012.8	2.0	19.9	11.6	81	45.6	5.6	-1.0	12.9	169	45	0	75	58	48	57	16	30	62
Feb.	1013.9	2.2	19.7	11.4	75	43.1	4.6	-0.6	14.1	196	59	1	87	62	51	60	20	45	67
Mar.	1016.6	2.3	18.4	10.9	74	41.7	2.8	-1.7	13.0	191	50	4	90	64	51	62	12	25	59
Apr.	1018.5	2.1	15.1	9.5	67	34.9	1.6	-3.9	11.9	195	69	0	80	72	56	69	2	23	53
May	1019.1	2.2	12.5	8.4	61	28.7	-1.2	-6.1	10.5	142	54	4	47	79	62	73	2	15	48
June	1019.6	1.8	10.2	7.2	54	22.4	-2.2	-6.7	9.7	115	52	15	44	83	67	75	0	7	59
July	1018.7	1.9	9.6	7.6	52	20.7	-2.8	-6.4	8.9	178	54	14	69	82	65	75	0	10	59
Aug.	1017.1	2.2	10.5	8.3	59	25.0	-2.1	-5.9	9.2	110	50	12	49	76	60	71	1	16	59
Sept.	1017.0	2.0	12.4	9.6	61	31.4	-0.6	-5.1	9.8	201	58	13	67	68	55	67	0	6	46
Oct.	1014.7	2.2	14.3	10.7	68	36.9	+0.1	-4.0	10.4	193	74	7	76	62	52	63	1	9	31
Nov.	1013.4	1.9	16.2	10.8	71	40.9	2.5	-4.1	11.4	206	70	6	73	60	52	61	2	21	44
Dec.	1012.2	1.9	18.4	11.2	77	43.7	4.4	+0.7	12.7	182	58	3	100	59	51	59	5	19	48
Annual	1016.1	2.1	14.8	9.8	81	45.6	-2.8	-6.7	10.9	206	691	0	100	69	56	65	—	—	80
Years	30	36	30	30	86	105	105	101	30	105	30	105	105	30	30	30	80	80	80

TABLE LXVIII (continued)

Month	Mean Austr. tank evapor. (mm)	Number of days									Mean cloudiness (oktas)		Meridian altitude of sun**	Sun-shine (h)	Wind					Mean daily glob. radiation (cal./cm²)
		precip.		min. temp.		thunder	light-ning	clear	fog	gale*	all types	low			most freq. dir.		speed (kt.)			
		>2.5 mm	>0.25 mm	<0 °C	<2.2 °C										09h00	15h00	mean	highest daily mean	highest gust	
Jan.	164	3	9	0.0	0.0	2	1.8	7	0.1	0	3.9	2.8	72°	7.8	S,SW	S	7.6	18	57	585
Feb.	128	4	8	0.0	0.0	2	2.3	7	0.3	0	3.8	2.9	62°	7.4	N,S	S	7.4	17	64	504
Mar.	104	4	9	0.0	0.0	2	1.8	6	1.1	<1	4.2	2.7	52°	6.5	N	S	6.7	15	57	393
Apr.	63	7	13	0.0	0.0	1	1.2	5	2.3	0	4.7	3.4	42°	5.0	N	S	6.1	17	58	270
May	39	4	14	0.0	0.3	<1	0.5	3	6.8	0	4.9	3.8	32°	4.1	N	N	6.5	19	63	180
June	29	7	16	0.5	3.1	<1	0.4	3	6.5	0	5.2	3.8	29°	3.4	N	N	6.2	20	54	155
July	28	5	17	0.3	4.4	<1	0.3	3	6.5	<1	5.0	3.7	32°	3.7	N	N	7.5	20	59	160
Aug.	38	7	17	0.1	2.1	1	0.9	3	3.7	0	4.8	3.7	42°	4.6	N	N	7.1	18	56	231
Sept.	60	5	15	0.0	1.2	1	1.3	3	1.3	0	4.7	3.0	52°	5.5	N,W	N,S	7.4	18	60	324
Oct.	85	7	14	0.0	0.0	2	1.8	4	0.3	1	4.9	3.0	62°	5.8	N	S	7.3	16	60	422
Nov.	114	5	13	0.0	0.0	2	2.3	4	0.3	0	4.8	3.7	72°	6.2	S,SW	S	7.4	18	62	492
Dec.	148	5	11	0.0	0.0	2	1.9	4	0.2	0	4.5	2.8	76°	7.0	S,SW	S	7.6	18	53	545
Annual	1000	63	156	0.9	11.1	15	16.5	52	29.4	1	4.6	3.3	—	5.6	N	S	7.0	20	64	355
Years	82	11	30	10	10	34	30	33	30	5	30	5	—	35	30	30	15	48	51	12

* At Essendon Airport, 6 km north-northwest of Melbourne.

** 21st of the month.

TABLE LXIX

CLIMATIC TABLE FOR LAUNCESTON, TASMANIA, AUSTRALIA
Latitude 41°27'S, longitude 147°10'E, elevation 81 m

Month	Press.** (mbar)	Mean daily temp. (°C)	Mean daily temp. range (°C)	Extreme temp. (°C)		Precipitation (mm)				Relative humidity (%)			Evapor.** (mm)
				highest	lowest	mean	highest	lowest	max. 24 h	daily index	09h00	15h00	
Jan.	1011	17.7	13.2	37.8	+1.1	41	90	8	43	60	60	46	130
Feb.	1014	18.2	13.3	38.3	+0.9	50	81	<1	31	63	66	47	102
Mar.	1015	16.1	13.4	34.4	−0.6	40	157	3	71	67	72	52	81
Apr.	1017	12.9	11.1	28.9	−2.3	62	150	0	104	72	81	59	51
May	1017	10.2	10.2	23.8	−4.4	73	232	1	38	77	87	66	31
June	1016	8.1	9.0	19.0	−5.6	71	167	13	36	75	90	71	20
July	1015	7.4	9.3	19.0	−6.1	86	194	15	56	77	90	70	23
Aug.	1014	8.5	9.9	20.0	−3.9	80	161	9	48	77	86	65	30
Sept.	1013	10.4	10.3	23.9	−4.4	65	146	15	36	75	77	60	46
Oct.	1012	12.3	11.2	31.6	−3.9	68	170	7	31	72	70	57	69
Nov.	1012	14.6	12.2	32.8	+0.3	56	112	5	43	65	63	51	92
Dec.	1010	16.5	12.7	36.1	−0.3	50	151	0	69	63	61	49	124
Annual	1014	12.7	11.2	38.3	−6.1	740	1018	467	104	69	75	58	799.6
Years	34	51	51	50	50	30	49	49	27	33	32	32	10

TABLE LXIX (*continued*)

Month	Number of days								Cloudiness			Wind				Probability of precipitation (%)		
	precipitation		min. temp.		clear*	cloudy*	fog	thunder*	all types		low*	prev. dir.		speed (kt.)*		<12.5 mm	<25 mm	<50 mm
	>0.25 mm	>2.5 mm	<2.2	<0°C					09h00	15h00		09h00	15h00	09h00	15h00			
Jan.	8	4	0	0	2	4	2	1	4.2	4.2	3.6	NW,W	NW,W	11	14	5	22	61
Feb.	7	3	0.1	0	4	4	2	1	4.5	4.1	3.4	NW	NW	10	12	24	45	78
Mar.	9	5	0.3	0.1	4	6	3	1	4.5	4.6	3.4	NW,C	NW	11	12	6	33	65
Apr.	11	6	1.6	0.4	1	7	3	<1	4.6	4.7	3.4	C	NW	10	12	6	18	53
May	13	6	3.4	1.2	2	7	6	<1	5.2	4.7	3.7	C,NW	NW	11	12	6	12	39
June	16	6	>7	6.7	3	7	4	<1	5.3	4.8	4.4	C	NW	10	8	0	8	20
July	18	9	>6	4.9	3	8	8	<1	5.5	5.0	4.2	C,NW	NW,C	11	11	0	4	18
Aug.	17	8	>5	4.2	2	8	4	<1	5.0	4.8	3.9	C,NW	NW	11	12	2	4	29
Sept.	15	8	5.3	1.2	3	6	3	<1	4.7	4.9	3.8	NW,C	NW	10	14	0	8	26
Oct.	14	8	2.2	0.3	2	6	2	<1	4.7	4.9	3.2	NW	NW	11	13	2	8	35
Nov.	11	6	0.1	0	3	6	2	1	4.6	4.6	3.2	NW	NW	10	13	6	18	57
Dec.	10	6	0	0	3	7	1	1	4.5	4.4	3.1	NW,C	NW	10	12	16	28	61
Annual	149	75	>31	19.0	32	76	40	6	4.8	4.6	3.6					—	—	—
Years	30	10	10	10	10	10	9	10	27	27	3	5	5	5	5	49	49	49

* At Low Head, 41°03'S, 146°49'E, 28 m.
** At Cressy, 41°42'S, 147°03'E, 158 m.

TABLE LXX

CLIMATIC TABLE FOR HOBART, TASMANIA, AUSTRALIA (INITIALLY HOBARTON, HOBART TOWN)
Latitude 42°53'S, longitude 147°20'E, elevation 53.9 m

Month	M.s.l. press. (mbar)	Mean daily temp. (°C)	Mean daily temp. range (°C)	Temp. extremes (°C) highest sun	Temp. extremes (°C) highest shade	Temp. extremes (°C) lowest shade	Temp. extremes (°C) lowest grass	Mean vapor press. (mbar)	Precipitation (mm) mean (*)	Precipitation (mm) mean (**)	Precipitation (mm) max. 24 h	Relat. humidity (%) 09h00	Relat. humidity (%) 15h00	Relat. humidity (%) mean	Relat. humidity (%) daily index	Probability of precipitation (%) <12.5 mm	Probability of precipitation (%) <25 mm	Probability of precipitation (%) <50 mm
Jan.	1010.9	16.3	9.7	71	40.6	4.5	−0.8	10.5	46	42	75	58	53	59	58	12	30	60
Feb.	1012.1	16.1	9.4	74	40.2	3.9	−2.1	11.6	43	47	56	62	56	63	61	21	40	69
Mar.	1014.5	15.1	9.0	66	37.3	1.8	−2.5	11.0	54	52	88	66	56	67	64	10	33	67
Apr.	1014.8	12.4	7.9	61	30.6	0.7	−3.9	9.8	59	63	133	72	61	72	66	8	23	58
May	1014.8	10.5	7.3	53	25.4	−1.6	−6.7	8.9	43	51	44	77	63	78	70	7	29	64
June	1016.0	8.3	6.4	50	20.7	−1.6	−6.1	7.9	57	66	147	79	70	80	73	5	13	55
July	1014.5	7.8	6.7	49	18.9	−2.4	−7.4	7.7	54	47	64	80	69	80	72	3	16	49
Aug.	1012.1	8.8	7.6	54	22.0	−1.7	−6.6	7.9	46	53	58	75	61	76	68	3	22	69
Sept.	1012.7	10.6	8.5	59	27.6	−0.6	−7.6	8.1	48	53	156	67	58	67	63	2	14	61
Oct.	1010.4	11.8	9.1	69	33.3	0	−4.6	8.8	64	72	66	63	56	63	61	2	10	48
Nov.	1009.1	13.6	9.3	68	36.8	1.7	−3.3	9.6	57	58	94	59	54	60	60	4	13	44
Dec.	1009.3	15.1	9.2	72	40.7	3.3	−2.7	10.4	64	64	85	58	54	58	60	14	28	61
Annual	1012.6	12.2	8.3	74	40.7	−2.4	−7.6	9.2	636	668	156	67	58	69	64	—	—	—
Years	30	30	30	57	91	91	73	30	30	30	78	56	53	54	30	92	92	92

368

TABLE LXX (*continued*)

Month	Mean Austr. tank evap. (mm)	Number of days						Mean cloudiness (oktas)		Mean sun-shine (h)	Wind						Extr. monthly rainfall			
		precip.		tunder-storm	light-ning	clear	fog	all types	low		most freq. dir.		mean speed (kt.)		highest daily mean	highest gust	highest***		lowest***	
		>2.5 mm	>0.25 mm								09h00	15h00	09h00	15h00						
Jan.	123	4	13	1	0.9	1.9	0.0	5.1	3.5	7.7	NNW	SSE	7.0	10	18	66	150	150	1	4
Feb.	94	4	10	1	1.0	2.3	0.0	5.0	3.3	7.1	NNW	SSE	6.3	9	22	58	232	131	2	3
Mar.	79	4	13	1	1.2	2.4	0.3	4.9	3.3	6.4	NW,N	SSE	5.9	8	18	69	193	255	1	7
Apr.	50	5	14	<1	0.7	1.7	0.2	5.2	3.5	5.0	NW,N	W,NW,SE	5.8	8	21	64	216	248	2	2
May	35	4	14	0	0.4	2.4	0.9	4.9	3.6	4.4	NNW	NW,N	5.5	6	17	69	162	214	3	4
June	23	4	16	<1	0.4	2.4	0.8	5.0	3.4	4.0	NNW	NW,N	5.4	6	21	65	207	238	6	7
July	24	4	17	<1	0.3	2.0	1.0	4.9	3.0	4.4	NW	NNW	5.6	6	20	68	153	153	8	4
Aug.	33	6	18	<1	0.4	2.1	0.4	4.9	3.7	5.1	NNW	NW,N	5.9	7	22	76	258	161	6	8
Sept.	50	6	17	<1	0.7	1.5	0.1	5.0	3.8	5.9	NNW	NW	6.9	9	19	73	181	200	10	10
Oct.	77	5	18	1	0.6	1.0	0.0	5.3	3.5	6.1	NNW	SW,SE	7.1	10	17	64	169	193	7	10
Nov.	96	6	16	1	0.7	1.3	0.1	5.1	3.4	7.2	NNW	S,SE	6.9	10	18	73	227	188	4	8
Dec.	111	6	14	1	0.5	1.1	0.0	5.4	3.2	7.3	NNW	SSE	6.6	10	20	61	229	196	3	4
Annual	795	58	180	6	7.8	22.1	3.8	5.0	3.4	5.9	NNW	W	6.3	8	22	76	1,102	1,102	341	392
Years	30	10	30	19	30	30	30	30	2–3	30	30	30	30	28	68	70	100	78	100	78

* 1911–40.

** 1931–60.

*** Second column excludes 22 initial years (sites not comparable).

Note: Between June and August there are on the average 1.7 days with minima <0°C and 10.7 days with minima <2.2°C. Minima <2.2°C occur also on 0.8 days in April–May and 1.3 in September, giving and annual total of 12.8 days.

TABLE LXXI

CLIMATIC TABLE FOR MAATSUYKER ISLAND, TASMANIA, AUSTRALIA
Latitude 43°42'S, longitude 146°18'E, elevation 95 m

Month	Mean sta. level press. (mbar)*		Temperature (0°C)				Precipitation (mm)				Relat. humidity (%)		
	09h00	15h00	mean	daily range	extreme max.	min.	mean	highest	lowest	max. 24 h	daily index	09h00	15h00
Jan.	1012	1012	13.3	6.2	34.4	5.0	77	163	19	32	81	83	81
Feb.	....	....	13.4	6.1	31.7	3.3	81	184	14	41	83	84	81
Mar.	....	....	13.1	5.7	30.0	4.4	89	207	25	37	84	85	82
Apr.	1016	1014	11.3	5.0	23.9	1.1	106	217	21	29	84	87	85
May	....	....	10.4	4.2	22.2	2.8	119	169	45	40	86	88	85
June	....	....	9.1	4.0	16.1	2.2	118	192	46	26	87	88	87
July	1014	1013	8.6	3.9	16.1	1.7	130	216	52	45	85	87	86
Aug.	....	....	8.8	4.7	19.4	0.6	130	213	49	28	85	88	83
Sept.	....	....	9.4	5.6	22.8	1.7	106	211	26	37	84	88	84
Oct.	1011	1010	10.4	5.7	22.8	1.1	108	188	39	53	84	84	85
Nov.	....	....	11.3	5.8	27.8	1.1	97	210	19	28	84	85	84
Dec.	....	....	12.7	6.3	27.8	4.4	90	315	12	57	83	86	82
Annual	....	....	11.1	5.6	34.4	0.6	1252	1568	697	57	84	86	84
Years	—	—	13	13	8	8	30	52	52	52	12	8	12

TABLE LXXI (*continued*)

Month	Evapor. (mm)**	Number of days with				Cloudiness all types (oktas)	Mean monthly sunshine** (h)	Wind				Probability of precipitation (%)		
		precipitation		thunder-storm	fog			most freq. dir.		mean speed (m/sec)		<12.5 mm	<25 mm	<50 mm
		>0.25 mm	>2.5 mm					09h00	15h00	09h00	15h00			
Jan.	101	19	10	1	1	6.2	198	W	W	20	22	0	2	26
Feb.	76	17	9	1	2	5.8	195	W	W	18	21	0	21	53
Mar.	64	20	8	1	2	5.8	175	W	W	17	18	0	2	19
Apr.	38	22	11	0	1	6.2	120	W	W	16	17	0	2	7
May	30	24	12	<1	1	6.6	96	W	W	20	20	0	0	7
June	20	22	15	0	<1	6.2	72	W,NW	W	20	17	0	0	7
July	23	24	15	1	0	6.3	73	W,NW	W	18	17	0	0	0
Aug.	25	26	15	<1	0	6.2	110	W,NW	W	20	17	0	0	2
Sept.	38	23	13	<1	0	6.3	123	W	W	23	22	0	0	12
Oct.	46	24	11	<1	1	6.1	145	W	W	18	21	0	0	5
Nov.	70	21	10	<1	2	6.1	170	W	W	18	20	0	2	19
Dec.	110	20	10	<1	5	6.2	183	W	W	22	23	2	7	23
Annual	641	262	139	6	15	6.2	1660	—	W	19	20	—	—	—
Years	—	30	8	4	4	8	—	—	5	5	5	5	43	43

Probability of precipitation annual: <12.5 mm = 5, <25 mm = 43, <50 mm = 43

* . . . = not available.

** Estimated by interpolation from maps.

TABLE LXXII

CLIMATIC TABLE FOR CHATEAU TONGARIRO, NEW ZEALAND
Latitude 39°12′S, longitude 175°32′E, elevation 1119 m

Month	Mean daily temp. (°C)	Mean daily temp. range (°C)	Temperature extremes[1] (°C)		Mean vapor press. (mbar)	Mean precip. (mm)	Max. precip.[1] (mm/24h)
			highest	lowest			
Jan.	11.5	10.0	29.4	− 1.7	10.7	206	107.2
Feb.	12.2	9.4	25.0	− 1.1	11.3	236	141.5
Mar.	10.6	8.9	24.4	− 2.8	10.4	173	181.9
Apr.	8.1	8.1	21.9	− 7.5	9.1	254	84.8
May	5.3	7.3	19.2	− 7.2	7.5	249	118.1
June	3.1	6.7	18.9	− 7.2	6.6	264	79.8
July	2.3	7.1	15.0	−13.6	6.3	246	78.7
Aug.	3.0	7.1	19.4	− 9.2	6.4	229	97.8
Sept.	4.6	8.3	19.4	− 8.9	6.9	226	68.6
Oct.	6.3	8.6	23.3	− 8.3	7.8	287	135.1
Nov.	8.1	9.1	22.7	− 5.8	8.8	254	97.5
Dec.	10.1	9.6	30.0	− 4.2	9.9	267	119.4
Annual	7.1	8.3	30.0	−13.6	8.5	2,890	181.9

Month	Number of days with			Mean cloudiness (oktas)	Prev. wind dir. 09h00
	precip.	ground frost	screen frost		
Jan.	15	3.2	0.4	4.8	WNW
Feb.	13	2.0	0.3	4.9	WNW
Mar.	13	3.9	0.9	4.8	WNW
Apr.	15	8.9	3.2	4.9	WNW
May	17	14.1	8.8	5.1	WNW
June	18	17.8	15.6	5.1	WNW
July	18	21.3	17.9	4.9	WNW
Aug.	18	20.0	16.0	5.0	WNW
Sept.	16	17.2	12.5	5.0	WNW
Oct.	19	12.5	7.4	5.4	WNW
Nov.	17	7.8	2.5	5.4	WNW
Dec.	17	4.7	1.0	5.3	WNW
Annual	196	133.4	86.5	5.0	WNW

[1] 1930–1960.

TABLE LXXIII

CLIMATIC TABLE FOR AUCKLAND (ALBERT PARK), NEW ZEALAND
Latitude 36°51′S, longitude 174°46′E, elevation 49 m

Month	Station press. (mbar) mean daily	Mean daily temp. (°C)	Mean daily temp. range (°C)	Temperatures extremes[1] (°C) highest	Temperatures extremes[1] (°C) lowest	Mean vapor press. (mbar)	Mean precip.[1] (mm)	Max. precip. (mm/24h)	Mean evapo- ration (mm)
Jan.	1014.3	19.2	7.3	32.2	7.3	16.2	84	97.0	119
Feb.	1016.0	19.6	7.1	32.4	8.6	17.0	104	162.3	89
Mar.	1017.5	18.4	7.0	29.9	5.6	16.1	71	116.1	76
Apr.	1018.0	16.4	6.6	27.2	4.1	14.8	109	131.6	43
May	1016.3	13.8	6.1	23.6	1.9	12.5	122	91.4	25
June	1015.2	11.8	5.9	21.0	—0.1	11.3	140	91.2	25
July	1015.9	10.8	6.1	19.2	0.7	10.6	140	85.8	25
Aug.	1016.2	11.3	6.4	19.4	1.1	10.9	109	77.0	31
Sept.	1017.5	12.6	6.6	21.7	1.2	11.5	97	54.6	43
Oct.	1016.3	14.3	6.5	23.9	2.2	12.6	107	63.8	71
Nov.	1015.0	15.9	6.9	27.2	5.0	13.6	81	80.0	99
Dec.	1014.5	17.7	7.2	31.8	6.1	14.9	79	136.9	114
Annual	1016.1	15.2	6.6	32.4	—0.1	13.5	1,242	162.3	762

Month	Number of days with precip.	thunder- storm	fog	gusts[2] (m/sec) ≥18	gusts[2] (m/sec) ≥27	ground frost	screen frost	Mean cloudi- ness (oktas)	Wind most freq. dir.	Wind mean speed[2] (m/sec)	Mean sun- shine (h)
Jan.	10	0.6	0.1	3.0	0.1	–	–	4.9	WSW	4.9	233
Feb.	10	0.5	0.1	2.7	–	–	–	4.9	WSW	4.4	194
Mar.	12	0.3	0.3	3.1	0.2	–	–	4.6	SW	4.3	190
Apr.	15	0.8	0.5	3.8	0.1	–	–	4.7	SW	4.4	155
May	19	0.6	1.3	5.3	0.2	0.1	–	4.7	SW	4.2	138
June	20	0.9	1.5	6.3	0.4	0.8	–	4.6	WSW	4.3	123
July	20	1.1	1.8	4.3	0.3	1.7	–	4.7	WSW	4.2	129
Aug.	19	0.8	1.4	5.1	0.3	0.6	–	4.5	WSW	4.6	151
Sept.	16	0.8	0.6	4.1	0.3	0.2	–	4.8	WSW	4.7	168
Oct.	17	0.8	0.2	4.9	0.1	–	–	5.2	WSW	5.4	181
Nov.	15	0.9	–	5.3	0.2	–	–	5.1	WSW	5.0	207
Dec.	12	0.6	–	4.1	0.2	–	–	4.9	WSW	4.8	224
Annual	185	8.7	7.8	52.0	2.4	3.4	0.0	4.8	WSW	4.6	2,094

[1] Various sites, 1868–1960.

[2] Mechanics Bay data.

TABLE LXXIV

CLIMATIC TABLE FOR TAURANGA, NEW ZEALAND
Latitude 37°40′S, longitude 176°12′E, elevation 4 m

Month	Mean daily sta. press. (mbar)	Mean daily temp. (°C)	Mean daily temp. range (°C)	Temperature extremes[1] (°C) highest	lowest	Mean vapor press. (mbar)	Mean precip. (mm)	Max. precip.[2] (mm/24h)	Mean evapor. (mm)
Jan.	1015.2	18.6	10.3	33.3	3.3	16.3	89	134.4	127
Feb.	1016.9	19.0	9.8	31.1	1.7	16.8	89	159.5	104
Mar.	1017.1	17.5	9.7	29.7	0.7	15.7	97	95.5	81
Apr.	1016.6	15.0	9.7	27.9	−0.6	14.2	127	239.0	48
May	1015.1	12.4	9.2	23.9	−5.3	11.9	124	115.8	28
June	1013.9	10.1	9.2	21.9	−4.6	10.2	140	163.8	23
July	1016.1	9.3	9.6	22.8	−4.2	9.6	127	134.6	23
Aug.	1015.6	10.1	9.2	19.6	−3.4	10.3	122	76.4	31
Sept.	1018.5	11.6	9.5	24.7	−4.6	11.1	97	156.2	56
Oct.	1015.1	13.3	9.1	25.6	−2.3	12.5	117	129.4	84
Nov.	1013.6	15.2	9.8	28.3	+0.6	13.9	84	66.8	102
Dec.	1013.2	17.0	9.8	30.6	−0.3	15.3	86	132.1	122
Annual	1015.6	14.1	9.6	33.3	−5.3	13.2	1,300	239.0	813

Month	Number of days with precip.	thunder-storm	fog	wind gusts (m/sec) ≥18	≥27	ground frost	screen frost	Mean cloudi-ness (oktas)	Mean sun-shine (h)	Wind most freq. dir.	mean speed (m/sec)
Jan.	10	0.5	1.1	2.9	–	0.3	–	4.6	255	W	3.8
Feb.	9	0.5	1.6	1.4	–	0.2	–	4.5	210	W	3.4
Mar.	11	0.7	2.6	3.0	0.1	0.6	–	4.4	207	W	3.5
Apr.	12	1.0	2.8	3.2	0.1	2.4	–	4.2	176	WSW	3.5
May	14	0.9	3.6	3.5	0.2	7.0	0.4	4.2	155	WSW	3.2
June	14	0.5	1.9	4.0	0.2	11.3	2.2	4.1	141	W	3.4
July	15	0.4	1.8	3.3	0.3	14.0	3.0	4.2	152	W	3.3
Aug.	15	1.0	1.4	5.1	0.1	11.7	1.2	4.3	164	W	3.5
Sept.	14	0.4	2.0	4.0	0.2	9.0	0.2	4.3	186	W	3.9
Oct.	14	1.0	1.2	4.1	0.1	4.0	–	4.8	201	W	4.2
Nov.	13	1.0	0.3	4.5	–	1.6	–	4.9	226	W	4.3
Dec.	12	0.7	0.2	3.6	0.1	0.5	–	4.7	243	W	4.2
Annual	153	8.6	20.5	42.6	1.4	62.6	7.0	4.4	2,316	W	3.7

[1] Various sites, 1913–1960.
[2] Various sites, 1898–1960.

TABLE LXXV

CLIMATIC TABLE FOR NEW PLYMOUTH, NEW ZEALAND
Latitude 39°04′S, longitude 174°05′E, elevation 49 m

Month	Mean daily sta. press. (mbar)	Mean daily temp. (°C)	Mean daily temp. range (°C)	Temperature extremes[1] (°C)		Mean vapor press. (mbar)	Mean precip. (mm)	Max. precip.[2] 24h (mm)	Mean evapor. (mm)
				highest	lowest				
Jan.	1015.0	17.1	7.5	29.1	4.6	15.6	119	123.4	122
Feb.	1016.6	17.6	7.7	30.0	5.1	16.4	104	185.2	94
Mar.	1016.8	16.6	7.5	27.5	1.6	15.2	94	104.1	84
Apr.	1016.2	14.7	7.2	28.4	1.8	13.4	127	99.1	58
May	1014.5	12.4	6.7	22.2	−0.4	11.6	135	112.5	43
June	1013.2	10.3	6.3	20.4	−1.6	10.0	163	95.2	36
July	1015.9	9.6	6.6	21.4	−1.4	9.5	157	93.0	33
Aug.	1015.1	10.1	6.7	18.6	−1.6	9.9	150	71.6	46
Sept.	1018.3	11.3	6.8	21.3	−0.2	11.1	122	105.9	58
Oct.	1014.6	12.7	6.6	23.9	0.6	12.3	147	113.8	79
Nov.	1013.3	14.1	6.9	25.8	1.3	13.3	114	69.9	99
Dec.	1012.8	15.6	6.9	27.2	2.3	14.5	122	120.6	114
Annual	1015.2	13.5	6.9	30.0	−1.6	12.7	1,554	185.2	869

Month	Number of days with							Mean cloudi-ness (oktas)	Mean sun-shine (h)	Wind	
	precip.	thunder-storm	fog	gusts (m/sec)[3]		ground frost	screen frost			most freq. dir.	mean speed[3] (m/sec)
				≥18	≥27						
Jan.	9	0.4	0.5	5.5	0.3	−	−	5.6	245	W	5.3
Feb.	10	0.2	0.2	4.1	0.2	−	−	5.4	194	W	5.2
Mar.	11	0.4	0.2	5.6	0.8	0.1	−	4.8	196	SE	4.6
Apr.	13	1.4	0.1	6.3	0.8	0.3	−	5.0	165	SE	5.0
May	15	1.8	0.3	8.3	0.9	0.3	−	5.1	145	SE	4.8
June	18	1.7	0.7	9.8	1.1	1.7	0.1	5.3	116	SE	5.0
July	17	1.6	0.7	8.0	1.3	2.6	0.2	5.6	135	SE	5.3
Aug.	17	1.3	0.3	8.4	1.0	1.8	0.1	5.3	154	SE	5.0
Sept.	14	0.6	0.7	7.1	0.7	0.7	0.1	5.3	168	W	5.6
Oct.	17	1.0	0.7	7.7	0.3	0.2	−	6.0	175	W	5.8
Nov.	14	0.7	0.3	7.6	0.2	−	−	5.7	201	W	5.6
Dec.	14	0.7	1.0	6.3	0.4	−	−	5.7	216	W	5.3
Annual	169	11.8	5.7	84.7	8.0	7.7	0.5	5.4	2,110	W	5.1

[1] 1921–1960.
[2] Various sites, 1862–1880, 1882–1884, 1888–1891, 1894–1960.
[3] New Plymouth Airport data.

TABLE LXXVI

CLIMATIC TABLE FOR NAPIER, NEW ZEALAND
Latitude 39°29′S, longitude 176°55′E, elevation 2 m

Month	Mean daily temp. (°C)	Mean daily temp. range (°C)	Temperature extremes[1]		Mean vapor press. (mbar)	Mean precip. (mm)	Max. precip.[2] (mm/24h)	Mean evapor. (mm)
			highest (°C)	lowest (°C)				
Jan.	18.7	9.3	35.8	4.9	14.5	66	110.2	140
Feb.	18.8	8.8	35.0	3.3	15.4	71	166.1	109
Mar.	17.0	8.9	32.8	0.0	14.5	56	204.0	86
Apr.	14.8	9.0	29.2	−0.3	12.7	69	169.2	56
May	11.8	8.8	25.1	−1.7	10.5	97	148.1	36
June	9.4	9.1	22.8	−2.5	8.8	79	144.0	25
July	8.6	8.8	23.1	−2.8	8.5	89	90.2	28
Aug.	9.6	8.6	21.9	−2.8	9.2	71	181.9	43
Sept.	11.3	9.3	26.7	−1.5	9.9	48	84.6	53
Oct.	13.7	9.1	30.9	−0.8	11.2	48	162.3	94
Nov.	15.7	9.3	31.5	1.9	12.3	48	61.5	117
Dec.	17.4	9.0	33.9	3.3	13.7	51	158.2	140
Annual	13.9	9.0	35.8	−2.8	11.8	792	204.0	925

Month	Number of days with							Mean cloudiness (oktas)	Mean sunshine (h)	Wind	
	precip.	thunderstorm	fog	gusts (m/sec)[3]		ground frost	screen frost			most freq. dir.	mean speed[3] (m/sec)
				≥18	≥27						
Jan.	8	0.7	0.1	2.6		−	−	4.0	250	E	3.4
Feb.	8	0.5	0.2	1.4	−	−	−	4.1	206	E	3.0
Mar.	8	0.2	0.5	1.4	−	−	−	4.0	205	WSW	2.9
Apr.	9	0.2	0.5	2.0	−	0.5	−	4.2	163	WSW	2.9
May	11	0.1	0.7	1.8	−	2.8	0.3	4.2	144	WSW	3.0
June	11	−	0.6	1.4	0.2	9.1	2.0	4.1	144	WSW	2.8
July	12	0.2	0.7	2.2	−	10.4	3.4	4.3	140	WSW	2.8
Aug.	12	0.1	0.6	1.4	0.2	7.7	1.3	4.1	162	WSW	3.0
Sept.	10	0.1	0.5	2.8	−	3.8	0.4	3.9	191	W	3.0
Oct.	10	0.7	0.3	1.4	−	1.0	0.1	4.2	209	W	2.9
Nov.	9	1.2	0.2	2.3	−	0.1	−	4.2	221	W	3.7
Dec.	8	0.5	0.1	3.0	−	0.1	−	4.1	246	E	3.5
Annual	116	4.5	5.0	23.8	0.4	35.5	7.5	4.1	2,281	WSW	3.1

[1] Various sites, 1869–1880, 1905–1960.
[2] Various sites, 1864–1880, 1888–1960.
[3] Napier Airport data.

TABLE LXXVII

CLIMATIC TABLE FOR NELSON (AIRPORT), NEW ZEALAND
Latitude 41°17'S, longitude 173°13'E, elevation 2 m

Month	Mean station press. (mbar)	Mean daily temp. (°C)	Mean daily temp. range (°C)	Temp. extremes[1] (°C) highest	Temp. extremes[1] (°C) lowest	Mean vapor press. (mbar)	Mean precip. (mm)	Max. precip.[2] (mm/24h)	Mean evapor. (mm)
Jan.	1013.5	16.8	9.6	33.1	2.8	14.1	69	88.9	117
Feb.	1015.1	16.9	9.4	28.9	3.1	14.7	58	95.0	91
Mar.	1016.0	15.4	9.7	28.3	−0.2	13.7	66	105.7	79
Apr.	1015.4	12.4	10.2	26.9	−2.8	11.9	76	108.2	48
May	1014.2	9.5	10.6	21.4	−3.3	9.9	86	86.9	28
June	1013.1	7.0	10.9	18.6	−6.6	7.8	79	115.6	20
July	1015.5	6.4	10.8	18.6	−6.1	7.6	74	122.7	20
Aug.	1014.6	7.4	10.2	20.0	−5.8	8.6	91	88.0	25
Sept.	1016.5	9.4	10.2	22.2	−3.1	9.9	74	69.6	43
Oct.	1013.4	11.4	9.2	24.0	−1.7	11.0	91	70.6	58
Nov.	1011.5	13.4	9.6	27.3	−1.0	11.9	69	77.7	91
Dec.	1011.3	15.3	9.4	30.6	1.2	13.1	79	96.0	117
Annual	1014.2	11.8	10.0	33.1	−6.6	11.2	912	122.7	744

Month	Number of days with precip.	Number of days with thunder-storm	Number of days with fog	gusts (m/sec) ⩾18	gusts (m/sec) ⩾27	ground frost	screen frost	Mean cloudi-ness (oktas)	Mean sun-shine (h)	Wind most freq. dir.	Wind mean speed (m/sec)
Jan.	9	0.4	1.1	3.2	–	0.1	–	4.6	260	NNE	3.8
Feb.	8	0.4	2.5	3.3	0.1	0.2	–	4.7	218	NNE	3.5
Mar.	10	0.4	3.3	3.1	0.1	0.9	0.1	4.6	218	SW	2.9
Apr.	10	0.7	3.6	3.9	0.1	5.3	0.4	4.3	189	SW	2.8
May	12	0.8	5.8	2.5	0.1	12.8	0.9	4.5	161	SW	2.0
June	10	0.6	5.2	2.2	–	18.1	11.3	4.1	151	SW	1.7
July	11	0.2	4.1	2.1	–	21.1	14.2	4.4	161	SW	1.8
Aug.	12	0.4	3.1	1.7	0.1	18.7	8.3	4.6	172	SW	2.2
Sept.	10	0.3	2.3	3.4	–	12.7	2.7	4.3	202	NNE	3.0
Oct.	13	0.4	2.3	2.8	–	4.6	0.6	4.9	210	NNE	3.4
Nov.	12	1.0	1.3	3.5	0.1	1.4	0.1	4.8	226	NNE	3.9
Dec.	10	0.7	1.1	4.5	0.1	0.7	–	4.9	246	NNE	3.9
Annual	127	6.3	35.7	36.2	0.7	96.6	42.6	4.6	2,414	NNE	2.9

[1] Two sites, 1920–1951, 1943–1960.
[2] Various sites, 1893–1960.

TABLE LXXVIII

CLIMATIC TABLE FOR WELLINGTON (KELBURN), NEW ZEALAND
Latitude 41°17′S, longitude 174°46′E, elevation 126 m

Month	Mean station press. (mbar)	Mean daily temp. (°C)	Mean daily temp. range (°C)	Temp. extremes[1] (°C) highest	lowest	Mean vapor press. (mbar)	Mean precip. (mm)	Max. precip.[1] (mm/24h)	Mean evapor. (mm)
Jan.	1012.6	16.2	7.2	29.4	4.1	14.6	74	113.5	107
Feb.	1014.9	16.4	7.0	31.1	4.7	15.1	91	160.5	82
Mar.	1016.7	15.4	6.8	27.2	3.9	14.2	79	144.8	74
Apr.	1017.0	13.5	6.1	27.3	2.1	12.7	94	125.7	46
May	1014.9	10.9	5.6	21.9	−0.7	11.2	119	144.8	31
June	1013.9	8.8	5.4	20.6	−1.2	9.4	122	86.6	25
July	1013.8	8.1	5.4	18.9	−1.9	8.8	130	82.6	18
Aug.	1014.6	8.8	5.8	20.0	−1.6	9.4	135	95.2	28
Sept.	1014.4	10.2	6.2	20.6	−0.6	10.3	97	96.0	43
Oct.	1012.8	11.7	6.4	24.2	1.1	11.5	122	105.4	58
Nov.	1011.6	13.3	6.8	26.9	1.7	12.7	81	67.3	79
Dec.	1011.8	15.1	7.1	29.1	3.4	13.7	107	152.4	94
Annual	1014.1	12.4	6.3	31.1	−1.9	12.0	1,250	160.5	686

Month	Number of days with precip.	thunderstorm	fog	gusts (m/sec) ≥18	≥27	ground frost	screen frost	Mean cloudiness (oktas)	Mean sunshine (h)	Wind most freq. dir.	mean speed[2] (m/sec)
Jan.	10	0.5	0.3	13.4	2.6	–	–	5.6	234	NW	5.2
Feb.	9	0.4	0.8	13.6	4.2	–	–	5.7	195	NW	5.0
Mar.	10	0.5	1.1	9.8	1.6	–	–	5.6	189	NW	4.7
Apr.	13	0.4	1.3	12.1	1.9	0.1	–	5.6	151	NW	5.0
May	15	0.5	1.5	13.1	2.1	1.3	–	5.5	118	NW	5.4
June	16	0.6	2.1	13.9	0.6	3.5	0.03	5.4	106	NW	5.1
July	18	0.3	1.8	11.4	1.4	5.6	–	5.6	108	NW	5.7
Aug.	16	0.3	1.5	9.8	1.5	4.3	0.03	5.6	139	NW	4.8
Sept.	14	0.4	0.5	11.6	1.8	2.3	–	5.6	170	NW	5.6
Oct.	14	0.3	0.5	13.3	3.8	0.6	–	5.9	183	NW	5.5
Nov.	12	0.6	0.5	15.6	3.9	0.2	–	5.8	197	NW	5.6
Dec.	12	0.5	0.5	13.8	3.2	–	–	5.8	222	NW	5.3
Annual	159	5.3	12.4	151.4	28.6	17.9	0.06	5.6	2,012	NW	5.2

[1] Various sites, 1862–1960.

[2] Rongotai data.

TABLE LXXIX

CLIMATIC TABLE FOR HOKITIKA (SOUTH), NEW ZEALAND
Latitude 42°43′S, longitude 170°57′E, elevation 4 m

Month	Mean station press. (mbar)	Mean daily temp. (°C)	Mean daily temp. range (°C)	Temp. extremes[1] (°C) highest	lowest	Mean vapor press. (mbar)	Mean precip. (mm)	Max. precip.[2] (mm/24h)	Mean evapor. (mm)
Jan.	1012.7	14.8	8.2	26.7	1.7	14.8	249	217.9	84
Feb.	1015.1	15.1	7.8	28.9	2.3	15.4	218	232.9	64
Mar.	1016.9	14.1	8.2	29.2	−0.6	14.3	213	193.3	46
Apr.	1016.8	11.9	8.5	23.6	−0.6	12.3	224	207.3	28
May	1015.1	9.6	9.0	21.9	−2.6	10.0	229	139.7	13
June	1013.5	7.2	9.3	18.2	−3.9	8.2	203	113.8	13
July	1013.6	6.9	9.3	18.3	−3.9	8.0	211	153.4	11
Aug.	1014.3	7.7	9.2	19.5	−3.6	8.7	236	133.9	23
Sept.	1014.4	9.1	8.6	19.8	−2.8	10.3	211	118.1	33
Oct.	1013.2	10.7	7.8	23.3	−2.1	11.2	277	137.7	58
Nov.	1011.8	12.1	7.7	23.4	−0.1	12.4	241	116.1	76
Dec.	1012.4	13.6	7.7	26.1	0.6	13.8	251	114.3	94
Annual	1014.1	11.1	8.4	29.2	−3.9	11.6	2,764	232.9	541

Month	Number of days with precip.	thunder-storm	fog	gusts (m/sec) ⩾18	⩾27	ground frost	screen frost	Mean cloudi-ness (oktas)	Mean sun-shine (h)	Wind most freq. dir.	mean speed (m/sec)
Jan.	14	1.2	2.6	2.1	0.2	−	−	5.6	208	W	3.0
Feb.	14	0.5	3.1	1.7	0.2	0 1	−	5.8	170	W	3.1
Mar.	16	1.4	2.5	1.9	0.5	0.5	0.1	5.6	166	E	2.5
Apr.	16	2.1	2.5	3.5	0.3	1.9	0.1	5.4	140	E	2.8
May	16	1.9	1.7	2.4	0.1	6.9	2.0	5.3	122	E	2.3
June	16	1.6	1.5	3.0	0.4	13.4	7.6	5.1	107	E	2.6
July	16	1.1	1.1	3.2	0.2	15.6	8.8	5.1	129	E	2.9
Aug.	16	1.7	1.1	3.8	0.4	12.3	5.7	5.3	138	E	2.9
Sept.	17	0.7	1.7	3.0	0.5	6.8	1.4	5.5	152	E	3.4
Oct.	19	1.7	1.5	3.5	0.1	2.8	0.3	6.1	160	E	3.5
Nov.	18	1.7	1.8	3.6	0.2	0.8	0.1	5.9	173	E	3.7
Dec.	17	0.9	2.3	3.0	0.1	0.4	−	6.0	190	W	3.6
Annual	195	16.5	23.4	34.7	3.2	61.5	26.1	5.6	1,855	E	3.0

[1] Various sites, 1866–1880; 1913–1960.
[2] Various sites, 1865–1880; 1913–1960.

TABLE LXXXIII

CLIMATIC TABLE FOR CHRISTCHURCH (GARDENS), NEW ZEALAND
Latitude 43°32′S, longitude 172°37′E, elevation 7 m

Month	Mean station press. (mbar)	Mean daily temp. (°C)	Mean daily temp. range (°C)	Temp. extremes[1] (°C)		Mean vapor press. (mbar)	Mean precip. (mm)	Max. precip.[2] (mm/ 24h)	Mean evapor. (mm)
				highest	lowest				
Jan.	1010.4	16.4	9.8	36.1	1.1	13.2	56	82.0	127
Feb.	1013.1	16.2	9.4	35.2	1.2	13.5	46	78.0	109
Mar.	1015.4	14.6	9.2	32.4	−0.9	13.0	43	80.3	84
Apr.	1014.9	12.0	9.3	28.4	−3.6	11.5	46	119.6	46
May	1013.6	8.7	9.1	26.5	−5.9	9.1	76	101.6	28
June	1012.3	6.3	8.8	20.7	−5.8	7.5	69	74.7	18
July	1013.4	5.7	8.7	21.1	−7.1	7.2	61	70.9	18
Aug.	1013.1	6.9	8.7	22.6	−5.0	8.0	58	82.6	33
Sept.	1012.9	9.4	9.6	27.3	−4.8	9.0	51	56.1	58
Oct.	1010.6	11.7	10.1	31.0	−3.3	9.9	51	68.8	76
Nov.	1008.9	13.6	10.7	32.2	−1.2	11.0	51	45.7	104
Dec.	1009.7	15.4	9.9	33.5	0.6	12.5	61	63.5	114
Annual	1012.4	11.4	9.4	36.1	−7.1	10.4	668	119.6	820

Month	Number of days with							Mean cloudi- ness (oktas)	Mean sun- shine (h)	Wind	
	precip.	thunder- storm	fog	gusts (m/sec)[3]		ground frost	screen frost			most freq. dir.	mean speed[3] (m/sec)
				≥18	≥27						
Jan.	10	0.8	0.2	5.7	0.4	0.2	–	5.1	215	E	4.8
Feb.	8	0.3	0.1	5.1	0.2	0.3	–	5.0	182	E	4.3
Mar.	9	0.3	0.2	4.9	0.4	1.2	–	5.0	174	E	3.8
Apr.	10	0.1	2.5	4.0	0.4	4.9	0.7	4.9	143	ENE	3.5
May	12	0.1	4.6	4.4	0.2	11.3	3.7	4.9	133	ENE	3.5
June	13	0.1	5.2	3.8	0.2	17.4	9.8	4.9	116	SW	3.2
July	14	0.1	4.4	3.8	0.2	19.0	11.3	4.9	126	SW	3.2
Aug.	11	0.1	3.1	2.7	–	17.0	8.2	4.7	145	SW	3.4
Sept.	10	0.2	1.2	4.8	0.5	10.0	2.5	4.8	169	ENE	4.2
Oct.	11	0.2	0.4	6.4	0.5	5.5	0.4	4.8	187	ENE	4.5
Nov.	10	0.5	0.5	5.9	0.2	2.6	0.1	4.8	206	ENE	4.4
Dec.	11	0.7	0.4	5.6	0.1	0.6	–	5.1	194	E	4.7
Annual	129	3.5	22.8	57.1	3.3	90.0	36.7	4.9	1,990	ENE	3.9

[1] 1864–1880, 1905–1960.
[2] 1864–1884, 1894–1960.
[3] Christchurch Airport data.

TABLE LXXXI

CLIMATIC TABLE FOR LAKE TEKAPO, NEW ZEALAND
Latitude 44°00′S, longitude 170°29′E, elevation 683 m

Month	Mean daily temp. (°C)	Mean daily temp. range (°C)	Temperature extremes[1] (°C)		Mean vapor press. (mbar)	Mean precip. (mm)	Max. precip.[1] (mm/24h)
			highest	lowest			
Jan.	15.5	13.2	33.3	− 3.2	9.9	53	50.8
Feb.	15.2	13.2	32.2	− 2.2	10.0	46	91.4
Mar.	13.2	12.5	30.0	− 5.1	10.0	41	61.2
Apr.	10.1	11.3	26.6	− 6.5	8.3	48	73.2
May	5.6	10.0	21.2	−11.1	6.4	46	68.1
June	3.3	9.2	17.8	−15.6	5.4	36	58.7
July	1.6	9.4	17.1	−13.1	4.9	41	55.9
Aug.	3.7	9.9	18.2	−13.8	5.7	41	66.0
Sept.	7.6	12.0	23.4	− 8.7	6.6	53	38.1
Oct.	9.6	11.7	24.7	− 6.1	7.3	56	120.6
Nov.	11.9	11.8	28.3	− 6.6	8.1	46	55.1
Dec.	13.8	12.3	30.2	− 5.1	9.7	58	69.8
Annual	9.3	11.4	33.3	−15.6	7.7	564	120.6

Month	Number of days with			Mean cloudiness (oktas)	Mean sunshine (h)	Wind most freq. dir. 09h00
	precip.	ground frost[2]	screen frost[3]			
Jan.	7	2.3	0.1	3.8	258	N
Feb.	6	2.1	0.2	4.1	215	N
Mar.	8	6.2	0.8	3.9	210	N
Apr.	9	16.0	2.6	4.2	165	N
May	10	19.5	12.1	4.2	132	N
June	7	22.1	18.1	4.0	108	N
July	8	27.4	25.0	4.1	118	N
Aug.	7	24.3	20.1	4.0	152	N
Sept.	6	17.1	10.0	3.8	180	N
Oct.	10	10.0	4.5	4.1	211	N
Nov.	10	4.9	0.6	4.2	218	N
Dec.	10	2.4	0.4	4.0	232	N
Annual	98	154.3	94.5	4.4	2,199	N

[1] Various sites, 1925–1960.

[2] New site averages, 1950–1960; old site: annual average 204 days.

[3] New site averages, 1950–1960; old site: annual average 111 days.

TABLE LXXXII

CLIMATIC TABLE FOR ALEXANDRA, NEW ZEALAND
Latitude 45°15′S, longitude 169°24′E, elevation 158 m

Month	Mean daily temp. (°C)	Mean daily temp. range (°C)	Temperature extremes[1] (°C)		Mean vapor press. (mbar)	Mean precip. (mm)	Max. precip.[2] (mm/24h)	Mean evapor. (mm)
			highest	lowest				
Jan.	16.8	12.5	37.2	1.4	11.0	46	68.6	109
Feb.	16.6	12.7	33.1	0.7	11.0	38	45.0	91
Mar.	14.4	12.8	32.1	− 0.8	10.5	31	38.6	74
Apr.	10.8	12.1	27.7	− 4.1	9.0	33	33.9	38
May	6.1	10.9	23.1	− 7.1	6.8	23	32.5	18
June	3.3	9.1	19.8	− 8.9	6.0	20	17.5	8
July	2.4	9.3	18.2	−11.7	5.5	18	19.8	5
Aug.	5.2	11.5	19.8	− 9.4	6.1	15	22.1	20
Sept.	8.6	12.4	22.3	− 5.6	7.1	20	33.8	41
Oct.	11.5	12.4	27.8	− 3.4	8.1	30	49.5	69
Nov.	13.8	12.5	31.7	− 1.8	9.1	28	40.6	91
Dec.	15.8	12.2	33.4	0.6	10.6	33	35.0	109
Annual	10.4	11.7	37.2	−11.7	8.4	335	68.6	676

Month	Number of days with					Mean cloudi-ness[3] (oktas)	Mean sun-shine (h)	Wind most freq. dir.
	precip.	thunder-storm	fog	ground frost	screen frost			
Jan.	9	0.7	–	1.0	–	4.5	237	SSW
Feb.	8	0.6	0.3	1.1	–	4.5	199	SSW
Mar.	8	0.1	1.0	5.1	0.3	4.6	196	SSW
Apr.	9	0.1	2.9	11.3	2.9	4.7	152	SSW
May	9	–	5.4	20.7	12.8	5.0	124	SSW
June	9	–	4.3	24.8	19.9	5.1	102	NE
July	8	0.1	4.7	27.6	24.1	5.2	111	NE
Aug.	7	0.1	1.6	25.4	18.2	4.6	152	NE
Sept.	7	–	0.5	14.2	7.8	4.2	175	SSW
Oct.	9	0.2	0.1	12.9	1.9	4.6	203	SSW
Nov.	9	0.7	–	6.0	0.2	4.6	209	SSW
Dec.	10	0.4	–	1.3	–	4.9	221	SSW
Annual	102	3.0	20.8	156.1	88.1	4.7	2,081	SSW

[1] 1928–1960.
[2] 1922–1960.
[3] Observations at 09h00, 12h00, 15h00.

TABLE LXXXIII

CLIMATIC TABLE FOR DUNEDIN (MUSSELBURGH), NEW ZEALAND
Latitude 45°55'S, longitude 170°31'E, elevation 2 m

Month	Mean station press. (mbar)	Mean daily temp. (°C)	Mean daily temp. range (°C)	Temp. extremes[1] (°C) highest	lowest	Mean vapor press. (mbar)	Mean precip. (mm)	Max. precip.[2] (mm/24h)
Jan.	1011.3	14.9	8.4	34.4	2.2	12.0	71	86.6
Feb.	1012.4	15.1	8.4	33.5	2.8	12.2	64	137.2
Mar.	1013.4	13.6	8.0	29.4	1.1	11.6	64	134.4
Apr.	1012.8	11.6	7.5	29.4	−1.2	10.4	64	229.1
May	1012.4	8.9	7.3	24.9	−2.2	8.7	66	103.1
June	1010.6	6.8	7.1	20.8	−4.4	7.5	74	94.5
July	1013.8	6.4	7.4	19.4	−5.0	7.0	64	97.5
Aug.	1012.3	7.3	7.8	21.1	−3.9	7.6	58	116.8
Sept.	1013.8	9.4	8.9	25.0	−1.7	8.3	56	113.3
Oct.	1010.3	11.2	8.2	28.3	−1.1	9.3	64	58.4
Nov.	1008.0	12.8	8.4	30.0	0.0	10.4	71	113.8
Dec.	1008.7	13.9	7.9	31.1	1.7	11.2	74	137.7
Annual	1011.6	11.0	7.9	34.4	−5.0	9.7	787	229.1

Month	Number of days with precip.	thunder-storm	fog	gusts (m/sec) ≥18	≥27	ground frost	screen frost	Mean cloudi-ness (oktas)	Mean sun-shine (h)	Wind most freq. dir.[4]	mean speed[4] (m/sec)
Jan.	14	1.0	0.8	5.6	0.2	0.5	–	5.3	194	NE	4.0
Feb.	12	0.5	1.3	5.5	0.1	0.5	–	5.0	166	NE	3.8
Mar.	13	0.2	0.9	4.3	0.2	1.3	0.1	5.2	152	NE	3.4
Apr.	13	–	1.7	4.3	0.3	4.1	0.2	5.2	122	NE	3.2
May	13	0.1	1.1	4.1	0.3	10.9	0.5	4.9	110	NE	2.6
June	14	–	2.0	3.6	0.2	15.9	3.3	5.1	94	W	2.8
July	14	0.2	2.1	3.1	0.1	18.0	4.1	4.7	110	W	2.2
Aug.	12	–	2.3	3.5	0.2	17.4	2.7	4.6	132	W	2.8
Sept.	12	–	1.3	5.5	0.2	13.4	0.6	4.5	147	NE	3.4
Oct.	14	0.2	1.4	6.6	0.2	5.1	0.2	5.3	166	NE	3.8
Nov.	15	0.3	0.4	5.8	0.4	1.9	–	5.2	174	NE	4.1
Dec.	16	0.8	1.3	5.8	0.4	0.6	–	5.7	167	NE	4.2
Annual	162	3.3	16.6	57.7	2.8	89.6	11.7	5.1	1,734	NE	3.4

[1] Various sites, 1865–1960.
[2] Various sites, 1852–1960.
[3] Taieri Airport data.
[4] Observations at 09h00.

TABLE LXXXIV

CLIMATIC TABLE FOR INVERCARGILL (AIRPORT), NEW ZEALAND
Latitude 46°25′S, longitude 168°19′E, elevation 0 m

Month	Mean station press. (mbar)	Mean daily temp. (°C)	Mean daily temp. range (°C)	Temp. extremes[1] (°C) highest	lowest	Mean vapor press. (mbar)	Mean precip. (mm)	Max. precip.[2] (mm/24h)	Mean evapor. (mm)
Jan.	1011.7	13.5	9.8	33.8	−0.7	12.6	91	63.5	109
Feb.	1012.5	13.5	10.0	31.1	−2.4	12.5	97	70.9	81
Mar.	1013.0	12.2	9.9	30.0	−2.4	11.7	109	58.2	64
Apr.	1012.5	10.2	9.4	26.1	−3.8	10.3	97	43.2	36
May	1011.8	7.2	9.7	21.7	−6.9	8.1	94	43.2	20
June	1010.6	5.6	7.9	19.4	−6.7	7.4	99	43.2	10
July	1013.6	5.0	9.0	20.6	−7.2	6.9	71	42.7	13
Aug.	1012.8	6.2	10.1	21.8	−6.4	7.4	76	31.8	20
Sept.	1014.8	8.2	10.2	23.3	−4.5	8.9	84	82.6	38
Oct.	1011.2	10.2	9.6	25.7	−2.8	9.9	86	51.3	61
Nov.	1007.6	11.4	9.9	28.3	−1.4	10.7	91	50.8	86
Dec.	1009.0	12.7	9.2	31.1	−1.1	11.6	91	58.9	94
Annual	1011.8	9.7	9.6	33.8	−7.2	9.8	1,087	82.6	633

Month	Number of days with precip.	thunder-storm	fog	gusts (m/sec) ≥18	≥27	ground frost	screen frost	Mean cloudiness (oktas)	Mean sunshine (h)	Wind most freq. dir.	mean speed (m/sec)
Jan.	17	1.3	3.8	10.1	1.2	2.4	0.2	6.1	198	W	4.9
Feb.	16	1.0	4.6	7.5	0.7	2.8	0.1	6.1	162	W	4.2
Mar.	17	0.6	6.7	9.0	0.7	4.8	0.6	6.1	146	W	4.6
Apr.	18	0.8	5.4	6.5	0.6	6.5	1.2	6.0	105	W	4.4
May	18	1.0	7.2	7.4	1.6	14.3	7.3	5.8	93	W	3.5
June	20	1.1	3.4	5.1	0.6	14.9	9.2	6.1	68	N	3.2
July	17	0.5	4.5	5.3	0.6	20.3	13.4	5.5	97	N	3.0
Aug.	15	0.2	6.0	4.6	0.5	20.2	11.3	5.5	121	N	3.9
Sept.	15	0.5	5.5	8.0	0.5	13.6	4.8	5.6	144	W	4.8
Oct.	17	0.5	3.8	9.3	0.6	7.4	1.5	6.0	166	W	4.9
Nov.	18	0.9	3.0	9.3	0.6	4.9	0.7	6.0	172	W	5.1
Dec.	18	0.9	2.8	8.8	0.8	2.1	−	6.1	189	W	5.4
Annual	206	9.3	56.7	90.5	9.6	114.2	50.3	5.9	1,661	W	4.3

[1] Various sites, 1905–1960.
[2] Various sites, 1890–1906, 1910–1960.

References Index

Geographical Index

Subject Index